Lecture Notes in Energy 12

For further volumes:
http://www.springer.com/series/8874

Breda Kegl • Marko Kegl • Stanislav Pehan

Green Diesel Engines

Biodiesel Usage in Diesel Engines

Breda Kegl
Marko Kegl
Stanislav Pehan
Faculty of Mechanical Engineering
University of Maribor
Maribor
Slovenia

ISSN 2195-1284 ISSN 2195-1292 (electronic)
ISBN 978-1-4471-6939-0 ISBN 978-1-4471-5325-2 (eBook)
DOI 10.1007/978-1-4471-5325-2
Springer London Heidelberg New York Dordrecht

Printed on acid-free paper

Springer is part of Springer Science+Business Media (www.springer.com)

Contents

Chapter 1
Introduction

Modern diesel engines power much of the world's equipment and, most notably, are prime movers commonly available today. Among personal and commercial vehicles, the diesel engines hold a significant market share worldwide. And their share and popularity are both increasing.

Traditionally, diesel engines run on mineral diesel, which is produced from crude oil. This fact comes with several consequences, giving rise to various concerns, which cannot be put aside. In fact, they have to be, and in many areas they already are, addressed decisively in order to ensure long-term and sustainable use of these excellent machines in the future.

The *first concern* is the limited crude oil reserves. Crude oil covers about 37 % of world's energy demands (Fig. 1.1) (Asif and Muneer 2007; Dorian et al. 2006; Kegl 2012; Kjärstad and Johnsson 2009).

World's ultimate oil reserves are estimated at 2 trillion barrels, of which 900 billion barrels have already been consumed (Bentley 2002). This is the total volume that would have been produced when production eventually ceases. The global daily consumption of oil equals about 75 million barrels. Various countries are at different stages of their reserve depletion curves. Many of them, such as USA, are past their midpoint and are in terminal decline, whereas others are close to midpoint such as UK and Norway. Luckily, the five major Gulf producers—Saudi Arabia, Iraq, Iran, Kuwait, and United Arab Emirates—are at an early stage of depletion. Anyhow, a quick calculation reveals that at present production rates oil supplies are predicted to last only about 40 years. This estimate might not be very accurate, but this changes nothing on the fact that we have to search for *alternative fuels*, on which the diesel engines will run in the future. Of course, these alternative fuels should preferably be environment-friendly and produced from renewable sources.

The *second concern* is the immense quantity of fuel, consumed by diesel engines all over the world. A vast majority of diesel engines are engaged in road transport, which accounts for about 81 % of total energy used for transportation (Fig. 1.2) (Chapman 2007). Besides the road transport, diesel engines are also engaged in

B. Kegl et al., *Green Diesel Engines*, Lecture Notes in Energy 12,
DOI 10.1007/978-1-4471-5325-2_1, © Springer-Verlag London 2013

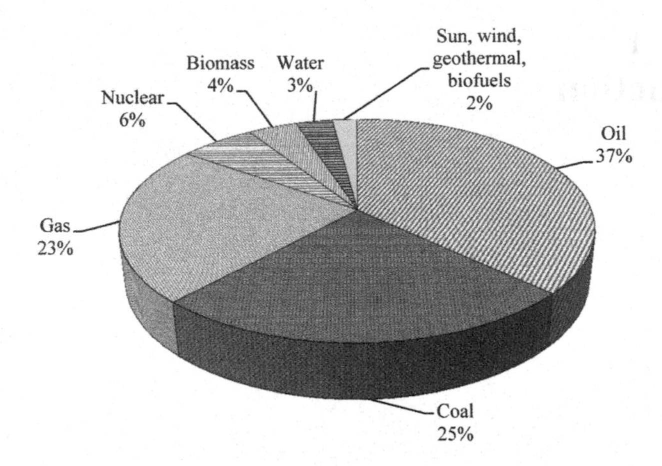

Fig. 1.1 Shares of global energy sources

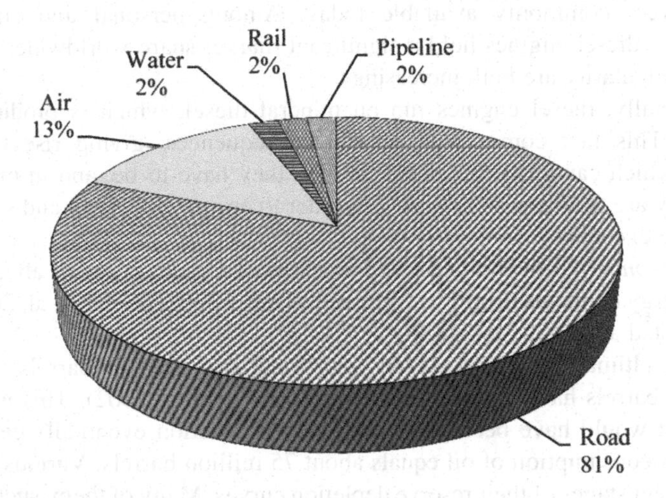

Fig. 1.2 Energy consumption shares in the transport sector

railway transport, naval transport, electricity generation, and so on. Anyhow, even if we manage to develop efficient alternative fuel production, the immense fuel volume, consumed daily by world's diesel engines, will still be a problem. For this reason, trying to *reduce* the diesel engine *fuel consumption* to the limits of possible should be worth of every effort.

The *third concern* is related to the chemical process of transformation of internal fuel energy into mechanical work or, more precisely, to the exhaust emissions of this process. A diesel engine burns the fuel under high pressure and emits the burning products into the environment. Therefore, diesel engines are one of the significant contributors to environmental pollution worldwide, with large increases expected in vehicle population and vehicle miles traveled. A diesel engine produces

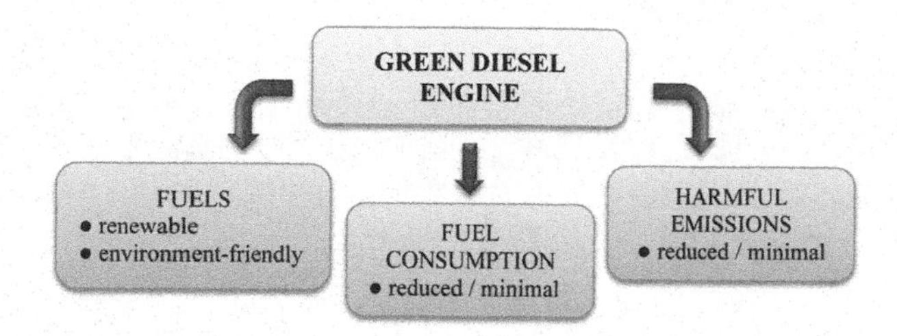

Fig. 1.3 Green diesel engine features

mainly CO_2, NO_x, CO, unburned HC, and PM/smoke emissions. These emissions contribute negatively to:

- Global climate changes
- General pollution of air, water, and soil
- Direct health effects (cancer, cardiovascular and respiratory problems, . . .)

For many years, diesel engine emissions have not been considered to be a major problem. Luckily, this has changed substantially. Nowadays, respectable efforts are put into the development of diesel engines and new supplemental technologies with the aim to *reduce* these *harmful emissions*. Special attention is focused on engine management, fuel injection and combustion control, exhaust gas recirculation, catalytic after treatment and filtering of exhaust emission, and alternative fuel usage.

The term *green diesel engine* is used here to denote a diesel engine that addresses all three major concerns to the extent possible. Thus, a green diesel engine should be able to run on alternative fuels, with minimal fuel consumption and emission (Fig. 1.3). This book is intended to promote further steps in the development of such diesel engines.

References

Asif, M., & Muneer, T. (2007). Energy supply, its demand and security issues for developed and emerging economies. *Renewable and Sustainable Energy Reviews, 11*, 1388–1413.

Bentley, R. W. (2002). Global oil & gas depletion: an overview. *Energy Police, 30*, 189–205.

Chapman, L. (2007). Transport and climate change: a review. *Journal of Transport Geography, 15*, 354–367.

Dorian, J. P., Franssen, H. T., & Simbeck, D. R. (2006). Global challenges in energy. *Energy Policy, 34*, 1984–1991.

Kegl, T. (2012). Transformation of heat energy into mechanical work at low environmental pollution. *Journal of Young Investigators, 24*(6), 81–87.

Kjärstad, J., & Johnsson, F. (2009). Resources and future supply of oil. *Energy Policy, 37*, 441–464.

Chapter 2
Diesel Engine Characteristics

Diesel engine is a compression ignition engine of a 2- or 4-stroke type. From the $p - V$ diagrams (Fig. 2.1), it can be seen that the duration of the whole diesel cycle is 360°CA for the *two-stroke engine* and 720°CA for the *four-stroke engine*. The whole cycle consists of the following *phases*: intake of air, compression of air, fuel injection, mixture formation, ignition, combustion, expansion, and exhaust. The intake phase begins with the intake valve opening and lasts till the intake valve closing. After that the intake air is compressed to a level corresponding to compression ratios from 14:1 to 25:1 (Bauer 1999) or even more. The compression ratio ε is a geometrical quantity, defined as

$$\varepsilon = \frac{V_{\max}}{V_{\min}} = \frac{V_{\mathrm{h}} + V_{\mathrm{c}}}{V_{\mathrm{c}}}, \tag{2.1}$$

where $V_{\max}$ and $V_{\min}$ denote maximal and minimal volume above the piston, V_{c} is the clearance or compression volume, and V_{h} denotes the piston displacement—the volume between the bottom dead center (BDC) and the top dead center (TDC) of the piston.

Towards the end of compression, fuel is injected into the precombustion (indirect injection) or combustion (direct injection) chamber under high pressure. The fuel in the spray mixes with the compressed hot air, evaporates, and then the mixture ignites by itself. It has to be pointed out that injection, atomization, spray development, mixture formation, ignition, combustion, and emission formation processes proceed largely simultaneously and interact with each other. During the combustion process heat is released and both the in-cylinder pressure and the in-cylinder temperature increase. At the end of the expansion phase, the exhaust valve opens and the exhaust phase begins. This phase and the whole cycle end as the exhaust valve closes.

In automotive application, diesel engines are practically always of the 4-stroke type (Bauer 1999), either naturally aspirated or turbocharged.

In a *naturally aspirated diesel engine* (Fig. 2.2), the pressure p_k in the intake tubes is smaller than the ambient pressure p_o. Since at the end of the exhaust process the pressure p_r of the residual gases is higher than p_k, this means that the

B. Kegl et al., *Green Diesel Engines*, Lecture Notes in Energy 12,
DOI 10.1007/978-1-4471-5325-2_2, © Springer-Verlag London 2013

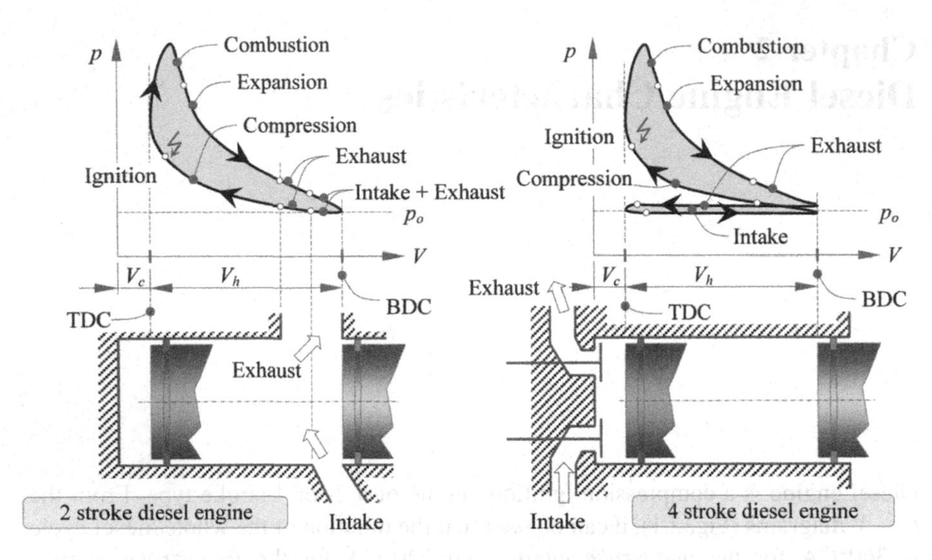

Fig. 2.1 The $p - V$ diagrams of two-stroke and four-stroke diesel engine

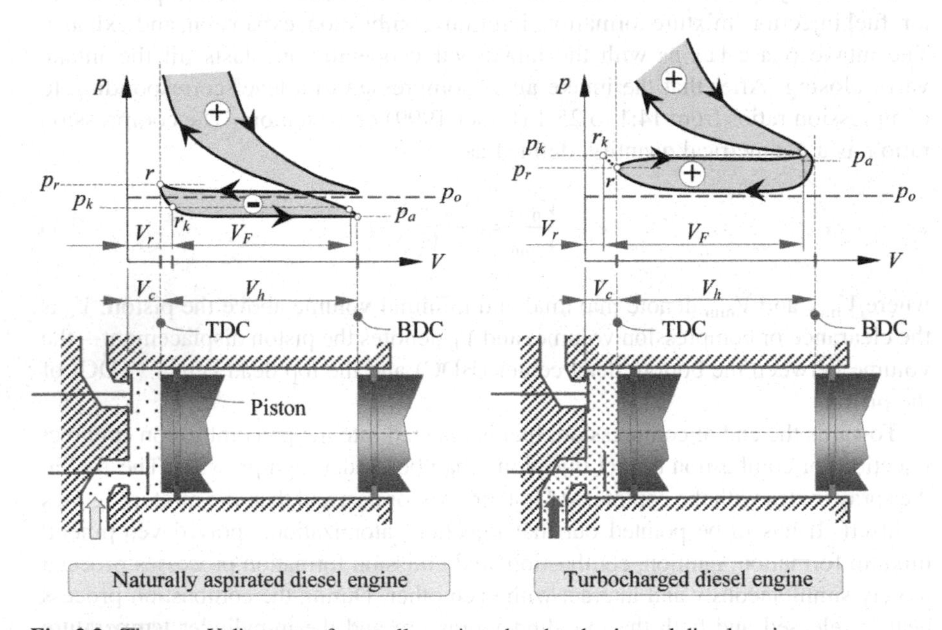

Fig. 2.2 The $p - V$ diagrams of naturally aspirated and turbocharged diesel engine

intake of air starts only after the piston travels a significant distance towards the bottom dead center. This means that the in-cylinder pressure p_a at the end of the intake process is always lower than p_k. In fact, in a naturally aspirated diesel engine the following relations are always valid: $p_r > p_k > p_a$ and $V_F < V_h$, where the symbol V_F denotes the actual volume of fresh intake air.

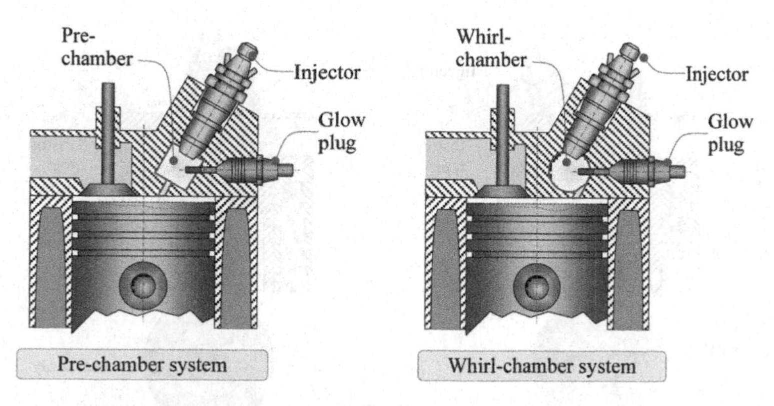

Fig. 2.3 Pre-chamber and whirl-chamber systems

In a *turbocharged diesel engine* (Fig. 2.2), the pressure in the intake tubes is higher than the pressure of the residual gasses, i.e., $p_k > p_a > p_r$. This means that in spite of the residual gases and the hydraulic losses, the volume V_{F} of fresh air may be higher than V_{h}. In other words, the volumetric efficiency η_{v}, defined as

$$\eta_{\mathrm{v}} = \frac{V_{\mathrm{F}}}{V_{\mathrm{h}}}, \tag{2.2}$$

can be higher than 1.0.

Fuel injection into the cylinder is realized by the fuel injection system, which can be of either an electronically (high pressure) or mechanically controlled type. More recent *electronically controlled injection systems* exhibit many advantages over the older mechanically controlled systems. In spite of that, one must admit that currently there are an enormous number of diesel engines with mechanically controlled systems still operating throughout the world. Moreover, *mechanically controlled injection systems* come with an attractive feature: they are very robust and relatively insensitive to fuel properties and quality. If the engine has to run on various biofuels, this can be a notable advantage.

Mechanically controlled fuel injection systems can be roughly classified as *direct* or *indirect injection systems*. Indirect systems inject the fuel into the pre-chamber or whirl chamber (Fig. 2.3), rather than directly into the cylinder. According to practical experiences in the past, one can roughly say that mechanically controlled indirect injection systems may deliver lower harmful emissions, especially of NO_x, compared to the direct injection systems. It is worth taking a brief look at this point.

At first, it has to be pointed out that in *indirect injection systems*, fuel is injected at a relatively low pressure up to 300 bar into the pre-chamber or whirl chamber when air is compressed through the piston movement to TDC. The injected fuel is mixed with the swirling air and the combustion starts in the pre-chamber (or whirl chamber), where the mixture is rich on fuel. Then, the air–fuel mixture is forced into the main combustion chamber and mixed with the residual compressed air. The combustion continues and completes in the main combustion chamber, where

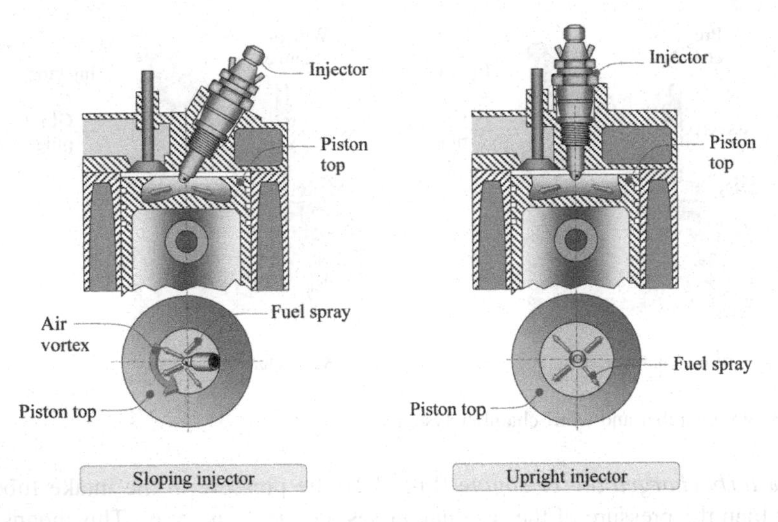

Fig. 2.4 Direct injection system—air distributed system

the mixture has a surplus of air. In this way, a two-phase combustion process is obtained: the first phase with a surplus of fuel, and at the second phase with a surplus of air. With an optimized pre-chamber or whirl chamber this can result in tempered release of energy at a low overall pressure level, soft combustion with low noise, low engine load, and low harmful emissions.

In *direct injection systems* fuel is injected with high injection pressure directly into the combustion chamber above the piston, using multi-hole nozzles. Here, fuel atomization, heating, evaporation, and mixture formation occur in very rapid succession. The air vortex is achieved by the special shape of the intake ports in the cylinder head. Furthermore, the design of the piston top with integrated combustion chamber contributes to the air movement at the end of the compression phase, when the injection process starts. Two fuel distribution methods are commonly used. The first one is the *air distributed method* (Fig. 2.4), where mixture formation is achieved by mixing fuel particles with air particles surrounding them.

The second method is the so-called *wall distribution method* (Fig. 2.5). In this, so-called *M fuel injection system* fuel is injected with a single-hole nozzle at relatively low pressure towards the walls of the combustion chamber. The fuel evaporates from the combustion chamber walls and is picked up by the swirling air. This method is characterized by extremely homogenous air–fuel mixtures, long combustion duration, low pressure rise, and soft combustion. However, the fuel consumption is higher compared to the air-distribution system.

The *features* of a diesel engine are defined by its *characteristics*. Diesel engine characteristics (Fig. 2.6) can roughly be classified into six groups as follows:

- *Fuel injection* characteristics
- *Fuel spray* characteristics
- *Combustion* characteristics

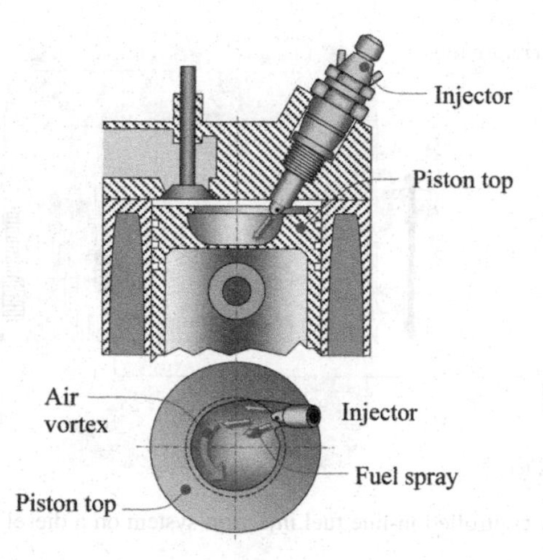

Fig. 2.5 Direct injection systems—wall distributed system

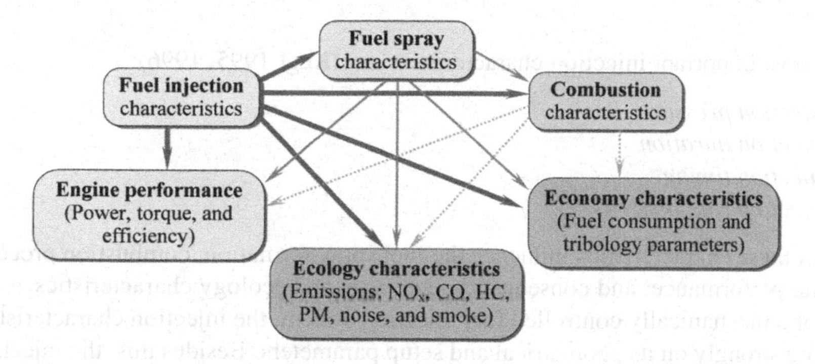

Fig. 2.6 Diesel engine characteristics

- *Engine performance* characteristics
- *Ecology* characteristics
- *Economy* characteristics

All these characteristics depend on the most basic parameters such as fuel type or injection system type and on various process characteristics such as the injection process, fuel spray development, atomization, mixture fuel/air formation, ignition and combustion, and so on (Merker et al. 2005). This means that even for a fixed engine type the diesel engine characteristics may be influenced by various geometrical and setup parameters. This holds true for a diesel engine with either mechanically or electronically controlled fuel injection system.

In *mechanically controlled injection systems*, like in-line, single, and distributor injection pumps, unit pumps, or unit injectors, the geometrical parameters play the most important role. Meanwhile, in *electronically controlled injection systems*, like common rail systems, engine behavior and performance are determined by setup parameters.

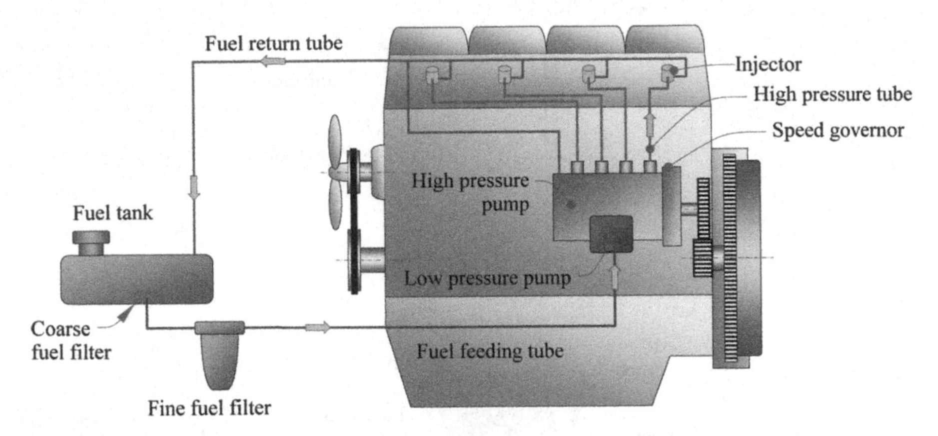

Fig. 2.7 Mechanically controlled in-line fuel injection system on a diesel engine

2.1 Injection Characteristics

The most important injection characteristics are (Kegl 1995, 1996):

- *Injection pressure*
- *Injection duration*
- *Injection timing*
- *Injection rate history*

All these characteristics influence the fuel spray formation, combustion process, engine performance, and consequently economy and ecology characteristics.

For a mechanically controlled fuel injection system, the injection characteristics depend strongly on its geometrical and setup parameters. Besides this, the injection characteristics can be influenced by the rest of the fuel path (Fig. 2.7), i.e., by the *coarse fuel filter*, *fine fuel filter*, *low pressure pump*, and so on.

The *mechanically controlled fuel injection system* consists basically of a high pressure fuel pump, high pressure tubes, and injectors (Kegl 1999). The *in-line fuel injection pump*, driven by a camshaft, has one pumping element for each cylinder. Pumping elements are mounted vertically in a straight line, side by side (Fig. 2.8). The lower half of the pump housing supports and encloses a horizontally positioned cam shaft, which has so many cam profiles as there are pumping elements. Each pumping element consists of a pump plunger which reciprocates in the barrel in dependence on camshaft profile. The camshaft profile converts the angular movement of the camshaft into a linear plunger motion by the roller cam follower and plunger return spring. Of course, the individual cam profiles are arranged according to the engine's firing order sequence. The top of each barrel is enclosed by its own delivery valve and optional snubber valve assembly.

The fuel enters the fuel ports at gallery pressure up to 1.5 bar, filling the space between the plunger and delivery valve. The plunger moves up in dependence on camshaft rotation following the cam profile and cuts off the feed/spill ports.

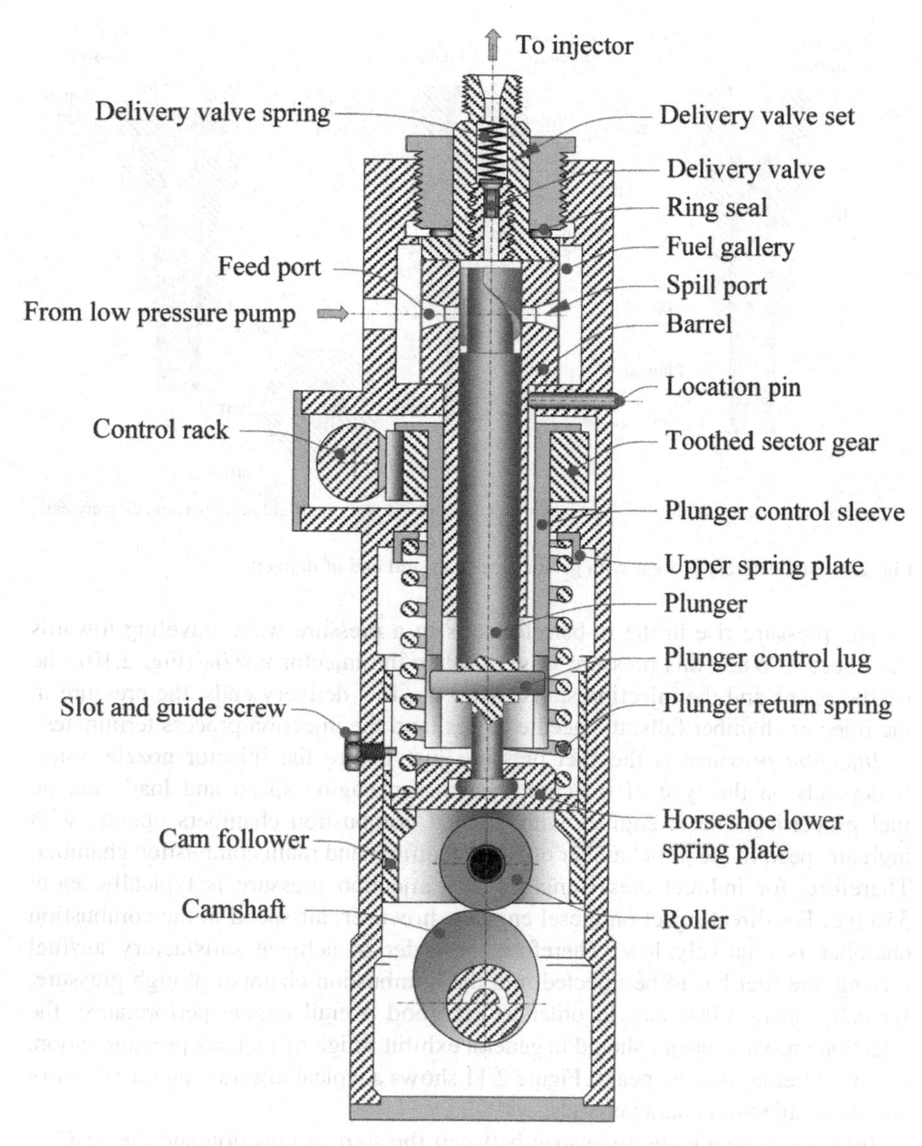

Fig. 2.8 Pumping elements

When all feed/spill ports are closed, the *geometrical fuel delivery begins* (Fig. 2.9). The in-barrel chamber pressure rises and eventually (dependent on the spring force and residual pressure in the high pressure tube) opens the delivery and the optional snubber valve. The pressure in the in-barrel chamber increases until the edge of the plunger helix unveils the feed/spill ports. Instantly, the in-barrel pressure collapses as fuel begins to escape down the vertical slot and exits through the feed/spill port. This is the moment of the *geometrical end of fuel delivery* (Fig. 2.9).

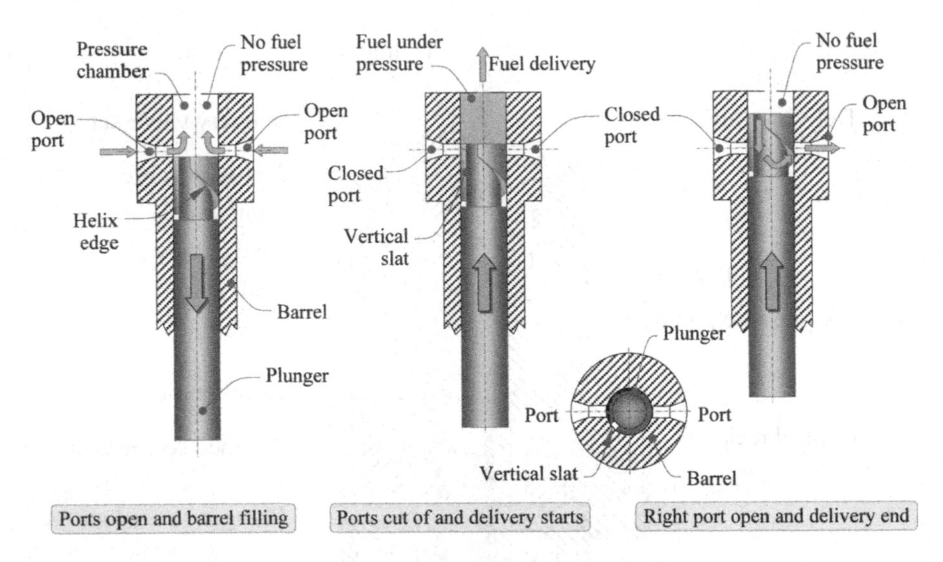

Fig. 2.9 Fuel delivery process with geometrical start and end of delivery

The pressure rise in the in-barrel results in a pressure wave traveling towards the injector. When this pressure wave reaches the injector nozzle (Fig. 2.10), the needle opens and the injection starts. After the fuel delivery ends, the pressure in the injector chamber falls, the needle closes, and the injection process terminates.

Injection pressure is the fuel pressure just before the injector nozzle holes. It depends on the type of injection system, on engine speed and load, and on fuel properties. Diesel engines with divided combustion chambers operate with high air speed in the pre-chamber or whirl chamber and main combustion chamber. Therefore, for indirect diesel engines, the injection pressure is typically about 350 bar. For direct injection diesel engines, however, air speed in the combustion chamber is relatively low. Therefore, in order to achieve satisfactory air/fuel mixing, the fuel has to be injected into the combustion chamber at high pressure, typically up to 1,000 bar. In order to get good overall engine performance, the injection pressure history should in general exhibit a high mean/peak pressure ration, i.e., no extreme pressure peaks. Figure 2.11 shows a typical injection pressure history and other injection characteristics.

Injection timing is the time span between the start of injection and the TDC of the engine piston. Injection timing has a strong influence on injection pressure, combustion process, and practically all engine emissions.

Injection duration is the time span from the beginning to the end of injection. In general, injection duration should be as short as possible. At higher engine speeds, injection duration should become longer and the mean injection pressure should be reduced (Desantes et al. 2004).

Injection rate represents the quantity of fuel injected per unit of time into the combustion chamber. Variations of injection rate history influence mixture preparation, combustion process, and harmful emissions. In a mechanical injection

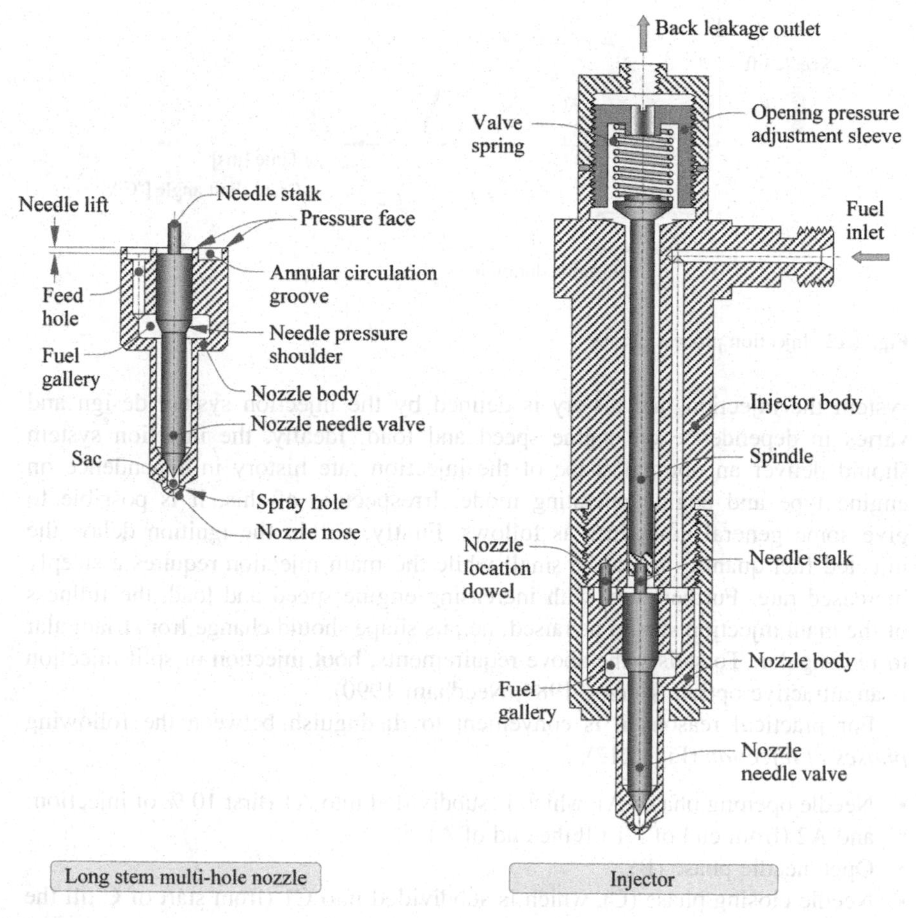

Fig. 2.10 Injector nozzle

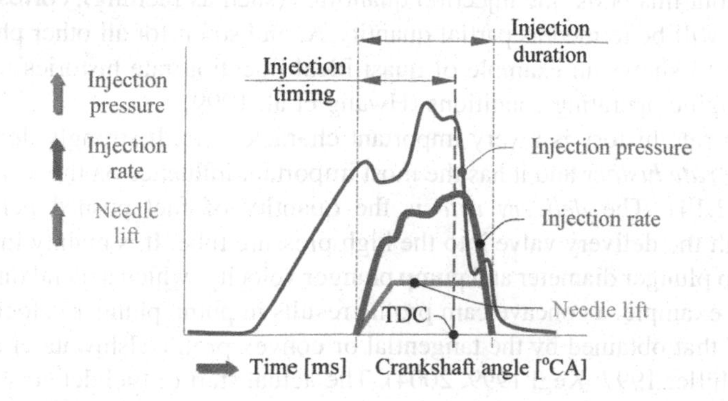

Fig. 2.11 Injection characteristics

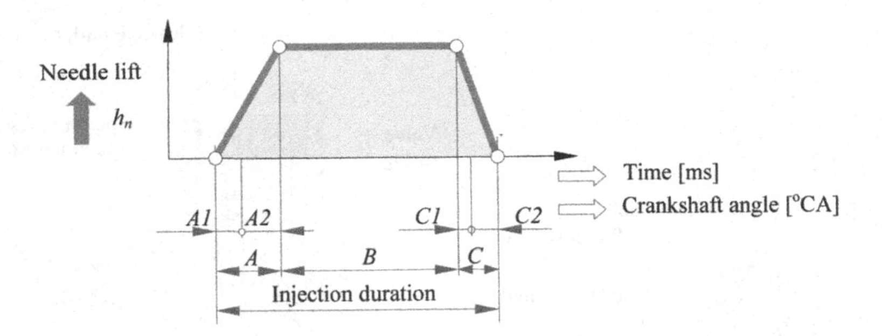

Fig. 2.12 Injection process phases

system the injection rate history is defined by the injection system design and varies in dependence on engine speed and load. Ideally, the injection system should deliver an optimal shape of the injection rate history in dependence on engine type and engine operating mode. Irrespective of this, it is possible to give some general guidelines as follows. Firstly, within the ignition delay, the injected fuel quantity has to be small while the main injection requires a steeply increased rate. Furthermore, with increasing engine speed and load, the fullness of the main injection should be raised, i.e., its shape should change from triangular to rectangular. To satisfy the above requirements, boot injection or split injection is an attractive option (Herzog 1989; Needham 1990).

For practical reasons it is convenient to distinguish between the following *phases of injection* (Fig. 2.12):

- Needle opening phase (A), which is subdivided into A1 (first 10 % of injection) and A2 (from end of A1 till the end of A)
- Open needle phase (B)
- Needle closing phase (C), which is subdivided into C1 (from start of C till the beginning of C2) and C2 (last 10 % of injection)

Throughout this book, the injection quantities (such as fuelling), corresponding to phase A, will be termed as partial quantity A, and so on for all other phases.

Figure 2.13 shows an example of quasi-ideal injection rate histories in dependence on engine operating conditions (Hwang et al. 1999).

Injection rate history is a very important characteristic. It strongly depends on the *delivery rate history* and it has the most important influence on the *heat release rate* (Fig. 2.14). The *delivery rate* is the quantity of fuel pushed per unit of time through the delivery valve into the high pressure tube. It is mainly influenced by the pump plunger diameter and pump plunger velocity, which depend on the cam profile. For example, a concave cam profile results in pump plunger velocity being higher than that obtained by the tangential or convex profile (Ishiwata et al. 1994; Kegl and Müller 1997; Kegl 1999, 2004). The actual start of fuel delivery depends on *injection pump timing*, which is usually given in crankshaft angle before TDC and indicates the moment when the pump plunger begins compressing the fuel. The delivery rate history is a well-controllable characteristic. The time from delivery

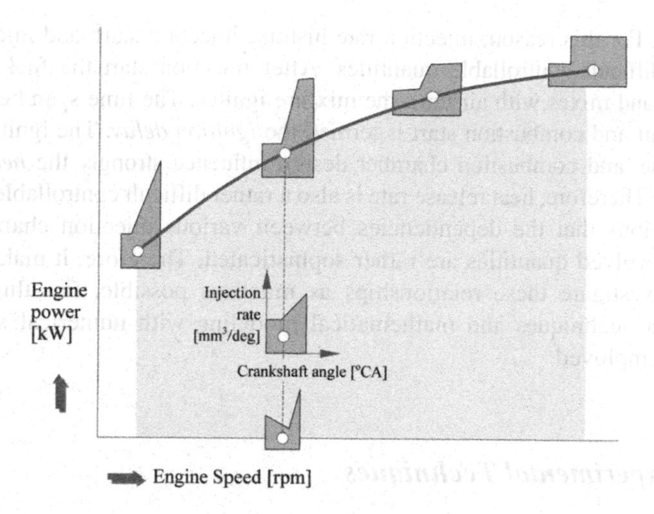

Fig. 2.13 Ideal injection rate shapes at various engine regimes

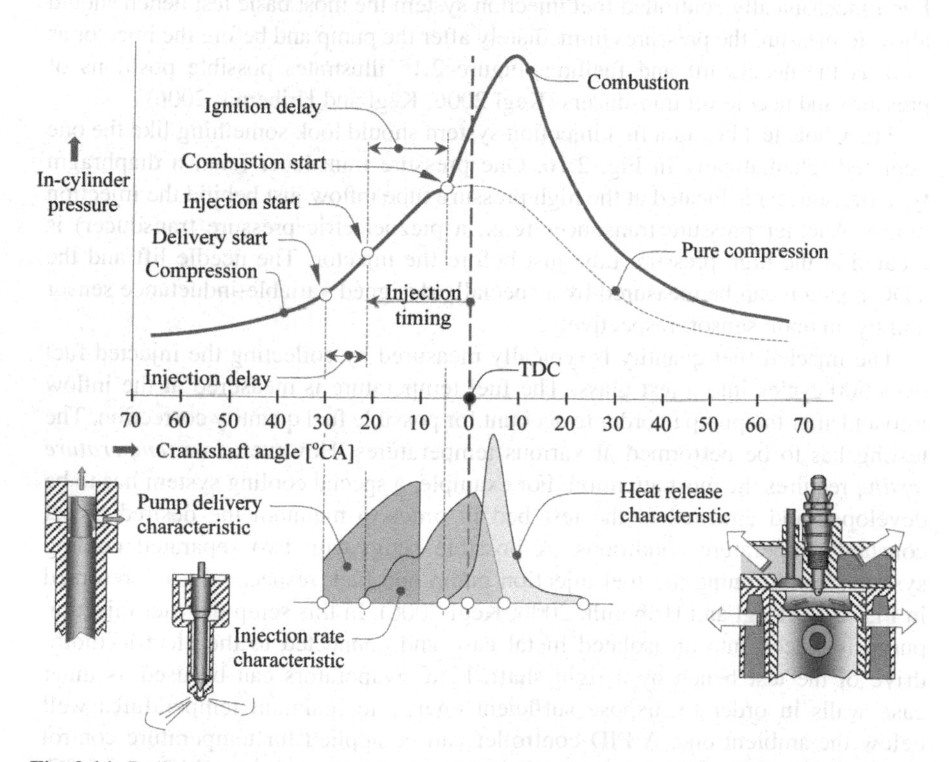

Fig. 2.14 Delivery rate history, injection rate history, heat release, and in-cylinder pressure

start to the *injection start* is the *injection delay*. The moment of injection start, given in crankshaft angle before TDC, is called the *injection timing*. Injection timing depends to a great extent on fuel properties and on geometrical parameters of the pump, high pressure tube, and injector. All these parameters influence the *injection rate history*. For this reason, injection rate history, injection start, and injection end are rather difficult controllable quantities. After injection start the fuel atomizes, evaporates, and mixes with air until the mixture ignites. The time span between the injection start and combustion start is termed the *ignition delay*. The ignition delay, injection rate, and combustion chamber design influence strongly the *heat release rate history*. Therefore, heat release rate is also a rather difficult controllable quantity.

It is obvious that the dependencies between various injection characteristics and other involved quantities are rather sophisticated. Therefore, it makes a good sense to investigate these relationships as much as possible. For this purpose experimental techniques and mathematical modeling with numerical simulation have to be employed.

2.1.1 Experimental Techniques

For a mechanically controlled fuel injection system the most basic test bench should allow to measure the pressures immediately after the pump and before the injector as well as the needle lift and fuelling. Figure 2.15 illustrates possible positions of pressure and needle lift transducers (Kegl 2006; Kegl and Hribernik 2006).

The whole test bed of a fuel injection system should look something like the one depicted schematically in Fig. 2.16. One pressure transducer (e.g., a diaphragm type transducer) is located at the high pressure tube inflow just behind the injection pump. Another pressure transducer (e.g., a piezoelectric pressure transducer) is located at the high pressure tube just before the injector. The needle lift and the TDC position can be measured by a specially designed variable-inductance sensor and by an optic sensor, respectively.

The injected fuel quantity is typically measured by collecting the injected fuel over 500 cycles into a test glass. The fuel temperature is measured at the inflow into and after the pump in order to account for possible fuel quantity correction. The testing has to be performed at various temperatures. However, *low temperature testing* requires the most attention. For example, a special cooling system has to be developed and attached to the test bed in order to maintain the desired preset constant temperature conditions. A possible setup with two separated cooling systems, conditioning the fuel injection pump and fuel, respectively, is presented in Fig. 2.17 (Kegl and Hribernik 2006; Kegl 2006). In this setup the fuel injection pump is placed into an isolated metal case and connected to the electric motor drive of the test bench by a rigid shaft. Plate evaporators can be used as inner case walls in order to dispose sufficient energy to maintain temperatures well below the ambient one. A PID controller can be applied for temperature control in order to ensure the tolerances of the set temperature to be within ±0.5 °C. Injectors should be mounted outside of the cooled case in order to enable simple measurement of the injected fuel amount. They can be connected to the injection pump by standard thermally isolated high pressure tubes. The fuel supply forms the second cooling system. A large fuel tank (e.g., 50 L) is placed into a refrigerator,

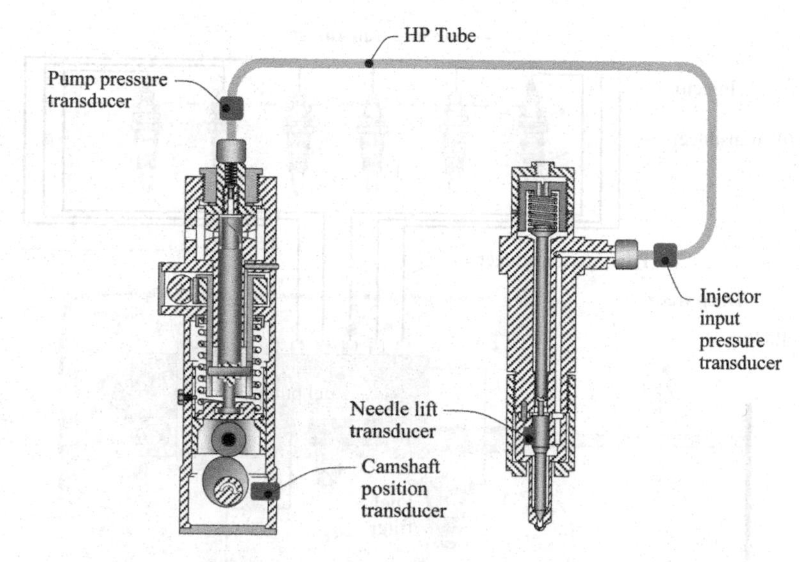

Fig. 2.15 Positions of transducers

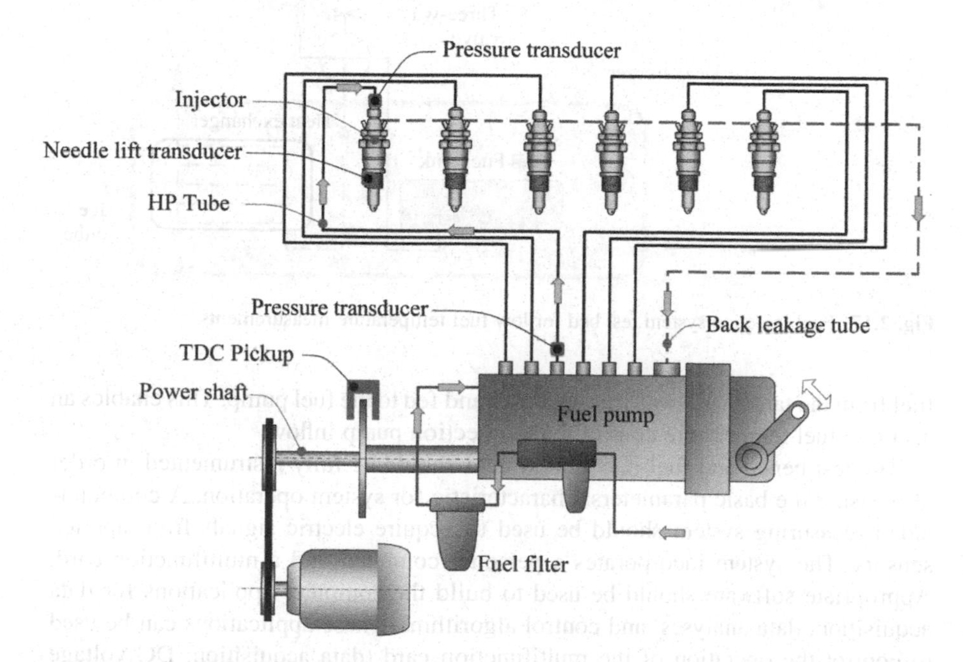

Fig. 2.16 Fuel injection system test bed

which allows the temperature to be controlled in the range between −30 °C and 40 °C. An ice-cooled heat exchanger, i.e., a pipe coil frozen within the ice cube, is placed next to the fuel tank. This exchanger cools the fuel flowing from the injection pump back to the fuel tank. This back flow can then be mixed with cool

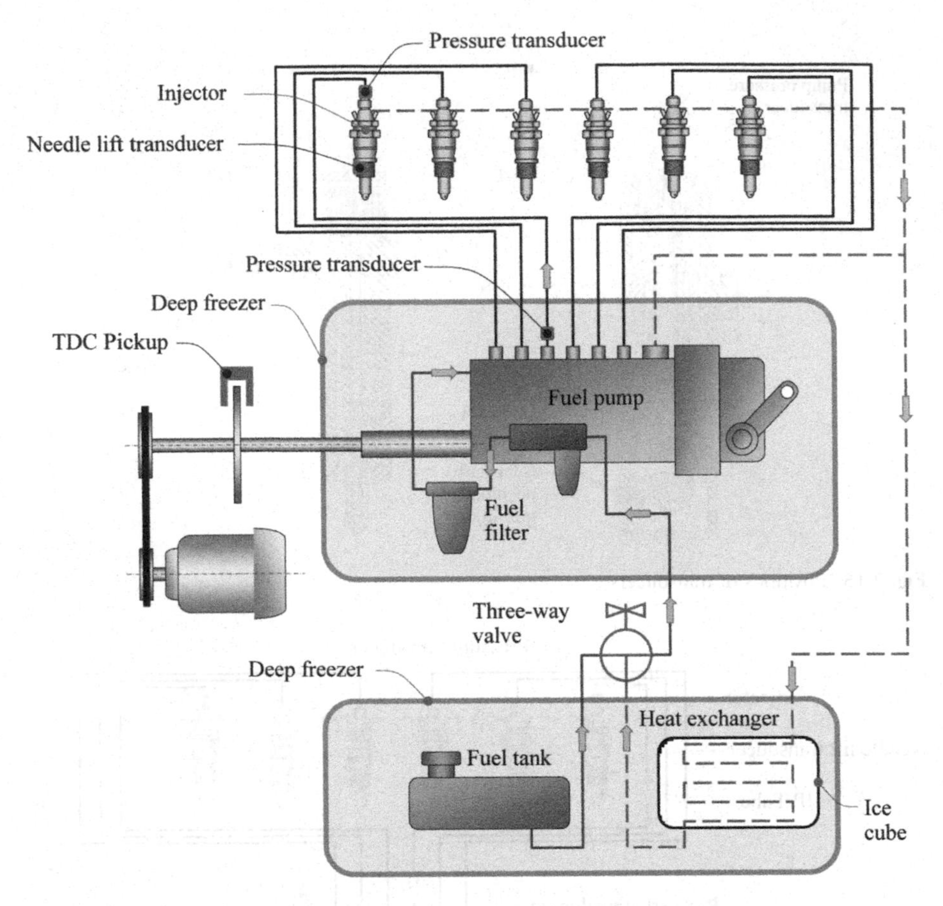

Fig. 2.17 Fuel injection system test bed for low fuel temperature measurements

fuel from the tank in a 3-way mixing valve and fed to the fuel pump. This enables an accurate fuel temperature control at the injection pump inflow.

The test bench and fuel injection system should be fully instrumented in order to measure the basic parameters, characteristic for system operation. A computer-aided measuring system should be used to acquire electric signals from applied sensors. The system incorporates a personal computer and a multifunction card. Appropriate software should be used to build the computer applications for data acquisition, data analyses, and control algorithms. These applications can be used to control the operation of the multifunction card (data acquisition, DC voltage output) and for data logging and post-processing.

Figure 2.18 illustrates the output from a typical test procedure, where p_{I} and p_{II} denote the measured pressures immediately after the pump and just before the injector. It is evident that the pressure before the injector delays and differs from the pressure after the pump. This is due to the compressibility of fuel, sound velocity of the fuel, spring forces, reflected pressure waves, etc. When the needle

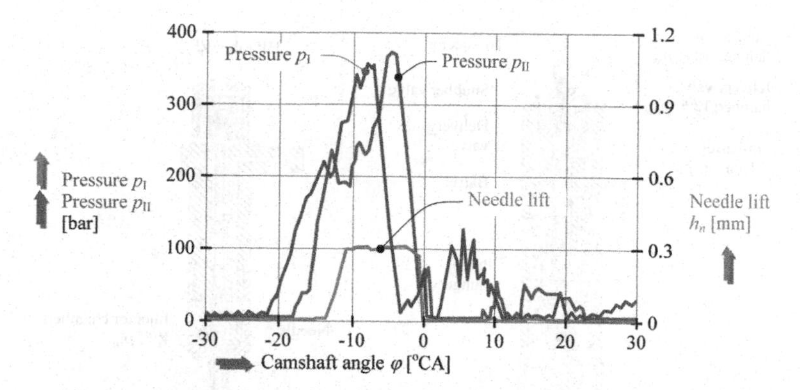

Fig. 2.18 Injection characteristics measured at the test bench

lifts, the pressure before the injector drops a little bit but starts to rise again once the injector is fully open. From the injection characteristics given in Fig. 2.18 the injection delay, injection timing, and injection duration can also be determined.

2.1.2 Mathematical Modeling and Simulation

In order to reduce the cost for experimental work, mathematical modeling of injection processes and numerical simulation may help substantially in the engine development process. Mathematical models of injection processes vary significantly in complexity, which leads to significant differences in accuracy, robustness, and computational efficiency. This book does not focus on mathematical modeling of injection processes. Therefore, only a relatively simple one-dimensional model will be presented in the following. The reason for this is that this model exhibits respectable computational efficiency, very good robustness, and quite reasonable accuracy. As such, it is also quite well suited to be employed in injection system optimization procedures.

The model (Kegl 1995) presented here simulates the injection process in a mechanically controlled in-line fuel injection system (Fig. 2.19) as a one-dimensional flow. The whole injection process is described by the following fifteen variables:

Injection rate $q\left[\frac{\text{mm}^3}{\text{ms}}\right]$,
Pressure in the in-barrel chamber p_k [MPa],
Pressure in the delivery valve chamber p_{dv} [MPa],
Pressure in the snubber valve chamber p_{sv} [MPa],
Pressure in the injector chamber p_{in} [MPa],
Volume of vapor cavities in the delivery valve chamber V_v^{dv} [mm^3],
Volume of vapor cavities in the snubber valve chamber V_v^{sv} [mm^3],

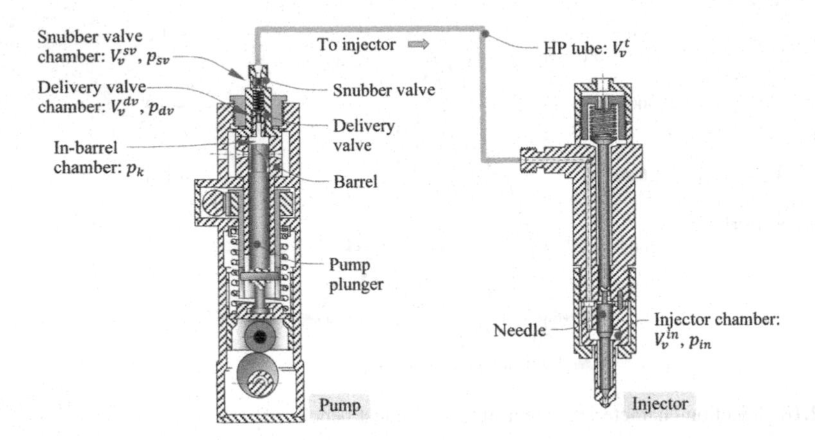

Fig. 2.19 Modeled fuel injection system

Volume of vapor cavities in the high pressure tube V_v^t [mm^3],
Volume of vapor cavities in the injector chamber V_v^{in} [mm^3],
Lift of the delivery valve h_{dv} [mm],
Velocity of the delivery valve v_{dv} $\left[\frac{\text{m}}{s}\right]$,
Lift of the snubber valve h_{sv} [mm],
Velocity of the snubber valve v_{sv} $\left[\frac{\text{m}}{s}\right]$,
Needle lift h_{n} [mm],
Needle velocity v_{n} $\left[\frac{\text{m}}{s}\right]$.

These quantities are related to each other and to time by a system of fifteen ordinary differential equations, involving an unknown parameter—the residual pressure p_0. Since this quantity is not known in advance, iterations are needed to solve this system of differential equations.

All of the used equations are either *equations of continuity* or *equations of motion*. The former ones are used to express the time derivatives of the injection rate; of the pressures in the in-barrel chamber, delivery valve, snubber valve, and injector chamber; and of the volume of vapor cavities in the delivery valve, snubber valve, and injector chamber. Equations of motion are used to express the time derivatives of the lift and velocity of the delivery valve, snubber valve, and the needle.

The *injection rate* $q = dq_{\text{inj}}/dt$ is given by

$$q = (\mu_{\text{in}} A_{\text{in}})\sqrt{\frac{2}{\rho}|p_{\text{in}} - p_a|}, \tag{2.3}$$

where q_{inj} is fuelling, $(\mu_{\text{in}} A_{\text{in}})$ denotes the actual (effective) cross-section area of the injector nozzles, ρ denotes the fuel density, p_{in} is the pressure in the injector chamber, and p_a is the pressure in the chamber into which the fuel is injected.

The time derivative of the *pressure* in the *in-barrel chamber* p_k is

$$\frac{dp_k}{dt} = \left[A_k v_k - (\mu_p A_p)\sqrt{\frac{2}{\rho}|p_k - p_n|} \times \mathrm{sgn}(p_k - p_n) - (\mu_{dv} A_{dv})\sqrt{\frac{2}{\rho}|p_k - p_{dv}|} \times \mathrm{sgn}(p_k - p_{dv}) - A_{dv} v_{dv}\right] \times \frac{E}{V_k} \tag{2.4}$$

where E is the modulus of fuel compressibility, V_k is the volume of the in-barrel chamber, A_k is the pump plunger cross-section area, v_k is pump plunger velocity, $(\mu_p A_p)$ is the effective flow area of feed/spill ports, p_k is the pressure in the in-barrel chamber, p_n is the pressure in the pump sump, $(\mu_{dv} A_{dv})$ is the effective flow area of the delivery valve, p_{dv} is the pressure in the delivery valve chamber, A_{dv} is the cross-section area of the delivery valve, and v_{dv} denotes the delivery valve velocity.

The time derivative of the *pressure* in the *delivery valve chamber* p_{dv} is

$$\frac{dp_{dv}}{dt} = \begin{cases} \left[(\mu_{dv} A_{dv})\sqrt{\frac{2}{\rho}|p_k - p_{dv}|} \times \mathrm{sgn}(p_k - p_{dv}) + A_{dv} v_{dv} - A_{sv} v_{sv} - (\mu_{sv} A_{sv})\sqrt{\frac{2}{\rho}|p_{dv} - p_{sv}|} \times \mathrm{sgn}(p_{dv} - p_{sv})\right] \times \frac{E}{V_{dv}}, & V_v^{dv} = 0 \\ 0, & V_v^{dv} \neq 0 \end{cases} \tag{2.5}$$

where V_{dv} denotes the volume of the delivery valve chamber, A_{sv} is the snubber valve cross-section area, v_{sv} is the snubber valve velocity, $(\mu_{sv} A_{sv})$ is the effective flow area of the snubber valve, p_{sv} is the pressure in the snubber valve chamber, and V_v^{dv} is the volume of vapor cavities in the delivery valve chamber.

The time derivative of the *pressure* in the *snubber valve chamber* p_{sv} is

$$\frac{dp_{sv}}{dt} = \begin{cases} \left[A_{sv} v_{sv} + (\mu_{sv} A_{sv})\sqrt{\frac{2}{\rho}|p_{dv} - p_{sv}|} \times \mathrm{sgn}(p_{dv} - p_{sv}) - A_t w_{\mathrm{I}}\right] \times \frac{E}{V_{sv}}, & V_v^{sv} = 0 \\ 0, & V_v^{sv} \neq 0 \end{cases} \tag{2.6}$$

where A_t denotes the cross-section area of the high pressure tube, w_{I} is the flow velocity at the point I–I (Fig. 2.20), V_{sv} is the volume in the snubber valve chamber, and V_v^{sv} is the volume of vapor cavities in the snubber valve chamber.

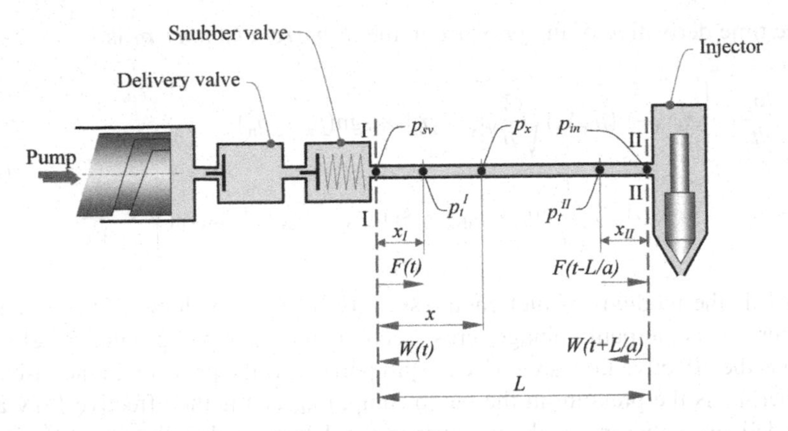

Fig. 2.20 Scheme of the in-line fuel injection system

The time derivative of the *pressure* in the *injector chamber* p_{in} is

$$\frac{dp_{\text{in}}}{dt} = \begin{cases} \begin{aligned}&[A_t w_{\text{II}} - \\ &(\mu_{\text{in}} A_{\text{in}})\sqrt{\frac{2}{\rho}|p_{\text{in}} - p_a|} \times \text{sgn}(p_{\text{in}} - p_a) - A_n^{\text{st}} v_n\Big],\end{aligned} & V_v^{\text{in}} = 0 \\ 0 & V_v^{\text{in}} \neq 0 \end{cases} \tag{2.7}$$

where w_{II} is the flow velocity at the point II–II (Fig. 2.20), $(\mu_{\text{in}} A_{\text{in}})$ is the effective flow area of the injector chamber, A_{n}^{st} is the cross-section area of the needle steam, v_{n} denotes needle velocity, V_{in} is the volume of the injector chamber, and V_v^{in} is the volume of vapor cavities in the injector chamber.

The time derivative of the *volume of vapor cavity* in the *delivery valve chamber* V_v^{dv} is

$$\frac{dV_v^{\text{dv}}}{dt} = \begin{cases} \begin{aligned}&-\Bigg[(\mu_{\text{dv}} A_{\text{dv}})\sqrt{\frac{2}{\rho}|p_k - p_{\text{dv}}|} \times \text{sgn}(p_k - p_{\text{dv}}) + A_{\text{dv}} v_{\text{dv}} - \\ &A_{\text{sv}} v_{\text{sv}} - (\mu_{\text{sv}} A_{\text{sv}})\sqrt{\frac{2}{\rho}|p_{\text{dv}} - p_{\text{sv}}|} \times \text{sgn}(p_{\text{dv}} - p_{\text{sv}})\Bigg],\end{aligned} & p_{\text{dv}} = p_v \\ 0, & \text{otherwise} \end{cases} \tag{2.8}$$

where p_v denotes the evaporation pressure.

The time derivative of the *volume of vapor cavity* in the *snubber valve chamber* V_v^{sv} is

$$\frac{dV_v^{sv}}{dt} = \begin{cases} -\left[A_{sv}v_{sv} + (\mu_{sv}A_{sv})\sqrt{\frac{2}{\rho}|p_{dv} - p_{sv}|} \times \text{sgn}(p_{dv} - p_{sv}) - A_t w_I\right], & p_{sv} = p_v \\ 0, & \text{otherwise} \end{cases} \quad (2.9)$$

The time derivative of the *volume of vapor cavity* in the *high pressure tube* V_v^t is

$$\frac{dV_v^t}{dt} = (w_{II} - w_I) \times A_t, \quad (2.10)$$

where A_t denotes the cross-section area of the high pressure tube.

The time derivative of the *volume of vapor cavity* in the *injector chamber* V_v^{in} is

$$\frac{dV_v^{in}}{dt} = \begin{cases} -\left[A_t w_{II} - (\mu_{in}A_{in})\sqrt{\frac{2}{\rho}|p_{in} - p_a|} \times \text{sgn}(p_{in} - p_a) - A_n^{st}v_n\right], & p_{in} = p_v \\ 0, & \text{otherwise} \end{cases} \quad (2.11)$$

The time derivatives of the *lift* h_{dv} and *velocity* v_{dv} of the *delivery valve* are

$$\frac{dh_{dv}}{dt} = \begin{cases} 0, & F_{dv} \geq 0 \text{ and } h_{dv} = h_{dv}^{max} \\ 0, & F_{dv} \leq 0 \text{ and } h_{dv} = 0 \\ v_{dv}, & \text{otherwise} \end{cases} \quad (2.12)$$

$$\frac{dv_{dv}}{dt} = \begin{cases} 0, & F_{dv} \geq 0 \text{ and } h_{dv} = h_{dv}^{max} \\ 0, & F_{dv} \leq 0 \text{ and } h_{dv} = 0 \\ \frac{F_{dv}}{m_{dv}}, & \text{otherwise} \end{cases} \quad (2.13)$$

The time derivatives of the *lift* h_{sv} and *velocity* v_{sv} of the *snubber valve* are

$$\frac{dh_{sv}}{dt} = \begin{cases} 0, & F_{sv} \geq 0 \text{ and } h_{sv} = h_{sv}^{max} \\ 0, & F_{sv} \leq 0 \text{ and } h_{sv} = 0 \\ v_{sv}, & \text{otherwise} \end{cases} \quad (2.14)$$

$$\frac{dv_{sv}}{dt} = \begin{cases} 0, & F_{sv} \geq 0 \text{ and } h_{sv} = h_{sv}^{max} \\ 0, & F_{sv} \leq 0 \text{ and } h_{sv} = 0 \\ \frac{F_{sv}}{m_{sv}}, & \text{otherwise} \end{cases} \quad (2.15)$$

where $F_{dv} = A_{dv} \times (p_k - p_{dv}) - F_0^{dv} - C_{dv}h_{dv}$ represents the sum of forces, which act on the delivery valve, h_{dv} denotes the delivery valve lift, h_{dv}^{max} is the maximum delivery valve lift, m_{dv} is the mass of delivery valve moving parts, F_0^{dv} is the preload spring force, and C_{dv} is the delivery valve spring stiffness, $F_{sv} = A_{sv} \times (p_{dv} - p_{sv}) - F_0^{sv} - C_{sv}h_{sv}$ represents the sum of forces, which act on the snubber valve, h_{sv} denotes the snubber valve lift, h_{dv}^{max} is the maximum snubber valve lift, m_{sv} is the mass of snubber valve moving parts, F_0^{sv} is the preload spring force, and C_{sv} is the snubber valve spring stiffness. The movement of the delivery valve is limited to the interval $[0, h_{dv}^{max}]$ and the movement of the snubber valve is limited to the interval $[0, h_{sv}^{max}]$.

The time derivatives of the *needle lift* h_n and *velocity* v_n are

$$\frac{dh_n}{dt} = \begin{cases} 0, & F_n \geq 0 \text{ and } h_n = h_n^{max} \\ 0, & F_n \leq 0 \text{ and } h_n = 0 \\ v_n, & \text{otherwise} \end{cases} \tag{2.16}$$

$$\frac{dv_n}{dt} = \begin{cases} 0, & F_n \geq 0 \text{ and } h_n = h_n^{max} \\ 0, & F_n \leq 0 \text{ and } h_n = 0 \\ \dfrac{F_n}{m_{in}}, & \text{otherwise} \end{cases} \tag{2.17}$$

where $F_n = (A_n^{st} - A_n^{se}) \times p_{in} + A_n^{se} \times p_{sac} - F_0^n - C_n h_n$ represents the sum of forces, which act on the needle, A_n^{se} is the cross-section area of the needle seat, p_{sac} is the pressure in the sac volume, h_n denotes the needle lift, h_n^{max} is the maximum needle lift, m_{in} is the mass of injector moving parts, F_0^n is the preload spring force, and C_n is the needle spring stiffness. The movement of needle is limited to the interval $[0, h_n^{max}]$.

The *pressure* in the *sac volume* p_{sac} is given by

$$p_{sac} = \frac{K^2}{1 + K^2} \times (p_{in} - p_a) + p_a, \tag{2.18}$$

where $K = \dfrac{(\mu_{in} A_{in})}{(\mu_{in} A_{in})_{max}}$.

In general, to determine the velocities w_I and w_{II} in the high pressure tube and to relate them to pressure waves (Fig. 2.20), the equations of momentum and continuity are used:

$$\frac{\partial p}{\partial x} + \rho \frac{\partial w}{\partial t} + \rho k w = 0, \tag{2.19}$$

$$\frac{\partial w}{\partial x} + \frac{1}{a^2 \rho} \times \frac{\partial p}{\partial t} = 0, \tag{2.20}$$

where ρ denotes the fuel density, k is the resistance factor, w is fuel velocity, a is the sound velocity, t is time, and x is a coordinate, measured along the HP tube.

By integrating these two equations, the pressure and velocity at x coordinate can be expressed as follows:

$$p_x = p_0 + F\left(t - \frac{x}{a}\right) \times e^{-\frac{kx}{a}} - W\left(t + \frac{x}{a}\right) \times e^{\frac{k(L-x)}{a}}, \tag{2.21}$$

$$w_x = \frac{1}{a\rho} \times \left[F\left(t - \frac{x}{a}\right) \times e^{-\frac{kx}{a}} + W\left(t + \frac{x}{a}\right) \times e^{-\frac{k(L-x)}{a}}\right], \tag{2.22}$$

where L is the length of the high pressure tube, a denotes the sound velocity in the fluid, and p_0 is the residual pressure in the high pressure system between two injections. The function F denotes the forward pressure wave, traveling from I–I to II–II (Fig. 2.20). The function W denotes the reflected pressure wave. Both pressure wave functions are defined later in the text. The pressures at the I–I and II–II cross section in the high pressure tube are marked as p_t^{I} and p_t^{II}. The numerical determination of the pressure at monitoring points I–I and II–II is necessary in order to verify the mathematical model with experiment.

The *resistance factor* k is derived on the basis of experimental work (Kegl 1995). Based on experimental data, ad hoc formulas for the determination of k can be developed. Typical parameters in these formulas are engine speed n, tube diameter d_t, tube length L, and maximum value of effective flow area of the injector chamber, for example,

$$k = 4.5 \times (60.36 + 3n) \times \left(\frac{0.002}{d_t}\right)^{2.75} \times \left(\frac{(\mu_{in}A_{in})_{max}}{4.272 \times 10^{-7}}\right)^{0.72} \times \left(\frac{0.5}{L}\right)^{0.8633}. \tag{2.23}$$

The forward pressure wave and fuel velocity at cross-section I–I can be expressed as

$$F(t) = p_{sv} - p_0 + W(t) \times e^{-\frac{kL}{a}}, \tag{2.24}$$

$$w_{\mathrm{I}} = \frac{1}{a\rho} \times \left(p_{sv} - p_0 + 2W(t) \times e^{-\frac{kL}{a}}\right), \tag{2.25}$$

and the reflected pressure wave and fuel velocity at cross-section II–II as

$$W\left(t + \frac{L}{a}\right) = p_0 - p_{in} + F\left(t - \frac{L}{a}\right) \times e^{-\frac{kL}{a}}, \tag{2.26}$$

$$w_{\rm II} = \frac{1}{a\rho} \times \left(p_0 - p_{\rm in} + F\left(t - \frac{L}{a}\right) \times e^{-\frac{kL}{a}}\right). \tag{2.27}$$

The *residual pressure* p_0 can be expressed in dependence on the delivery and snubber valve parameters as follows:

$$p_0 = p_{s(h_n=0)} - \frac{A_{\rm dv} h_{{\rm dv}(h_n=0)} E}{V_s}, \tag{2.28}$$

where $h_{{\rm dv}(h_n=0)}$ is the delivery valve lift and $p_{s(h_n=0)}$ is the pressure in the high pressure system at the time when the needle closes ($h_n = 0$). Due to the presence of the snubber valve,
$p_{s(h_n=0)}$ can be calculated as

$$p_{s(h_n=0)} = \frac{\sum_{i=1}^{m} p_{\rm t}^{z_i} + p_{\rm dv} + p_{\rm sv+p_{in}}}{m+3}, \tag{2.29}$$

where the quantities $p_{\rm t}^{z_i}, i = 1 \ldots m$ represent the pressures in the high pressure tube, calculated at m equidistant points, each point being positioned at the distance z_i, measured from the cross-section I–I. The pressure $p_{\rm t}^{z_i}$ can be expressed as

$$p_{\rm t}^{z_i} = p_0 + F\left(t - \frac{z_i}{a}\right) \times e^{-\frac{kz_i}{a}} - W\left(t + \frac{L - z_i}{a}\right) \times e^{-\frac{k(L-z_i)}{a}}, i = 1 \ldots m. \tag{2.30}$$

On the basis of the residual pressure p_0 and evaporization pressure p_v, the total volume of vapor cavities in the high pressure system, at the beginning of injection, can be calculated as

$$V_v^0 = \begin{cases} \frac{V_s}{E}(p_v - p_0), & p_v > p_0 \\ 0, & p_v \le p_0 \end{cases} \tag{2.31}$$

Furthermore, it can be assumed that initially the vapor in the system is distributed proportionally to the system dead volumes. Therefore, it follows that the initial vapor volumes in the delivery valve chamber $V_v^{{\rm dv},0}$, snubber valve chamber $V_v^{{\rm sv},0}$, high pressure tube $V_v^{t,0}$, and injector chamber $V_v^{{\rm in},0}$ are as follows:

$$V_v^{{\rm dv},0} = \frac{V_{\rm dv}}{V_s} V_v^0, \quad V_v^{{\rm sv},0} = \frac{V_{\rm sv}}{V_s} V_v^0, \quad V_v^{t,0} = \frac{V_{\rm t}}{V_s} V_v^0, \quad V_v^{{\rm in},0} = \frac{V_{\rm in}}{V_s} V_v^0. \tag{2.32}$$

In dependence on instantaneous vapor volume, pressure, and temperature, the instant values of fuel density, modulus of compressibility, and sound velocity can be computed (Kegl 1995, 2006).

The presented system of 15 first-order ordinary differential equations for the simulation of fuel injection process could be integrated straightforwardly by using any standard method, if the residual pressure would be known a priori. Since this is not the case, iterations are required starting with an assumed value of the residual pressure. In the first iteration, time integration can be done with an estimated residual pressure, being somewhat higher than the evaporation pressure. After the first integration is completed, the new residual pressure is calculated as discussed in the text. This value is then used as an estimation of the residual pressure for the next iteration and so on. The procedure is terminated when the difference of the calculated residual pressure in two successive iterations is smaller than some prescribed value (Kegl 1995).

The model presented in this section was implemented in a software package named BKIN. Therefore the term BKIN will be used throughout this book to reference this model.

2.2 Fuel Spray Characteristics

The most important diesel fuel spray characteristics may be classified as

- *Macroscopic quantities* such as:
  - Spray tip penetration
  - Cone angle
- *Microscopic quantities* such as:
  - Droplet size

All fuel spray characteristics (Fig. 2.21) influence the combustion process and consequently the economy and ecology characteristics and engine performance (Hiroyasu and Arai 1990; Kegl 2004; Soid and Zainal 2011).

Fuel spray tip penetration L_{p} is defined as the maximal distance measured from the injector to the spray tip. It represents the maximum penetration length achieved by the droplets in the center of the spray.

Spray cone angle θ is defined as the angle between two straight lines originating from the orifice exit of the nozzle and being tangent to the spray outline. This angle usually ranges from 5 to 30°.

Droplet size is usually measured on an average basis by the *medium diameter of the droplets* d_{32}, called the Sauter mean diameter. This quantity can be used to estimate the quality of atomization of the fuel.

Fuel spray penetration is determined by the equilibrium of two factors: the linear momentum of the injected fuel and the resistance of the working fluid (either gas or liquid) in the control volume. Due to friction, the kinetic energy of the fuel is transferred progressively to the working fluid. This energy transfer decreases continuously the kinetic energy of the droplets until their movement depends solely

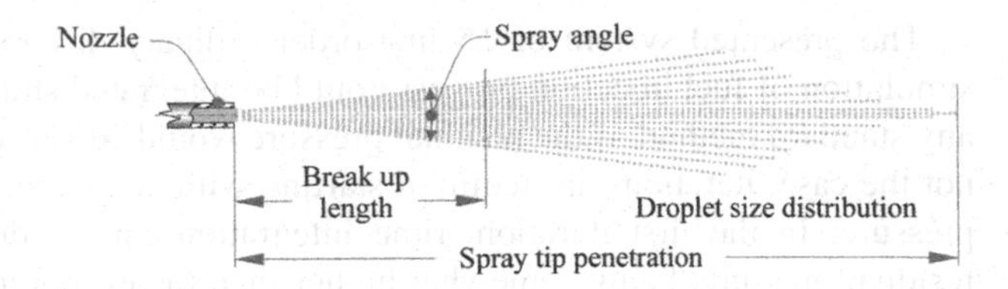

Fig. 2.21 Fuel spray characteristics

on the movement of the working fluid. Several studies show that spray penetration overcomes that of a single droplet. This is because the front droplets accelerate the surrounding fluid, causing the following droplets to have less aerodynamic resistance (Hiroyasu and Arai 1990; Gao et al. 2009a, b). One must emphasize that diesel fuel sprays tend to be of a compact type, which causes them to have large penetrations.

Diesel fuel spray penetration depends to a great extent on injection pressure, fuel properties, and nozzle geometry.

By increasing the injection pressure the fuel penetration velocity is increased. This means increased fuel momentum and larger spray penetration. Fuel properties like density, viscosity, and surface tension also affect spray penetration significantly. However, when making raw estimates, fuel density is often used as the only influencing property. In this context it is worth noting that fuel density of a given fuel may vary, for example, due to variations in fuel temperature. An increase of fuel temperature typically reduces the fuel density, which results in shorter spray penetration (Hiroyasu and Arai 1990; Gao et al. 2009a, b).

The *cone angle* is mainly affected by the geometric characteristics of the nozzle, the fuel and air density, and the Reynolds number of the fuel. Furthermore, the cone angle increases by increasing the injection pressure and by decreasing the working fluid temperature (Desantes et al. 2006; Hiroyasu and Arai 1990).

The *diameters of the droplets* depend on injection pressure, on working fluid temperature, and on fuel properties (Pogorevc et al. 2008; Desantes et al. 2006; Zhang et al. 2012). The diameters of the droplets tend to become smaller as the injection pressure raises. Furthermore, the working fluid temperature and fuel properties influence the evaporation rate, which also affects the droplet size. Namely, by increasing the temperature the rate of evaporation increases. Consequently, the droplets with small diameters tend to evaporate completely within a quite short time interval. On the other side, the droplets with greater diameters maintain a stable geometry for some time until they also evaporate completely.

In a fuel spray, fuel droplets evaporate as they travel away from the nozzle. The maximal distance, reached by the droplets before they all evaporate, is called the *liquid length*. After the liquid length is reached, the evaporated fuel continues to penetrate the surrounding gas and its range is denoted as the *vapor length*. It was found out that the liquid length tends to stabilize after a short spray development time and then remains approximately constant. On the other hand, in a typical diesel injection timeframe (a few milliseconds) the vapor length does not reach a steady state.

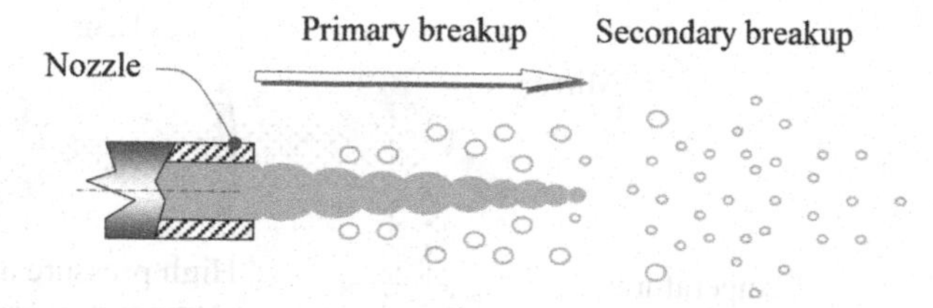

Fig. 2.22 Fuel spray breakup

Liquid spray formation is a rather sophisticated physical process, starting from the breakup of the liquid core into droplets, shortly after the nozzle exit, called the *primary breakup*. In the second stage the *formed droplets break up into smaller droplets*, which is called the *secondary breakup* (Fig. 2.22).

2.2.1 Experimental Techniques

The techniques used in the past for droplet size measurement can be classified into *mechanical methods* (droplet capture, cascade impaction, frozen drop and wax methods, and sedimentation technique), *electrical methods* (Wickse Dukler technique, the charged wire probe, and the hot wire anemometer), and *optical methods* (photography, holography, laser diffraction, laser anemometry, and various other techniques based on light scattering) (Soid and Zainal 2011; Leng et al. 2010; Myong et al. 2008; Andreassi et al. 2007; Payri et al. 2005). *Optical techniques* were commonly used for macroscopic and microscopic fuel sprays and combustion characterization. However, these techniques can often be difficult to implement and they are relatively expensive.

The macroscopic parameters such as the *cone angle* and the *tip penetration* can be determined through *direct visualization methods* (Fig. 2.23), where diesel spray injection can be studied in a high pressure and temperature cell. Spray propagation can be followed by using, for example, two intensified charge-coupled device cameras with double framing options. Two-dimensional flash light shadowgraph, laser elastic scattering, and chemiluminescence can be used to investigate the fuel propagation (Soid and Zainal 2011; Payri et al. 2005; Myong et al. 2008).

The microscopic parameters such as the *droplet size* can nowadays be measured by using either *PIV* or *PDPA techniques*. Figure 2.24 shows the experimental apparatus for the LIF-PIV experiment (Soid and Zainal 2011; Moon et al. 2010; Andreassi et al. 2007). A double pulsed Nd:YAG laser is utilized to obtain the fluoresced images of tracer droplets. A light sheet is generated using a sheet generator equipped with a cylindrical lens inside it. When the tracer droplets are fluoresced by the laser light, images are captured by a charge-coupled device (CCD) camera. A long-pass filter, which transmits the wavelengths longer than 560 nm, has to be installed in front of the CCD camera to capture the fluoresced signal of the tracer droplets. The pressures inside the constant volume vessel and the tracer reservoir can be controlled by pressure regulators. Two delay generators

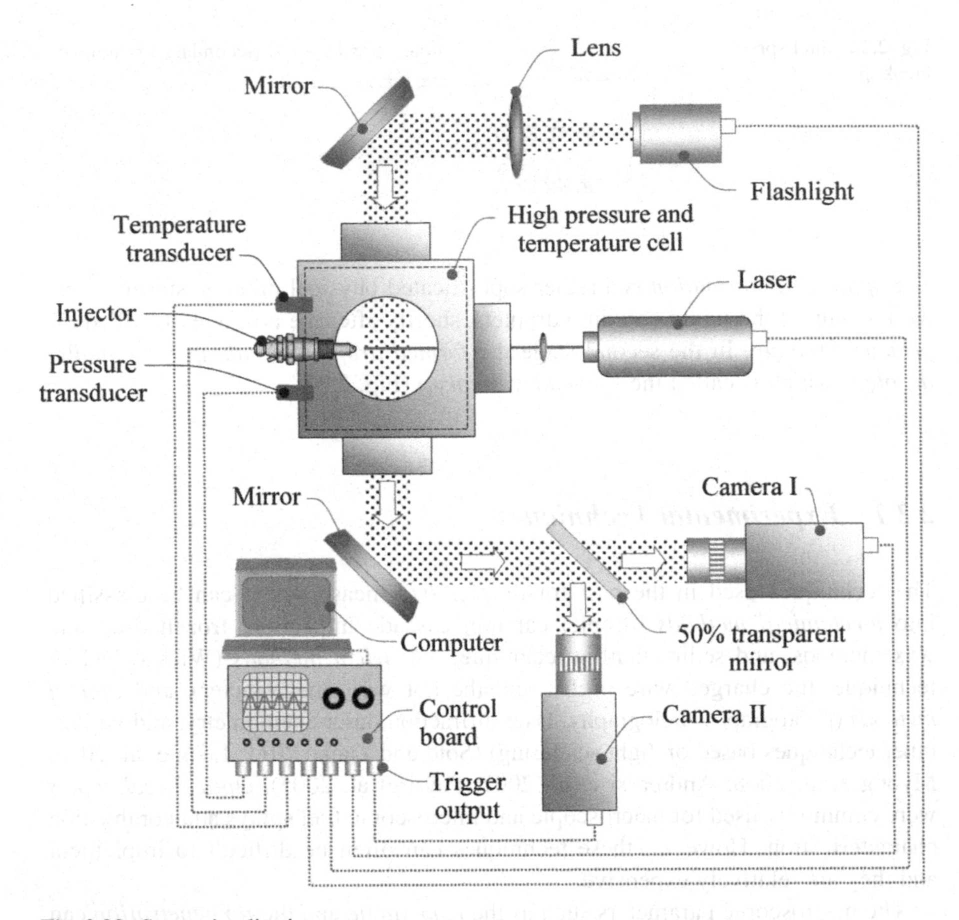

Fig. 2.23 Spray visualization setup

have to be employed to synchronize the tracer injection, fuel injection, laser firing, and image capturing timings. The image capturing timing and exposure time are determined by the laser shot timings and shot durations.

2.2.2 *Mathematical Modeling and Simulation*

Diesel fuel sprays have always been a challenge for fluid modelers since these sprays typically consist of a very large number of droplets. Each droplet has unique properties and is subject to complex interactions that are a function of those properties. Till now, most of the strategies, which were formulated over the years to address this problem, fall into *Eulerian–Eulerian* or *Eulerian–Lagrangian* type formulations.

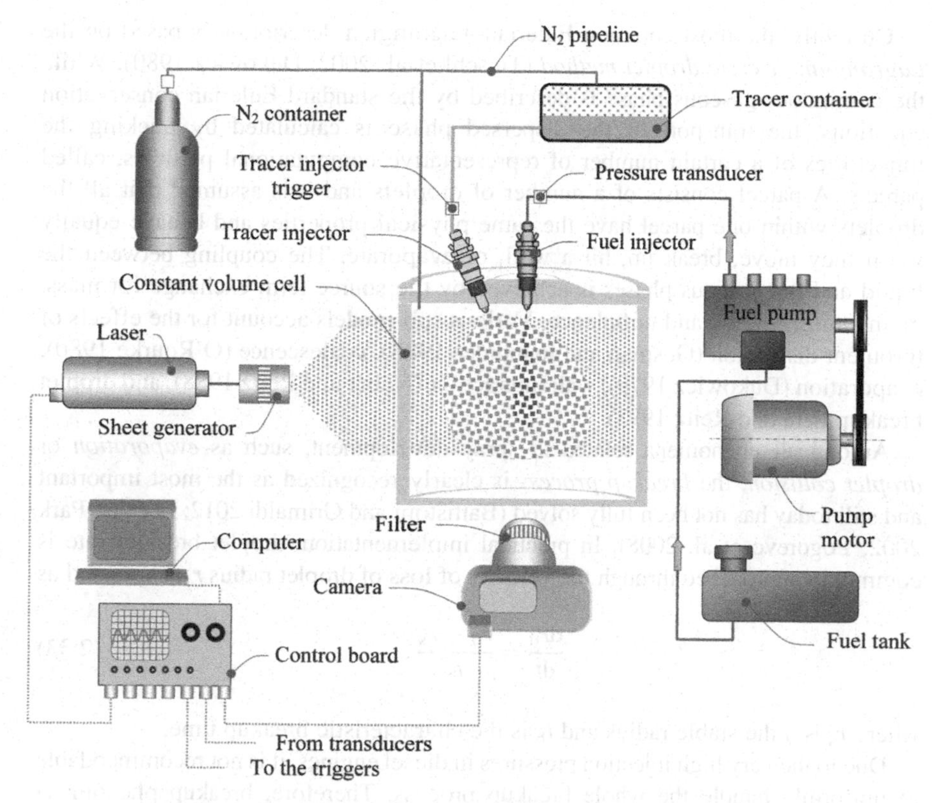

Fig. 2.24 Experimental apparatus for LIF-PIV

The *Eulerian–Eulerian* formulation uses for the spray a two-fluid model and solves the Eulerian field equations for both the liquid and the gaseous phase in order to obtain the liquid phase penetration. In the literature, the relative simplicity and grid independence are often listed as the main advantages of this approach. However, this approach cannot model easily the interaction between phases. Furthermore, the gas and the liquid velocities are assumed to be the same and the turbulence in the liquid phase is assumed to follow the gas phase turbulence (Iyer et al. 2002; Tatschl et al. 2002; Gidaspow 1994).

The *Eulerian–Lagrangian* formulation uses the Lagrangian approach for the liquid phase by modeling droplets as discrete particles. The positions of these particles are tracked in space and time by solving the Newton's equations of motion. The gaseous phase is treated in an Eulerian manner. In the literature, relative easy modeling of the interaction between phases by using sub-models for atomization, drop dispersion, collision, coalescence, breakup, and vaporization are often listed as the main advantages of this approach. The limitations are related to grid dependence, computational time, and memory requirements (Dukowicz 1980; Sazhina et al. 2000).

Currently, the most common Eulerian–Lagrangian description is based on the *Lagrangian discrete droplet method* (Tatschl et al. 2002; Dukowicz 1980). While the continuous gaseous phase is described by the standard Eulerian conservation equations, the transport of the dispersed phase is calculated by tracking the trajectories of a certain number of representative computational particles, called parcels. A parcel consists of a number of droplets and it is assumed that all the droplets within one parcel have the same physical properties and behave equally when they move, break up, hit a wall, or evaporate. The coupling between the liquid and the gaseous phases is achieved by the source term exchange for mass, momentum, energy, and turbulence. Various sub-models account for the effects of turbulent dispersion (Gosman and Ioannides 1983), coalescence (O'Rourke 1980), evaporation (Dukowicz 1979), wall interaction (Naber and Reitz 1988), and droplet breakup (Liu and Reitz 1993).

Among all phenomena related to spray development, such as *evaporation* or *droplet collision*, the *breakup process* is clearly recognized as the most important and still today has not been fully solved (Battistoni and Grimaldi 2012; Lee and Park 2002; Pogorevc et al. 2008). In practical implementation, droplet breakup rate is commonly approached through the velocity of loss of droplet radius r_d, expressed as

$$\frac{\mathrm{d}r_\mathrm{d}}{\mathrm{d}t} = \frac{r_\mathrm{d} - r_\mathrm{s}}{t_\mathrm{b}}, \tag{2.33}$$

where r_s is a the stable radius and t_b is the characteristic breakup time.

Due to the very high injection pressures in diesel engines, it is not recommendable to uniformly handle the whole breakup process. Therefore, breakup phenomena occurring near the nozzle exit, where the initial liquid column is disintegrated into ligaments or droplets, are accounted for by *primary breakup* models, whereas prediction of further breakup process, starting from initial drops, is handled by *secondary breakup* models (Fig. 2.25) (Battistoni and Grimaldi 2012; Lee and Park 2002; Pogorevc et al. 2008).

In the KH–RT hybrid model, it is assumed that primary breakup mainly occurs by KH instability while RT instability causes the secondary breakup as shown in Fig. 2.25. In region A, a droplet is detached from the intact liquid core by the KH instability. Once the droplet is detached from the intact sheet, the secondary breakup occurs by the competing KH and RT instabilities in the region B.

The *primary diesel breakup model* (KH model—Kelvin–Helmholtz model) is based on the rate approach and aims at capturing the combination of the aerodynamic mechanism, modeled through the wave approach, and the turbulence mechanism with included cavitation effects. The liquid breakup length L, (Fig. 2.25) is calculated by

$$L = C_3 \times d_\mathrm{nh} \times We^{0.32}, \tag{2.34}$$

where C_3 is the constant, d_nh is the nozzle hole diameter, and We is the Weber number of the fuel.

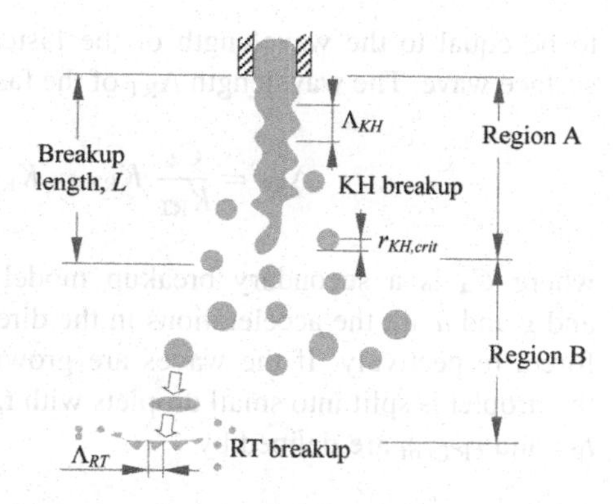

Fig. 2.25 Primary and secondary breakup models

During the breakup, the droplet radius r_{d} reduces to the critical radius $r_{\mathrm{KH,crit}}$ with uniform rate. In this case, the critical droplet radius $r_{\mathrm{KH,crit}}$ and the primary breakup time t_{KH} are given by

$$r_{\mathrm{KH},crit} = C_1 \times \Lambda_{\mathrm{KH}} \; \Lambda_{\mathrm{KH}} = \Lambda_{\mathrm{KH}}(r_{\mathrm{d}}, Z, T, We)$$
$$t_{\mathrm{KH}} = \frac{3.726 \times C_2 \times r_{\mathrm{d}}}{\Lambda_{\mathrm{KH}} \times \Omega_{\mathrm{KH}}} \; \Omega_{\mathrm{KH}} = \Omega_{\mathrm{KH}}(r_{\mathrm{d}}, Z, T, We, \sigma, \rho_f) \tag{2.35}$$

where C_1 and C_2 are the primary breakup model constants, Ω_{KH} and Λ_{KH} are the frequency and the corresponding wavelength of the fastest growing KH wave, r_{d} is the droplet radius, ρ_f is the fuel density, σ_f is the fuel surface tension, Z is the Ohnesorge number, T is the Taylor number, and Re is the Reynolds number:

$$We = \frac{\rho_a \times r_{\mathrm{d}} \times v_{\mathrm{rel}}^2}{\sigma_f},$$
$$T = Z\sqrt{We},$$
$$Z = \frac{\sqrt{We}}{Re}, \tag{2.36}$$
$$Re = \frac{\rho_f}{\mu_f} \times r_{\mathrm{d}} \times v_{\mathrm{rel.}}$$

Here, v_{rel} is the relative velocity between the droplet and ambient gas, μ_f is the fuel viscosity, and ρ_a is the ambient gas density.

Because the primary breakup occurs with uniform radius reduction rate, the new droplet radius r_{new} can be calculated as

$$\frac{r_{\mathrm{d}} - r_{\mathrm{new}}}{dt} = \frac{r_{\mathrm{d}} - r_{\mathrm{KH,\,crit}}}{t_{\mathrm{KH}}}. \tag{2.37}$$

The *secondary diesel breakup model* (RT model—Rayleight–Taylor model) is based on the wave model. Here, the initial size of the droplet diameter is supposed to be equal to the wavelength of the fastest growing or most probable unstable surface wave. The wavelength Λ_{RT} of the fastest growing wave can be calculated as

$$\Lambda_{\mathrm{RT}} = \frac{C_4}{K_{\mathrm{RT}}} \quad K_{\mathrm{RT}} = K_{\mathrm{RT}}(\rho_a, \rho_f \, \sigma_f, g, a), \tag{2.38}$$

where C_4 is a secondary breakup model constant, K_{RT} is the wave number, and g and a are the accelerations in the direction of travel due to gravity and drag force, respectively. If the waves are growing longer than the breakup time t_{RT}, the droplet is split into small droplets with radius $r_{\mathrm{RT,crit}}$. In the secondary breakup, t_{RT} and $r_{\mathrm{RT,crit}}$ are defined by

$$\begin{aligned} r_{\mathrm{RT,crit}} &= \frac{C_4}{K_{\mathrm{RT}}}, \\ t_{\mathrm{RT}} &= \frac{C_5}{\Omega_{\mathrm{RT}}} \quad \Omega_{\mathrm{RT}} = \Omega_{\mathrm{RT}}(\rho_a, \rho_f \, \sigma_f, g, a), \end{aligned} \tag{2.39}$$

where C_5 is a secondary breakup model constant, and Ω_{RT} is the frequency of fastest growing wave.

Usually, the fuel injected into the cylinder is initially assumed to form a liquid column that travels at a speed equal to the fuel injection speed until the fuel breakup time elapses. If one assumes that the flow through each nozzle is quasi-steady, incompressible, and one dimensional, the mass flow rate $\dot{m}_{\mathrm{inj}}$ of fuel injected through the nozzle is given by (Rad et al. 2010; Heywood 1988):

$$\dot{m}_{\mathrm{inj}} = C_{\mathrm{D}} A_{\mathrm{n}} \sqrt{2\rho_f \Delta p}, \tag{2.40}$$

where C_{D} is the discharge coefficient, A_{n} is the needle cross section area, ρ_f is fuel density, and Δp is the pressure difference between injection pressure p_{inj} and ambient pressure p_a (in-cylinder pressure).

For the conditions in which no cavitation occurs, the discharge coefficient C_{D} can be calculated from

$$C_{\mathrm{D}} = \frac{1}{\sqrt{C_i + f \dfrac{l_{\mathrm{nh}}}{d_{\mathrm{nh}}} + 1}}, \tag{2.41}$$

where c_i is the inlet loss coefficient, f is the friction coefficient, l_{nh} is the nozzle length, and d_{nh} is nozzle hole diameter. The friction coefficient can be calculated as

$$f = 0.3164 \times Re^{-0.25}. \tag{2.42}$$

Under the conditions with cavitation, the discharge coefficient C_D should be computed as

$$C_D = \sqrt{\frac{K}{2.6874 - 11.4 \dfrac{r_{nh}}{d_{nh}}}}, \tag{2.43}$$

where r_{nh} is the nozzle filet radius and K is the cavitation number, calculated as

$$K = \frac{p_{inj} - p_v}{\Delta p}, \tag{2.44}$$

where p_v is the saturated vapor pressure of the fuel.

Spray characteristics can also be calculated by using more or less empirical equations, developed from the experimental results. In this way, the first estimate of the fuel spray can be obtained.

Spray penetration before breakup can be calculated as (Rad et al. 2010; Heywood 1988):

$$L_p = v_{inj} t \; 0 < t < t_b, \tag{2.45}$$

where t_b is the breakup time and the fuel injection velocity v_{inj} at the nozzle tip can be expressed as

$$v_{inj} = C_D \sqrt{\frac{2\Delta p}{\rho_f}}. \tag{2.46}$$

The *spray penetration after breakup* can be obtained as

$$L_p = 2.95 \sqrt{\left(\frac{\Delta p}{\rho_a}\right)^{0.5} d_{nh} t} \;\; t \geq t_b, \tag{2.47}$$

where ρ_a is the air density.

The breakup time can be calculated as

$$t_b = 4.351 \frac{\rho_f \, d_{nh}}{C_D^2 \sqrt{\rho_a \Delta p}}. \tag{2.48}$$

The *spray penetration length* L_p can also be estimated independently of the breakup time by the Lustgarten equation as (Kegl et al. 2008):

$$L_p = 2 \times d_{nh}^{0.46} \times \left(\sqrt{\frac{2}{\rho_f}(p_{inj} - p_a)}\right)^{0.54} \times \left(\frac{\rho_f}{\rho_a}\right)^{0.23} \times t^{0.54}. \tag{2.49}$$

The *spray angle* can be obtained as (Rad et al. 2010; Heywood 1988):

$$\theta = \tan^{-1}\left(\frac{4\pi\left(\sqrt{\frac{3\rho_a}{\rho_f}}\right)}{18 + 1.68\frac{l_{\mathrm{nh}}}{d_{\mathrm{nh}}}}\right). \tag{2.50}$$

The *droplet size distribution* within a particular range is usually expressed in terms of the *Sauter mean diameter* d_{32}. The Sauter mean diameter is defined as the volume of the droplet divided by its surface area. It can be computed by using the following empirical formulas (Rad et al. 2010; Heywood 1988):

$$d_{32} = \max\left(d_{32,\mathrm{LS}}; d_{32,\mathrm{HS}}\right), \tag{2.51}$$

$$\frac{d_{32,\mathrm{LS}}}{d_{\mathrm{nh}}} = 4.12\, Re^{0.12} We^{-0.75}\left(\frac{\mu_f}{\mu_a}\right)^{0.54}\left(\frac{\rho_f}{\rho_a}\right)^{0.18}, \tag{2.52}$$

$$\frac{d_{32,\mathrm{HS}}}{d_{\mathrm{nh}}} = 0.38\, Re^{0.12} We^{-0.75}\left(\frac{\mu_f}{\mu_a}\right)^{0.37}\left(\frac{\rho_f}{\rho_a}\right)^{-0.47}, \tag{2.53}$$

where μ_f is the dynamic viscosity of fuel, μ_a is the dynamic viscosity of air, Re the Reynolds number, and We the Weber number. Furthermore, d_{32} can also be estimated by many other simplified expressions. One of these is the well-known Hiroyasu and Kadota formula (Aigal 2003; Kegl et al. 2008):

$$d_{32} = 2.39 \times 10^3 \left(p_{\mathrm{inj}} - p_a\right)^{-0.135} \times \rho_a \times q_{\mathrm{inj}}^{0.131}, \tag{2.54}$$

where q_{inj} [mm^3/stroke] is the fuelling.

The number of drops in a particular zone can be determined, if the d_{32} and the mass of injected fuel are known. Using an energy and mass balance for a single droplet in each zone, ordinary differential equations can be set up for the rate of change of the droplet temperature and diameter, which can be subsequently solved using a suitable numerical method.

2.3 Engine Performance, Ecology, and Economy Characteristics

Among all engine performances, the *engine power* and *engine torque* are the most important. The most important ecology characteristics are related to all *harmful emissions*, such as NO_x, CO, unburned hydrocarbons (HC), particulate materials (PM), smoke, and noise. Among the economy characteristics the *specific fuel consumption* and various *tribology parameters* related to wear, deposits, and lubrication phenomena will be considered.

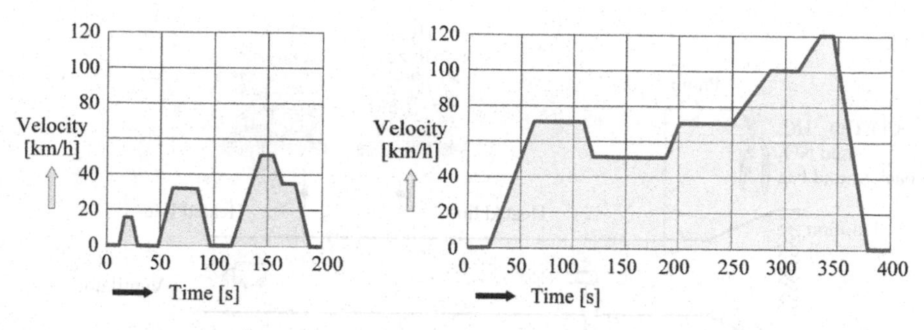

Fig. 2.26 ECE + EUDC test

2.3.1 Experimental Techniques

Nowadays, the experimental techniques related to engine power and torque are very well established and can be regarded as relatively straightforward. Therefore, in this chapter emphasis will be put onto experimental techniques related to ecology and economy.

In order to determine engine ecology characteristics, an engine is typically run under several prescribed operating conditions, termed the *engine cycle*. In practice, various steady-state and transient engine cycles are used. For example, a *transient test cycle FTP-75* for cars and light duty trucks is used for emission certification testing of cars and light duty trucks in the USA. In Japan, the urban *driving cycle JC08* is used for emission and fuel economy measurements of light duty vehicles. Furthermore, a combined *chassis dynamometer test ECE+EUDC* (Fig. 2.26) is used for emission certification of light duty vehicles (passenger cars and light commercial vehicles) in Europe.

In *transient testing* (Fig. 2.27), a constant volume sampling system is typically used to obtain the mass emission of exhaust gas components. A constant volume sampling system dilutes the entire exhaust gas with clean air to produce a constant flow. A fixed proportion of the entire diluted exhaust gas is collected in a sample bag to obtain a diluted exhaust gas that is representative for the average concentration while the engine runs. The diluted exhaust gas can also be measured continuously to obtain the average concentration during run. Mass emissions of exhaust components are obtained from the entire diluted exhaust gas flow and the average concentrations of the pollutant in the diluted gas.

A steady-state test ESC cycle for truck and bus engines is used for emission certification of heavy duty diesel engines in Europe. In the *steady-state test ESC cycle*, the importance of an individual operation mode is determined by the corresponding weighting factor. The weighting factors for all modes in percent are given in Fig. 2.28.

In *steady-state testing* (Fig. 2.29), the exhaust gas from the engine is directly collected without dilution. The mass emissions of the pollutants are obtained during the test from the quantity of exhaust gas flow, which is obtained from the air intake

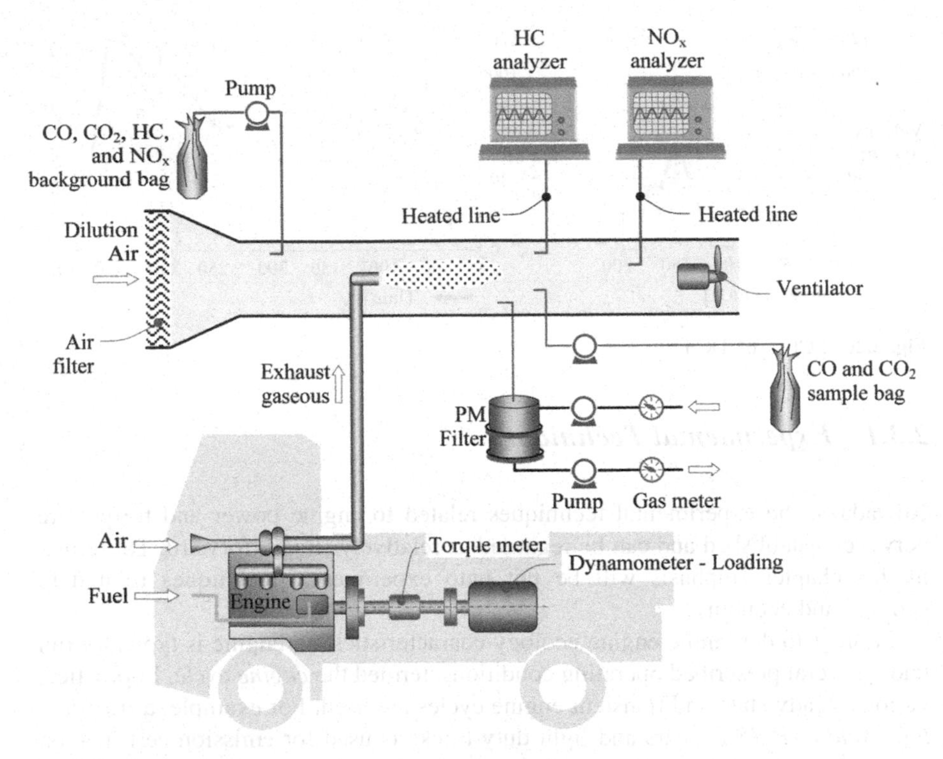

Fig. 2.27 System configuration for engine transient testing

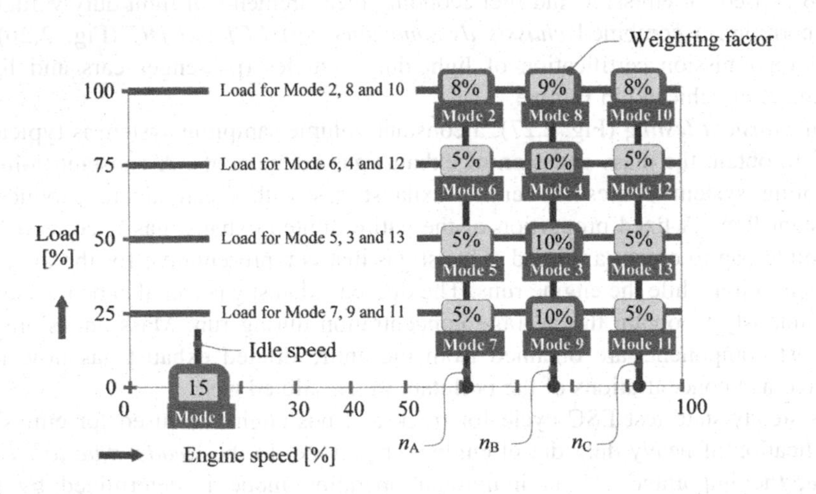

Fig. 2.28 ESC test, 13 mode cycle, weighting factors

flow and fuel flow, and the average concentrations of exhaust gas components. Because this direct measurement does not dilute the exhaust gas, this method is better than dilution measurements for measuring low concentration exhaust components.

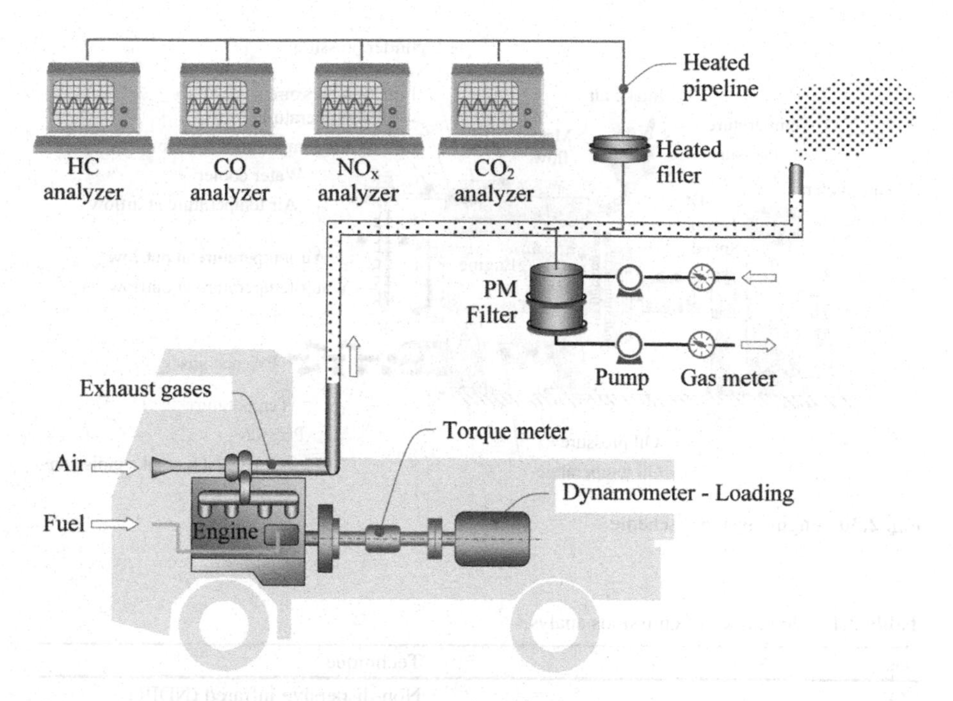

Fig. 2.29 System configuration of engine steady-state testing

A schematic diagram of the *engine test bed* is presented in Fig. 2.30 (Kegl 2008). The engine test bed consists of an engine and dynamometer, air flow rate meter, fuel consumption dynamic measuring system, analyzers of HC, NO_x, O_2, and CO, and smoke meter. By using a data acquisition system, the instantaneous pressure in the fuel high pressure tube, the instantaneous pressure in the cylinder, and the temperatures of fuel, ambient air, intake air, cooling water at inflow and outflow of the engine, oil, and exhaust gases are measured also (Fig. 2.30).

Adequate software has to be used to build the computer applications for data acquisition, data analyses, and control algorithms. These applications are used to control the operation of the data acquisition and for data logging and post-processing.

The usual techniques of emissions analysis are presented in Table 2.1 (Plint and Martyr 1995).

The *engine tribology characteristics* can be investigated by examining pump plunger surface, carbon deposits on injector and in combustion chamber, and nozzle discharge coefficient (Pehan et al. 2009). The *fuel lubrication* ability can be investigated by examining the alterations in pump plunger surfaces by using an electron microscope and by examining the *roughness parameters*. The *carbon deposits* in the combustion chambers can be examined using *endoscopic inspection*. The *deposits* in the *injector nozzle hole* can be estimated indirectly through *discharge coefficient* measurements.

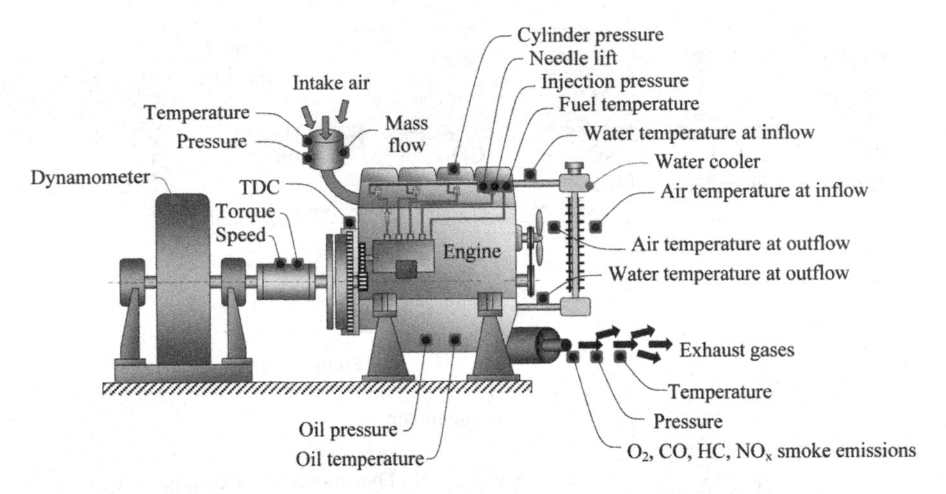

Fig. 2.30 Engine test bed scheme

Table 2.1 Techniques of emissions analysis

Gas	Technique
CO	Non-dispersive infrared (NDIR)
CO_2	Non-dispersive infrared (NDIR)
NO_x	Chemiluminescence
Unburned hydrocarbons, HC	Flame ionization detector (FID), Fast FID
Hydrocarbon species, CH_4	Fourier transform infrared (FTIR)
Particulate matter	Gravimetric analysis
Smoke	Filter method

The critical components for tribological assessment are presented schematically in Fig. 2.31.

To determine the alterations on *pump plunger surfaces* due to usage of various fuels, an *electron microscope* can be used. The surfaces have to be examined before and after fuel usage. To determine the influence of fuel usage on the *roughness* of the *pump plunger active surface*, the relevant roughness parameters have to be examined. The arithmetic roughness average R_a, the quadratic roughness average R_q, the maximum peak to valley height R_y, and the average peak to valley height R_z can be calculated on the basis of mechanical scanning of the pump plunger skirt surface and its head by a suitable device, for example, by the rotating head device PURV 3–100. It should be noted that a desired sliding surface (Fig. 2.32) exhibits a quasi-planar plateau separated by randomly spaced narrow grooves.

Carbon deposits on the injector and in the combustion chamber can be examined by *endoscopic inspection*. For this purpose an adequate videoscope system such as the OLYMPUS of type IPLEX SA with optical adapter IV76-AT120D/NF can be used.

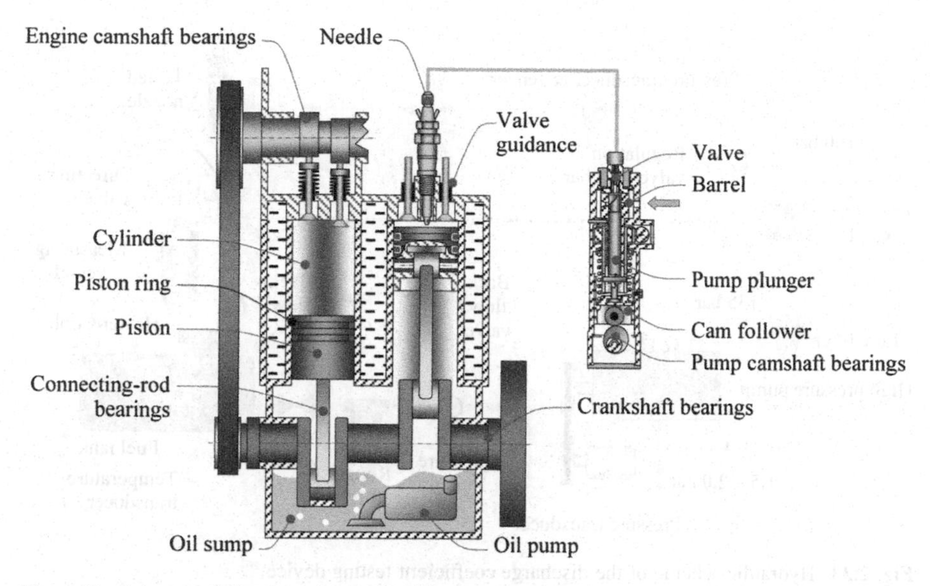

Fig. 2.31 Scheme of diesel engine with critical components for tribological assessment

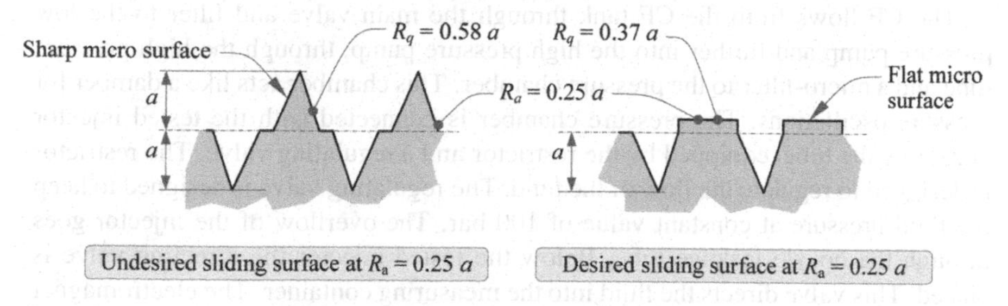

Fig. 2.32 Undesired and the desired sliding surfaces with equal roughness R_a

The fuel *flow coefficient* measuring device is intended to measure the fuel flow coefficients at steady-state conditions. The commonly used procedure is known as the Bosch procedure for measuring flow coefficients. The hydraulic scheme of the testing device is presented in Fig. 2.33.

The testing device consists of the calibrating fuel (CF) tank, filters, low pressure pump and high pressure pump (each pump is driven by its own electric motor), pressure chamber (to damp the pressure waves), restrictor, pressure regulating valve, directing valve (to direct the fuel flow into the desired measuring container), and nozzle holder for proper positioning of the testing nozzle. The needle position in the nozzle is calibrated by using a micrometer. Two pressure transducers for measuring the pressures before and after the high pressure pump are used. During the measurement, the temperature of the CF is regulated by a heat exchanger that is mounted into the fuel tank. For the calculation of fuel density, the fuel temperature in the measuring container is obtained by a thermoelement.

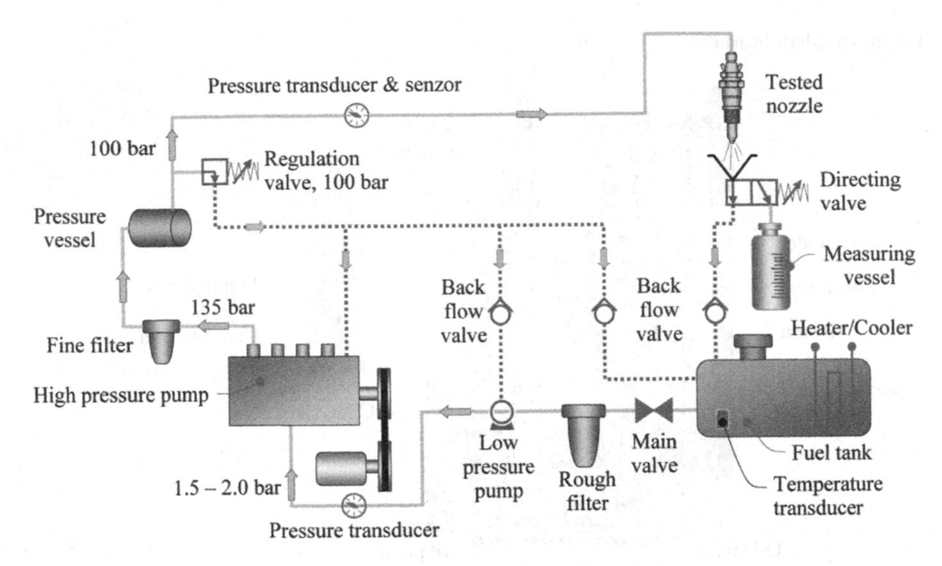

Fig. 2.33 Hydraulic scheme of the discharge coefficient testing device

The CF flows from the CF tank through the main valve and filter to the low pressure pump and further into the high pressure pump, through the high pressure tube and a micro-filter to the pressure chamber. This chamber acts like a damper for pressure oscillations. The pressure chamber is connected with the tested injector nozzle by the tube, equipped by the restrictor and a regulating valve. The restrictor is designed to regulate the flow of the fluid. The regulating valve is designed to keep the fluid pressure at constant value of 100 bar. The overflow of the injector goes through the nozzle leakage tube. Below the tested injector the directing valve is placed. This valve directs the fluid into the measuring container. The electromagnet of the directing valve is charged by the time switch, which can be precisely adjusted for the required duration of measurement. From the measured fuel quantity, the discharge coefficient can be calculated.

At the beginning of the measuring procedure, the required needle lift is set up by using a micrometer (Pehan et al. 2009). The fluid flows into the measuring container for some specified time interval (e.g., 30 s, 60 s,. . .) and at some specified pressure difference (e.g., 99 bar). By using the restrictor and the regulation valve, the pressure difference Δp is set to the required value. By using the time switch, the timer is started and the directing valve is activated, directing the fluid into the measuring container. After the prescribed time interval, the fluid flows through the injector back to the CF tank. On the basis of the fluid quantity in the measuring container and the time of measurement, the actual and the theoretical volume flows are calculated. The nozzle discharge coefficient μ is defined as the ratio between the measured (actual) volume flow $\dot{V}_{\mathrm{m}}$ and theoretical volume flow $\dot{V}_{\mathrm{t}}$, injected through the nozzle, i.e., $\mu = \frac{\dot{V}_{\mathrm{m}}}{\dot{V}_{\mathrm{t}}}$. According to the Bernoulli equation, the theoretical outflow velocity through one nozzle hole can be derived from the nozzle hole diameter d_{nh}, the pressure difference Δp, and fuel density ρ_f as $\dot{V}_{\mathrm{t}} = \frac{\pi \times d_{\mathrm{nh}}^2}{4}\sqrt{\frac{2 \times \Delta p}{\rho_f}}$. In general, the nozzle discharge coefficient has an important influence on the fuel spray characteristics and on injection characteristics (Kegl 2006).

2.3.2 *Mathematical Modeling and Simulation*

Computational fluid dynamics (CFD) modeling is becoming an attractive alternative for engine analysis in place of full experimental test stand study in recent years. It is broadly used by engine researchers to explore in-cylinder flow fields, heat transfer, combustion characteristics, and emission formation processes (Ismail et al. 2011; Gunabalan et al. 2010). The modeling of an internal combustion engine represents one of the most challenging fluid mechanics problems due to the compressible nature of the flow with large density variations, low Mach number, and turbulent, unsteady, cyclic, nonstationary, and nonuniform flow (Ismail et al. 2011). For the simulation of in-cylinder turbulent flow, a widely used Reynolds-averaged Navier–Stokes approach with RNG k–ε turbulence model is utilized as it accounts for high swirling flows (e.g., AVL FIRE and KIVA software).

In diesel engines, the combustion process can be categorized into two main phases, namely, the *premixed combustion* and *mixing-controlled combustion*. Premixed combustion is a rapid combustion process which heavily depends on the rate of atomization, breakup, vaporization, and mixing of the fuel droplets as well as chemical reactions. On the other hand, mixing-controlled combustion depends on the rate of turbulent mixing of fuel and air to form combustible mixture. A *non-premixed model* with the probability density function approach is used to model diesel combustion process for turbulent diffusion flames with fast chemistry (Ismail et al. 2011; Gunabalan et al. 2010).

Accurate and efficient numerical analysis is a key factor for successful determination of combustion characteristics and their improvements. For the steady-state performance prediction of diesel engines, there are several commercial packages (e.g., AVL BOOST) that are widely accepted, rigorously tested, and verified in practical applications.

The underlying mathematical models are based on the first law of thermodynamics. Basically, for the high pressure cycle this law states that the change of the internal energy in the cylinder $\left(\frac{d(m_c \times u)}{d\alpha}\right)$ is equal to the sum of piston work $\left(-p_c \times \frac{dV}{d\alpha}\right)$, the conversion of chemical energy to the thermal energy $\left(\frac{dQ_f}{d\alpha}\right)$, heat transfer $\left(-\sum \frac{dQ_w}{d\alpha}\right)$, and the enthalpy flow due to blow-by $\left(-h_{\mathrm{BB}} \times \frac{dm_{\mathrm{BB}}}{d\alpha}\right)$. For an internal combustion piston engine, it can be written as (Kegl 2011):

$$\frac{d(m_c \times u)}{d\alpha} = -p_c \times \frac{dV}{d\alpha} + \frac{dQ_f}{d\alpha} - \sum \frac{dQ_w}{d\alpha} - h_{\mathrm{BB}} \times \frac{dm_{\mathrm{BB}}}{d\alpha}, \tag{2.55}$$

where the symbol α denotes the angle of the crankshaft rotation, m_c is the mass of the mixture in the cylinder, u is the specific internal energy, p_c is the in-cylinder pressure, V is the cylinder volume, Q_f is the fuel energy, Q_w the heat transfer through the liner, and h_{BB} and m_{BB} are the enthalpy and mass of the mixture that escapes through the gap between the piston and the liner, respectively.

With the in-cylinder gas pressure p_c and the working displacement V_{D} of one piston, the indicated mean effective pressure p_i can be determined over the whole cycle duration as follows: $p_i = \frac{1}{V_{\mathrm{D}}} \int_{\mathrm{CD}} p_c dV$. Together with the gas equation:

$$p_c = \left(\frac{1}{V}\right) m_c R_o T_c, \tag{2.56}$$

where T_c is the in-cylinder gas temperature, equation (2.55) can be solved by using, for example, the Runge–Kutta method. Once the in-cylinder gas temperature is known, the in-cylinder gas pressure can be obtained from the gas equation.

To approximate the heat release characteristics of an engine, the Vibe function can be used:

$$x = 1 - e^{-ay(m+1)} \tag{2.57}$$

where x is the fraction of the fuel mass, which was burned since the start of combustion, $y = (\alpha - \alpha_o)/\Delta\alpha_c$, a is the Vibe parameter (a is usually equals 6.9 for complete combustion), m is the shape parameter (m usually equals 0.85), α is the angle of the crankshaft rotation, α_o is the angle of the combustion start, and $\Delta\alpha_c$ is the angle of combustion duration.

The actual heat release can be determined as

$$\frac{dQ_f}{d\alpha} = Q_{\mathrm{in}} \left(\frac{dx}{d\alpha}\right), \tag{2.58}$$

where Q_{in} is the total heat input.

Furthermore, for the modeling of the heat transfer in the cylinder, the heat transfer coefficient α_w can be determined by using the heat transfer model Woschni, given by

$$\alpha_w = 130 D^{-0.2} p_c^{0.8} T_c^{-0.53} \left(C_1 c_{\mathrm{m}} + C_2 \left(\left(V_{\mathrm{D}} \times T_{c,1} \right) \Big/ \left(p_{c,1} \times V_{c,1} \right) \left(p_c - p_{c,o} \right) \right) \right)^{0.8}, \tag{2.59}$$

where D is the cylinder bore, c_{m} is the mean piston velocity, $p_{c,o}$ is the pressure of pure compression, $p_{c,1}$ and $T_{c,1}$ are the in-cylinder pressure and temperature (at the moment when the intake valve closes), and C_1 and C_2 are given constants.

The coefficient α_p of the heat transfer through the intake, $\alpha_{p,i}$, and exhaust, $\alpha_{p,e}$, ports can be described by the Zapf formulas for the intake and exhaust side, respectively, as

$$\alpha_{p,i} = \left(C_7 + C_8 T_u - C_9 T_u^2\right) \times T_u^{0.33} \times \dot{m}^{0.68} \times d_{vi}^{-1.68}(1 - 0.765(h_v/d_{vi})),$$
$$\alpha_{p,e} = \left(C_4 + C_5 T_u - C_6 T_u^2\right) \times T_u^{0.44} \times \dot{m}^{0.5} \times d_{vi}^{-1.5}(1 - 0.797(h_v/d_{vi})), \tag{2.60}$$

where C_4, C_5, C_6, C_7, C_8, and C_9 are given constants, $\dot{m}$ is the mass flow, T_u is the temperature on the port input side, and d_{vi} and h_v are the valve seat diameter and the valve lift, respectively.

2.4 Discussion

Injection and *fuel spray characteristics* relate strongly to *engine performance*, *ecology*, *economy*, and *tribology characteristics*. All the most important injection characteristics like injection pressure, injection timing, injection duration, injection rate, and fuelling history, as well as fuel spray characteristics like spray tip penetration, spray angle, and Sauter mean diameter, determine mixture formation, in-cylinder pressure and temperature, self-ignition, and heat release. The engine combustion process is further strongly related to the engine power, torque, emissions, specific fuel consumption, carbon deposits, wear of engine parts, and so on. Unfortunately, the possibilities to control the combustion process directly are quite limited. In a mechanically controlled fuel injection system, even the injection process cannot be controlled directly. However, we can control injection indirectly through the fuel delivery process. For this reason, it is very important to know as good as possible the relationships among all injection, spray, and engine characteristics. Some of these relations are valid to a great extent for indirect and direct injection systems.

In general, low *injection pressure* enlarges the fuel droplet diameters and increases the *ignition delay* period during the combustion. This situation leads to an increase of the *in-cylinder pressure*, *NO_x*, and *CO emissions*. An increase of injection pressure typically improves the atomization at the nozzle outlet, resulting in a more distributed vapor phase, hence resulting in better mixing (Bruneaux 2001). When injection pressure is increased, fuel droplet diameters will become smaller and the mixing of the fuel with air during the ignition period improves, which leads to lower *smoke* level and *CO emission*. However, if the injection pressure is too high, the ignition delay becomes shorter. This worsens the mixing process and the combustion efficiency decreases. Consequently, smoke levels may rise (Celikten 2003). It has to be pointed out that the objective should be to increase the *mean injection pressure* to some reasonable maximum level. The maximum injection pressure is decisive for the mechanical loading of the fuel injection pump's components and drive. In a mechanically controlled fuel injection system, the injection pressure increases together with increasing engine speed and load. During the injection process, the maximum injection pressure can be more than double of the mean pressure. Therefore, the maximum fuel injection pressure is also limited by the *strength* of engine materials, *wear* of various elements of the injection system, and engine system *cost*.

At higher mean injection pressure the *injection duration* becomes shorter, if the fuelling is kept constant. In order to keep specific fuel consumption, emissions of smoke, unburned HC, and NO_x at acceptable levels, the injection duration must be adjusted properly to the operating regime and the *start of injection*. A long injection period, caused by low *injection rate*, makes poor *spray tip penetration*. Moreover, if the injection period is too long, it would inevitably produce excessive *smoke* and *particulate* emissions. In this case a pilot injection might be of some benefit (Hwang et al. 1999).

The *injection timing* has a considerable influence on the *start of combustion* of the air–fuel mixture. In general, if the injection timing is advanced, the *in-cylinder temperature* increases, thus leading to an increase of *NO_x emissions*. The retarded injection timing leads to incomplete combustion and to higher unburned *HC emissions* (Jayashankara and Ganesan 2010).

Many studies confirm that the *injection rate history* affects *ignition* and *combustion* characteristics and the temporal history of *smoke* and *NO_x* formation in direct injection diesel engines (Juneja et al. 2004; Desantes et al. 1999; Hwang et al. 1999). Fuel injection quantity should be controlled by engine parameters such as engine speed, load, and air motion in the combustion chamber. In general, the fuel amount at the beginning of injection should be relatively small. After that it should be increased as the piston reaches the TDC and decreased after TDC to prevent too rapid increase of the combustion pressure. Therefore, injection rate modulation devices are being developed mainly to reduce this high initial heat release rate. A modulation device limits the fuel injected during the ignition delay period by a separated pilot injection, prior to the main injection, or by initial injection rate control. If the initial rate of injection is much lower during the main injection, less fuel is injected and mixed by the time the first element of fuel has evaporated to form a suitable mixture and autoignition occurs. Hence the initial heat release rate is low and consequently the combustion noise level is reduced. The main injection follows with most of the fuel injected. Ideally, the degree of separation between the initial and main injections should be longer when the ignition delay is long, for example, when engine is running at low load and low-speed conditions or under transient loads in urban traffic.

In order to improve the efficiency of diesel/biodiesel engine development, the injection, fuel spray, and engine characteristics unavoidably have to be investigated experimentally and numerically. The more or less sophisticated mathematical models addressed briefly in this section have been tested at many operating regimes and with various fuels. On the basis on the *comparison* of the *numerically* and *experimentally obtained results* one can say that these mathematical models may perform quite well. To illustrate this, some results are presented in the following.

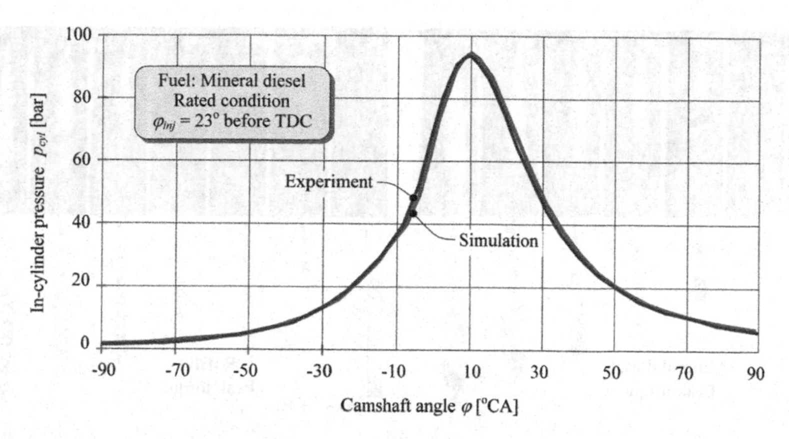

Fig. 2.34 Comparison of experimental and simulated in-cylinder pressure—AVL Boost

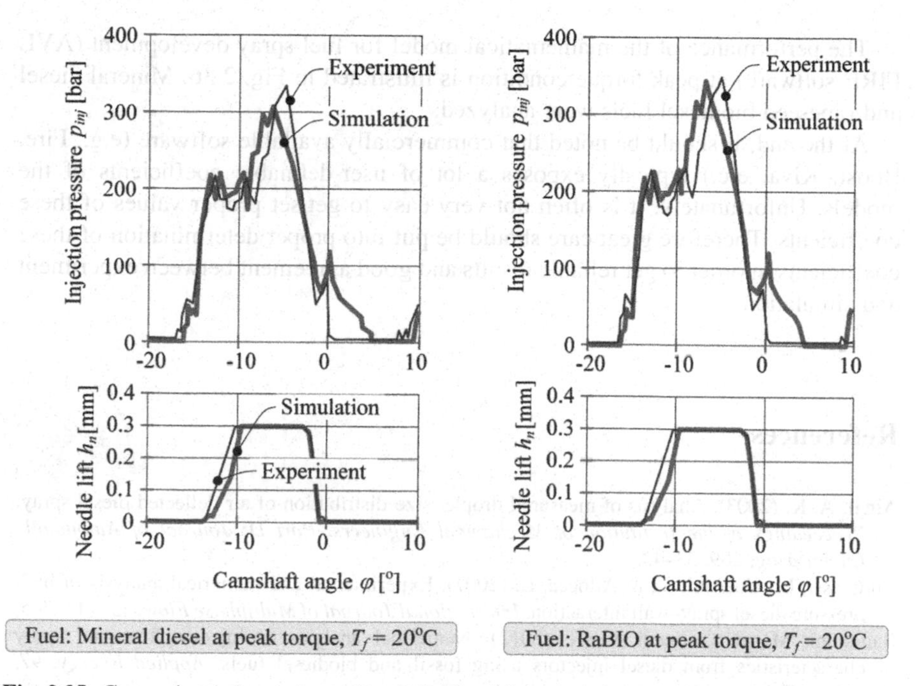

Fig. 2.35 Comparison of experimental and simulated injection characteristics—BKIN

A comparison of the in-cylinder pressure history of a diesel engine at rated conditions is shown in Fig. 2.34 (AVL Boost software).

The performance of the BKIN software is illustrated in Fig. 2.35. The given histories are for the diesel fuel M injection system of a MAN bus engine at peak torque condition. The used fuels are mineral diesel and rapeseed biodiesel.

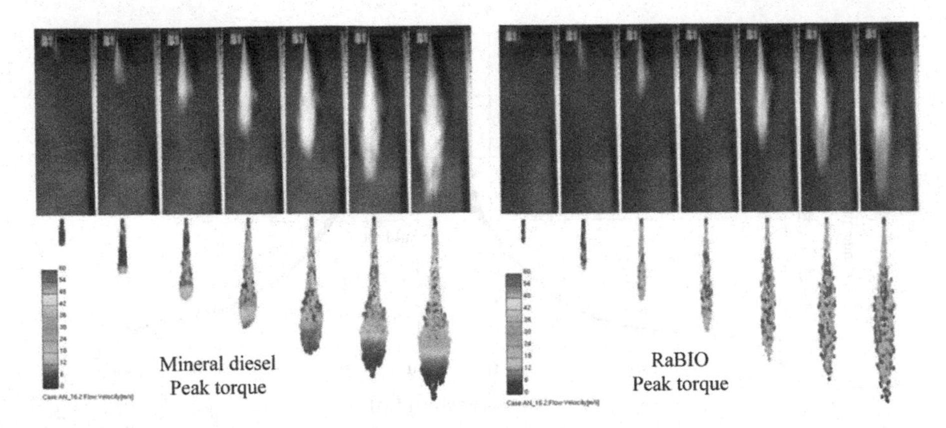

Fig. 2.36 Comparison of experimental and simulated fuel spray development—AVL FIRE

The performance of the mathematical model for fuel spray development (AVL FIRE software) at peak torque condition is illustrated in Fig. 2.36. Mineral diesel and rapeseed biodiesel fuels were analyzed.

At the end, it should be noted that commercially available software (e.g., Fire, Boost, Kiva, etc.) typically exposes a lot of user-definable coefficients of the models. Unfortunately, it is often not very easy to get/set proper values of these coefficients. Therefore great care should be put into proper determination of these coefficients in order to get reliable results and good agreement between experiment and simulation.

References

Aigal, A. K. (2003). Analysis of measured droplet size distribution of air deflected diesel spray. *Proceedings of the Institution of Mechanical Engineers, Part D: Journal of Automobile Engineering, 209*, 33–43.

Andreassi, L., Ubertini, S., & Allocca, L. (2007). Experimental and numerical analysis of high pressure diesel spray–wall interaction. *International Journal of Multiphase Flow, 33*, 742–765.

Battistoni, M., & Grimaldi, C. N. (2012). Numerical analysis of injector flow and spray characteristics from diesel injectors using fossil and biodiesel fuels. *Applied Energy, 97*, 656–666.

Bauer, H. (1999). *Diesel-engine management*. Stuttgart: SAE.

Bruneaux, G. (2001). Liquid and vapour spray structure in high-pressure common rail diesel injection. *Atomization and Sprays, 11*, 533–556.

Celikten, I. (2003). An experimental investigation of the effect of the injection pressure on engine performance and exhaust emission in indirect injection diesel engines. *Applied Thermal Engineering, 23*, 2051–2060.

Desantes, J. M., Arrègle, J., López, J. J., & García, J. M. (2006). Turbulent gas jets and diesel-like sprays in a crossflow: A study on axis deflection and air entrainment. *Fuel, 85*(14–15), 2120–2132.

Desantes JM, Arregle J, Rodriguez PJ (1999) Computational model for simulation of diesel injection systems, *SAE Paper* 1999-01-0915

Desantes, J. M., Benajes, J., Molina, S., & Gonzalez, C. A. (2004). The modification of the fuel injection rate in heavy-duty diesel engines. Part 1: Effects on engine performance and emissions. *Applied Thermal Engineering, 24*, 2701–2714.

Dukowicz, J.K. (1979). *Quasi-steady droplet phase change in the presence of convection*, Los Alamos Report LA-7997-MS

Dukowicz, J. K. (1980). A particle-fluid numerical model for liquid sprays. *Journal of Computational Physics, 35*, 229–253.

Gao, J., Matsumoto, Y., & Nishida, K. (2009a). Experimental study on spray and mixture properties of the group-hole nozzle for direct-injection diesel engines, part I: a comparative analysis with the single-hole nozzle. *Atomization and Sprays, 19*, 321–37.

Gao, J., Matsumoto, Y., & Nishida, K. (2009b). Experimental study on spray and mixture properties of the group-hole nozzle for direct-injection diesel engines, part II: effects of included angle and interval between orifices. *Atomization and Sprays, 19*, 339–55.

Gidaspow, D. (1994). *Multiphase flow and fluidization continuum and kinetic theory descriptions*. Boston, MA: Academic.

Gosman, A. D., & Ioannides, E. (1983). Aspects of computer simulation of liquid-fueled combusters. *Journal of Energy, 7*, 482–490.

Gunabalan, A., Tamilporai, P., & Ramaprabhu, R. (2010). Effect of injection timing and EGR on DI diesel engine performance and emission using CFD. *Journal of Applied Sciences, 10*(22), 2823–2830.

Herzog, P. (1989). The ideal rate of injection for swirl supported diesel engines. IMechE, Diesel Fuel Injection Systems Seminar, Birmingham

Heywood, J. B. (1988). *Internal combustion engines fundamentals*. New York, NY: McGraw Hill.

Hiroyasu H, Arai M (1990) Structures of fuel sprays in diesel engines. *SAE Paper* 900475

Hwang, J.W., Kal, H.J., Kim, M.H., Parkm, J.K., Shenghua, L., Martychenko, A.A., Chae, J.O. (1999). Effect of fuel injection rate on pollutant emissions in DI diesel engine. *SAE Paper* 1999-01-0195

Ishiwata, H., et al. (1994) Recent progress in rate shaping technology for diesel in line pumps. *SAE Paper* 940194

Ismail, H. M., Hg, H. K., & Gan, S. (2011). Evaluation of non-premixed combustion and fuel spray models for in-cylinder diesel engine simulation. *Applied Energy, 90*(1), 271–279.

Iyer, V. A., Abraham, J., & Magi, V. (2002). Exploring injected droplet size effects on steady liquid penetration in a diesel spray with a two-fluid model. *International Journal of Heat and Mass Transfer, 45*(3), 519–531.

Jayashankara, B., & Ganesan, V. (2010). Effect of fuel injection timing and intake pressure on the performance of a DI diesel engine—A parametric study using CFD. *Energy Conversion and Management, 51*, 1835–1848.

Juneja H, Ra Y, Reitz RD (2004) Optimization of injection rate shape using active control of fuel injection. *SAE Paper* 2004-01-0530

Kegl, B. (1995). Optimal design of conventional in-line fuel injection equipment. *Proceedings of the Institution of Mechanical Engineers, Part D: Journal of Automobile Engineering, 209*, 135–141.

Kegl, B. (1996). Successive optimal design procedure applied on conventional fuel injection equipment. *Journal of Mechanical Design, 118*(4), 490–493.

Kegl, B. (1999). A procedure for upgrading an electronic control diesel fuel injection system by considering several engine operating regimes simultaneously. *Journal of Mechanical Design, 121*(1), 159–165.

Kegl, B. (2004). Injection system optimization by considering fuel spray characteristics. *Journal of Mechanical Design, 126*, 703–710.

Kegl, B. (2006). Experimental investigation of optimal timing of the diesel engine injection pump using biodiesel fuel. *Energy Fuels, 20*, 1460–1470.

Kegl, B. (2008). Effects of biodiesel on emissions of a bus diesel engine. *Bioresource Technology, 99*, 863–873.

Kegl, B. (2011). Influence of biodiesel on engine combustion and emission characteristics. *Applied Energy, 88*, 1803–1812.

Kegl, B., & Hribernik, A. (2006). Experimental analysis of injection characteristics using biodiesel fuel. *Energy fuels, 20*(5), 2239–2248.

Kegl, B., Kegl, M., & Pehan, S. (2008). Optimization of a fuel injection system for diesel and biodiesel usage. *Energy & Fuels, 22*, 1046–1054.

Kegl, B., & Müller, E. (1997). Design of cam profile using Bézier's curve. *Journal of Mechanical Engineering, 43*(7–8), 281–288.

Lee, C. S., & Park, S. W. (2002). An experimental and numerical study on fuel atomization characteristics of high-pressure diesel injection sprays. *Fuel, 81*, 2417–2423.

Leng, X., Feng, L., Tian, J., Dua, B., Long, W., & Tian, H. (2010). A study of the mixture formation process for a third-generation conical spray applied in HCCI diesel combustion. *Fuel, 89*, 392–398.

Liu AB, Reitz RD (1993) Modeling the effects of drop drag and breakup on fuel sprays, *SAE Paper* 930072

Merker, G.P., Schwarz, C., Stiesch, G., Otto, F. (2005). *Simulating combustion.* Springer

Moon, S., Matsumoto, Y., Nishida, K., & Gao, J. (2010). Gas entrainment characteristics of diesel spray injected by a group-hole nozzle. *Fuel, 89*, 3287–3299.

Myong, K. J., Suzuki, H., Senda, J., & Fujimoto, H. (2008). Spray inner structure of evaporating multi-component fuel. *Fuel, 87*, 202–210.

Naber JD, Reitz RD (1988) Modeling engine spray/wall impingement, *SAE Paper* 880107

Needham J (1990) Injection timing and rate control—a solution for low emissions. *SAE Paper* 900854

O'Rourke PJ (1980) Modeling of drop interaction in thick sprays and a comparison with experiments, IMechE—Stratified Charge Automotive Engines Conference

Payri, R., García, J. M., Salvador, F. J., & Gimeno, J. (2005). Using spray momentum flux measurements to understand the influence of diesel nozzle geometry on spray characteristics. *Fuel, 84*, 551–561.

Pehan, S., Svoljšak-Jerman, M., Kegl, M., & Kegl, B. (2009). Biodiesel influence on tribology characteristics of a diesel engine. *Fuel, 88*, 970–979.

Plint, M., & Martyr, A. (1995). *Engine testing theory and practice*. Manchester: Butterworth-Heinemann.

Pogorevc, P., Kegl, B., & Škerget, L. (2008). Diesel and biodiesel fuel spray simulations. *Energy & Fuels, 22*, 1266–1274.

Rad, G. J., Gorjiinst, M., Keshavarz, M., Safari, H., & Jazayeri, S. A. (2010). An investigation on injection characteristics of direct injected heavy duty diesel engine by means of multi-zone spray modeling. *Oil and Gas Science and Technology—RevIFP Energies nouvelles, 65*(6), 893–901.

Sazhina, E. M., Sazhin, S. S., Heikal, M. R., Babushok, V. I., & Johns, R. J. R. (2000). A detailed modelling of the spray ignition process in diesel engines. *Combustion Science and Technology, 160*(1–6), 317–344.

Soid, S. N., & Zainal, Z. A. (2011). Spray and combustion characterization for internal combustion engines using optical measuring techniques—A review. *Energy, 36*, 724–741.

Tatschl R, Künsberg Sarre C, Berg E (2002) IC-Engine spray modeling – status and outlook, International Multidimensional Engine Modeling User's Group Meeting at the SAE Congress 2002.

Zhang, G., Qiao, X., Miao, X., Hong, J., & Zhen, J. (2012). Effects of highly dispersed spray nozzle on fuel injection characteristics and emissions of heavy-duty diesel engine. *Fuel, 102*, 666–673.

Chapter 3
Guidelines for Improving Diesel Engine Characteristics

Diesel engine characteristics depend significantly on the engine type. But, even for a given engine type, the engine characteristics can still be varied in a wide range in dependence on *engine management*, *exhaust gas after treatment*, and usage of *alternative fuels* (Fino et al. 2003; Gray and Frost 1998; Maiboom et al. 2008; Peng et al. 2008; Stanislaus et al. 2010; Twigg 2007) (Fig. 3.1). Engine management and alternative fuels usage offer a possibility to *reduce* the *formation* of harmful emissions. On the other hand, exhaust gas after treatment techniques enable a *reduction* of harmful emissions *already produced* by the engine.

3.1 Engine Management

Diesel engine management is mostly related to the fuel injection process. Modern fuel injection systems must in general fulfill the following requirements:

- *High pressure capability* and *injection pressure control*
- *Flexible timing control*
- *Injection rate control*

Increased *injection pressure* generally contributes to decreased fuel droplet size and improved combustion, resulting in a reduction of smoke emission. Low injection pressure, on the other hand, is required to reduce noise at idling and in the very low load range. In other words, optimum injection pressure must be determined in accordance with engine load and speed. Regarding the fuel consumption, it was found out that some improvement might be achieved by increasing the injection pressure. More precisely, the fuel should be injected with low pressure at the initial phase of injection and later with high pressure. According to above requirements, a fuel injection system should have a wide *pressure controllability*.

Injection timing influences to a great extent the combustion process and is especially important for the control of nitrogen oxidant (NO_x) and particulate matter (PM) emissions. Flexible injection–timing control, which considers both,

B. Kegl et al., *Green Diesel Engines*, Lecture Notes in Energy 12,
DOI 10.1007/978-1-4471-5325-2_3, © Springer-Verlag London 2013

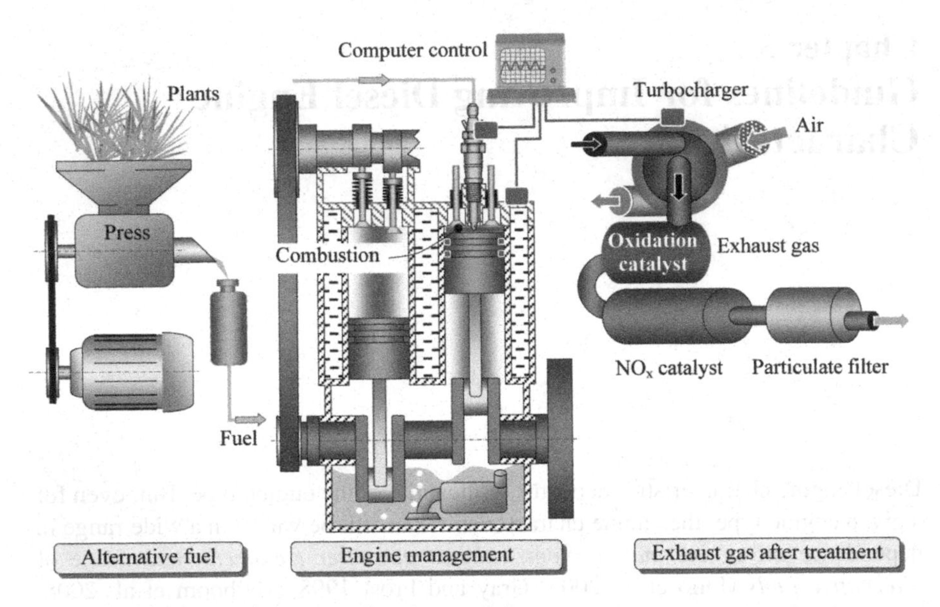

Fig. 3.1 Techniques to improve engine characteristics

engine speed and load, is therefore a highly desired feature of the injection system. The *injection rate* history variations can also be used to achieve reasonable NO_x–PM trade-offs. Besides of the NO_x and PM emission, the injection rate history also affects fuel consumption.

Once the injection characteristics are optimally matched to each other, they typically result in a fuel spray with small droplet size (good atomization), long tip penetration, and narrow spray angle. These spray characteristics play an important role in the efforts to improve engine performance and to reduce fuel consumption and harmful emissions. This is especially true for the fuel atomization (Kook et al. 2008).

To reduce harmful emissions in a wide engine operating range, multiple fuel injections can be used. Multiple fuel injections along with high fuel injection pressures represent an effective way of diesel engine combustion improvements (Kim et al. 2008; Wang et al. 2009). These are the main reasons for the development of electronically controlled high pressure fuel injection systems like the common rail system shown in Fig. 3.2. The common rail system basically consists of a supply pump, common rail, injectors, an ECU that controls these components, and sensors (Fig. 3.2).

The fuel pressure in the common rail is controlled by a solenoid valve that controls the fuel volume from the supply pump. This pressure is detected by a sensor in the common rail and set accordingly to the engine speed and load. This enables high pressure injection also at low speed regimes, allowing fuel to be well atomized at all operating regimes. Consequently, it is possible to reduce significantly the black smoke, being characteristic for diesel engines.

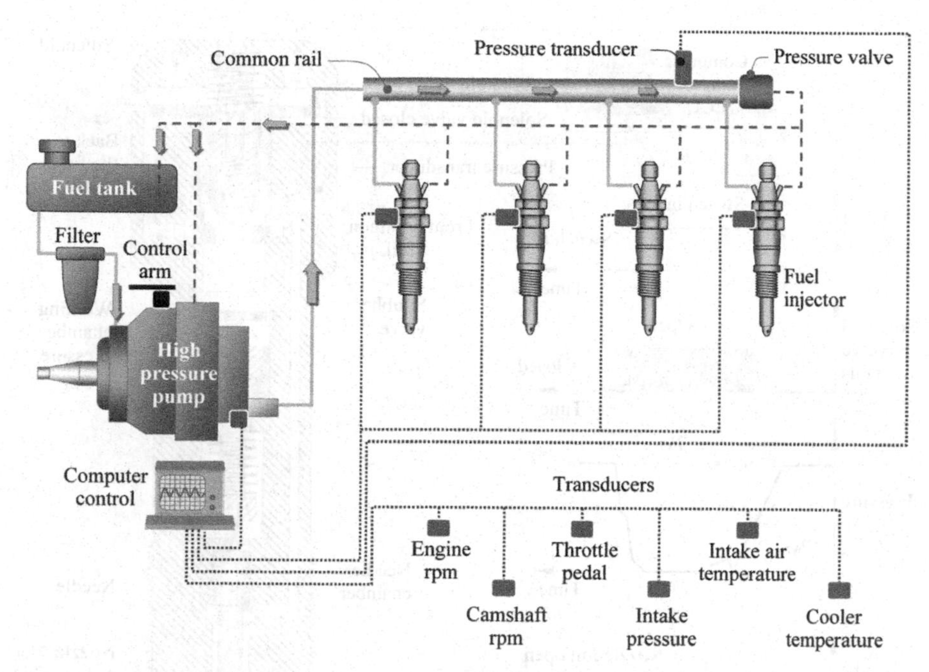

Fig. 3.2 Common rail diesel fuel injection system

The injection process in a common rail system can be roughly described as follows. The *injection volume* and *timing* are controlled by opening and closing of the solenoid valve (Fig. 3.3). When the solenoid valve is closed, the pressures in the working chamber and in the needle chamber are equal to the *common rail fuel pressure* and the needle is in the "closed" position. When the solenoid valve opens, the pressure in the working chamber falls and the common rail pressure in the needle chamber lifts the needle. This starts the injection. As the solenoid valve closes, the working chamber is pressurized again. This causes the needle to return to the "closed" position and terminates the injection. That means that the whole process is controlled by the solenoid valve, which, in turn, is controlled by the ECU. This means, for example, that by adding a power pulse to the solenoid valve prior to the main injection, one can easily get a pilot injection. This can be used to shorten the ignition delay and suppresses premixed combustion to address NO_x, combustion noise, and vibration reduction and to improve startability and fuel efficiency.

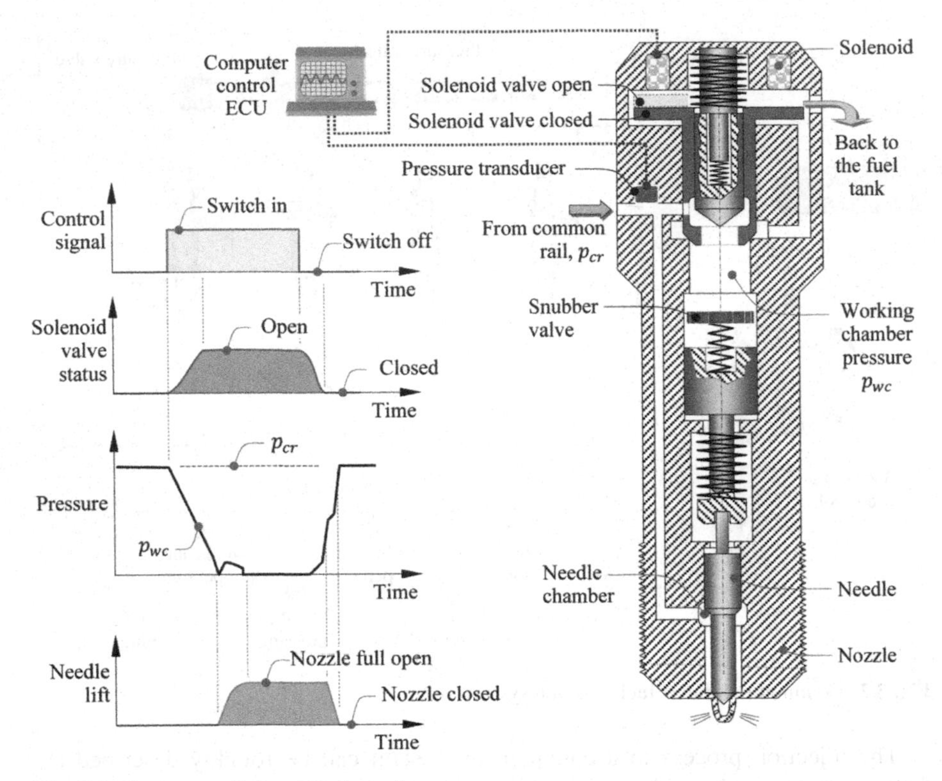

Fig. 3.3 Common rail diesel fuel nozzle

3.2 Exhaust Gas After Treatment

The exhaust of a diesel engine combustion process mainly contains the following harmful emissions: HC, CO, NO_x, and particulate matter (PM). Therefore, diesel engines should be equipped with *HC* and *CO oxidation control systems*, *lean–NOx control systems*, and *PM control*. In practice, these systems are typically *catalytic convertors* and *particulate filters* (Fig. 3.4) (Fino et al. 2003; Gray and Frost 1998; Peng et al. 2008; Stanislaus et al.2010; Twigg 2007). Additionally, *exhaust gas recirculation* may be successfully employed to reduce harmful emissions.

The most common of these devices are *catalytic convertors*. Physically, a catalytic convertor is an open monolithic structure that allows easy flow of gases, with the active catalyst coating applied to the channel walls of the monolith (Fig. 3.5).

The catalyst is typically platinum, palladium, rhodium, or some alloy. These catalysts can eliminate HC, CO, and NO_x emissions, and partially reduce particulates (Gray and Frost 1998; Sitshebo et al. 2009; Twigg 2007). The convertor oxidizes carbon monoxide and hydrocarbons to carbon dioxide and water and simultaneously reduces nitrogen oxides to nitrogen.

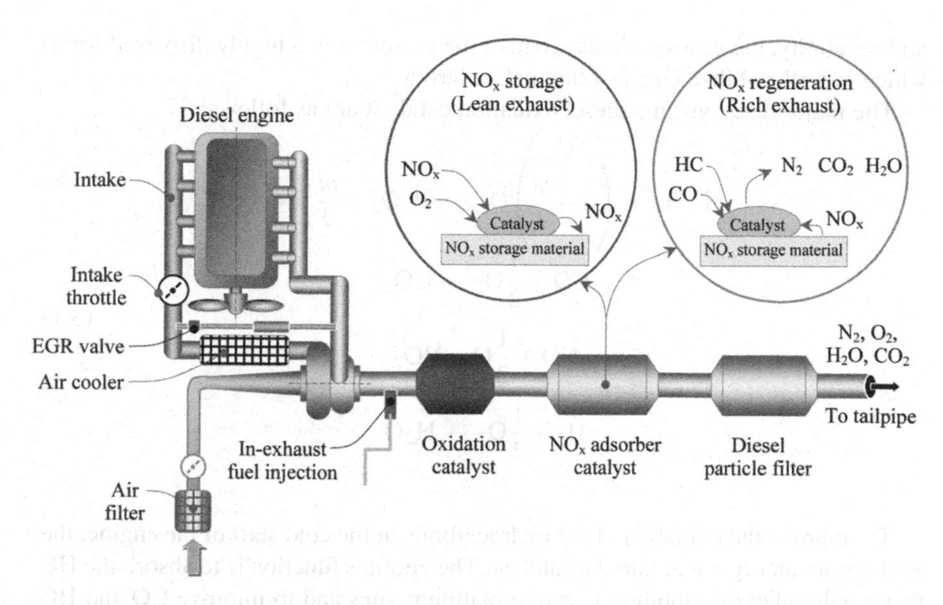

Fig. 3.4 Exhaust gas after treatment

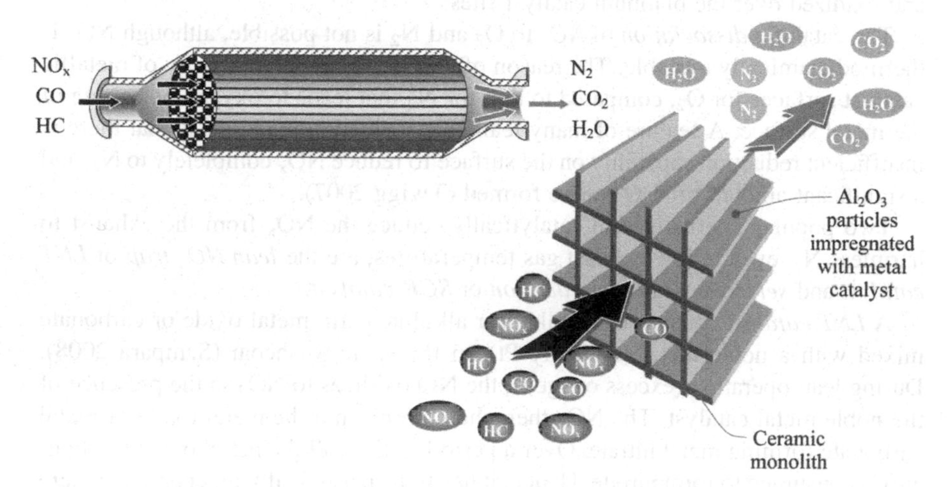

Fig. 3.5 Catalytic converter

Diesel catalytic converters are *oxidation catalyst* and *SCR catalyst* (Alkemade and Schumann 2006; Sitshebo et al. 2009; Forzatti et al. 2010). The diesel *oxidation catalyst* removes up to 90 % of carbon monoxide and hydrocarbons. The *SCR catalyst* removes NO_x.

The primary function of the *oxidation catalyst* is to oxidize the HC and CO produced during advanced combustion modes. For this purpose, the platinum-based oxidation catalysts are typically used (Twigg 2007). To achieve the performance and durability, the catalyst should contain the platinum in a highly dispersed form, which is well stabilized against thermal sintering.

The major reactions in a diesel oxidation catalyst are as follows:

$$\begin{aligned} &C_nH_m + \left(n + \frac{m}{4}\right) O_2 \rightarrow n\,CO_2 + \frac{m}{2}\,H_2O \\ &CO + \frac{1}{2} O_2 \rightarrow CO_2 \\ &NO + \frac{1}{2} O_2 \rightleftharpoons NO_2 \\ &H_2 + \frac{1}{2} O_2 \rightarrow H_2O \end{aligned} \tag{3.1}$$

To improve the oxidation of the hydrocarbons at the cold start of the engine, the zeolites are incorporated into the catalyst. The zeolites function is to absorb the HC that would otherwise inhibit the active platinum sites and to improve CO and HC oxidation performance. At higher temperatures, the HC is desorbed from the zeolite and oxidized over the platinum catalyst sites.

The catalytic *dissociation* of NO_x to O_2 and N_2 is not possible, although NO_x is thermodynamically unstable. The reason of this lies in the high affinity of metallic catalyst surfaces for O_2, compared to that for N_2 that leads to oxygen poisoning of the metal surface. A feature of many lean NO_x reduction reactions is that there is insufficient reduction capability on the surface to reduce NO_x completely to N_2, and a significant amount of N_2O can be formed (Twigg 2007).

Two popular alternatives to catalytically reduce the NO_x from the exhaust to harmless N_2, even at low exhaust gas temperatures, are the *lean* NO_x *trap* or *LNT catalyst* and *selective catalytic reduction* or *SCR catalyst*.

A *LNT catalyst* consists of an alkali or alkaline earth metal oxide or carbonate mixed with a noble metal (typically Pt) on the same washcoat (Sampara 2008). During lean operation (excess oxygen), the NO oxidizes to NO_2 in the presence of the noble metal catalyst. The NO_2 then chemisorbs onto the metal oxide or metal carbonate forming metal nitrate. Over a period of time all the metal oxide (carbonate) is consumed to form nitrate. Hence, it has to be periodically regenerated, where the nitrate decomposes back to metal oxide (carbonate).

The desirable reactions during lean operation are

$$\begin{aligned} 2NO + O_2 &\rightarrow 2NO_2 \\ NO_2 + mCO_3(mO) &\rightarrow mNO_3 + CO_2 \end{aligned} \tag{3.2}$$

Undesirable reactions during lean operation are

$$\begin{aligned} 2SO_2 + O_2 &\rightarrow 2SO_3 \\ SO_3 + mCO_3(mO) &\rightarrow mSO_4 + CO_2 \end{aligned} \tag{3.3}$$

and the desirable reactions during rich operation are

$$\begin{aligned} 2mNO_3 &\rightarrow 2mO + 2NO + O_2 \\ 2NO + 2CO(HC) &\rightarrow N_2 + 2CO_2 \\ mO + CO_2 &\rightarrow mCO_3 \end{aligned} \tag{3.4}$$

This regeneration is generally obtained by creating a rich environment with high levels of HC or CO or H_2. The NO formed due to the decomposition of the nitrate reacts with the reductant to form N_2. The undesirable reactions under LNT operation are the oxidation of SO_2 to form metal sulfates. NO oxidation to NO_2 occurs on the noble metal between 250 and 450 °C. This reaction is kinetically limited below 250 °C and thermodynamically limited beyond 450 °C. Also, the stability of the nitrate formed is severely limited beyond this temperature. There are two important problems with using an LNT for NO_x reduction. First, the saturated LNT with metal nitrate (mNO_3) should be periodically regenerated by providing a reducing environment which contains high levels of HC or CO or both. This is typically achieved by running the engine rich or by having a secondary fuel injection late in the exhaust stroke. This involves additional fuel penalties and complications with respect to control and design. Second, the sulfur in the fuel generally reacts with the metal carbonate or metal oxide to form metal sulfates (mSO_4). Reducing these back to its original form needs high temperatures of the order of 600 °C.

A number of various strategies have been intensively investigated to find an appropriate *SCR catalyst* of NO_x to N_2 (Kamasamudram et al. 2010; Sampara 2008; Sitshebo et al. 2009). Some of the practical techniques to reduce NO_x can be realized by using:

- Soot particulate
- Ammonia (urea)
- Hydrocarbons over zeolite-based catalysts
- Hydrocarbons over metal oxide catalysts
- Hydrocarbons over multistaged catalysts
- Hydrocarbons over noble metal catalysts

Among these options, the use of ammonia and hydrocarbons over zeolite-based catalyst are the most popular. Zeolite-based catalysts have received much attention due to their high activity and relatively wide temperature window since 1990.

SCR of NO with ammonia (NH_3) under lean conditions is a widely commercialized technology for NO_x removal from stationary sources. Vanadium-based catalysts are commonly used for this application. Sources of NH_3 can be compressed gas, or compounds such as urea ($(NH_2)_2CO$), which readily decompose to give NH_3:

$$(NH_2)_2CO + HO_2 \rightarrow 2NH_3 + CO_2 \tag{3.5}$$

This hydrolysis reaction occurs at temperatures beyond 160 °C. Over an SCR catalyst, NH_3 reacts with NO_x according to the following reactions:

$$\begin{aligned} 4NH_3 + 4NO + O_2 &\rightarrow 4N_2 + 6H_2O \\ 2NH_3 + NO + NO_2 &\rightarrow 2N_2 + 3H_2O \\ 4NH_3 + 2NO_2 + O_2 &\rightarrow 3N_2 + 6H_2O \end{aligned} \tag{3.6}$$

Among the above reactions, the reaction of NH_3 with NO and NO_2 is the most facile and therefore occurs at lower reaction temperatures. Thus, an external source which can provide a $\frac{NO}{NO_2} = \frac{1}{1}$ ratio would be ideal to achieve best SCR performance.

Metal catalysts such as platinum (Pt), copper (Cu), iridium (Ir), and more recently silver (Ag), may be mixed along with the zeolite on the same washcoat. The activity of these catalysts is closely related to the type of zeolite and their structure. As a rule of thumb, zeolite structures with lower acidity lead to smaller carbonaceous deposits, leading to higher NO_x conversions. Furthermore, HCs are trapped in the zeolite structures to increase the local HC concentration.

NO is oxidized to NO_2 at the catalyst surface which contains noble metal. The formed NO_2 then reacts with the trapped HC to produce N_2. The main obstacle to the use of this system is the hydrothermal stability of the zeolite. Furthermore, this method becomes problematic at high temperatures, where HC desorption is significant, which reduces local HC concentration at the catalyst surface.

Particulate filter is a device designed to remove diesel particulate matter or aerosolized diesel exhaust pollution particles. It looks similar to a traditional exhaust silencer, but is mounted closer to the engine. Inside of it there is a complex honeycomb ceramic structure, designed to filter the exhaust gases while minimizing the flow resistance (which would otherwise limit performance) (Fig. 3.6). By forcing the exhaust gases through the walls between the channels of the filter, particulate matter is deposited on the walls.

Diesel particulate filters are ceramic filters, fitted into the exhaust, which trap about 90 % of the exhaust gas particles. The particles are mainly a mixture of soot and ash. Soot is carbon based and is formed during combustion of the fuel; meanwhile ash is metal based and forms when engine oil is burnt.

An example of a particulate filter with a regenerative effect is shown in Fig. 3.7. The exhaust gases enter the filter and go through the porous ceramics piece that holds back the particles. As the particulate matter fills up the filter, the pressure on the intake side of the filter raises. This leads to an increased portion of the residual gases in the cylinder, which affects the fuel–air mixture exchange negatively. When the pressure behind the exhaust valve reaches some critical value, the filter needs to be regenerated (the particles have to be removed by burning). For this regeneration process several systems with automatic electronic control have been developed. Fig. 3.7 shows a system with a burner that needs to be supplied with additional fuel and air. The flame from the burner destroys the particulate matter.

Exhaust gas recirculation (EGR), schematically shown in Fig. 3.8, is widely used to reduce NO_x emission from diesel engines. Effectively, it reduces the combustion flame temperature due to its dilution, thermal, and chemical effects (Peng et al. 2008).

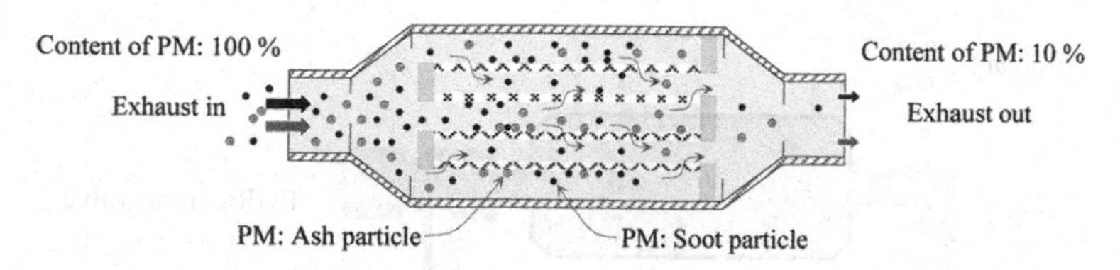

Fig. 3.6 Particulate filter

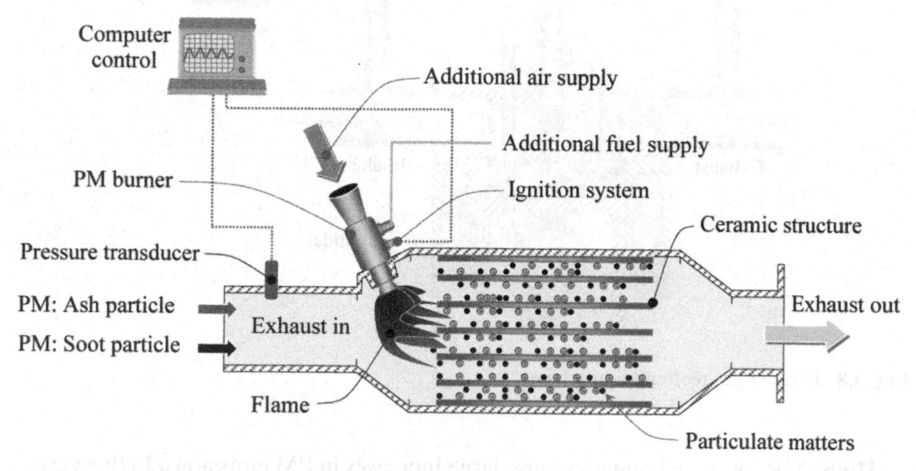

Fig. 3.7 Particulate filter with regenerative effect

The composition of the exhaust gas depends on the engine operating conditions. In the case of a cold start, the exhaust gas comprises a great deal of unburned hydrocarbons, fuel vapor, and products of partial oxidation reaction. In contrast, during normal conditions the exhaust gas contains much more products of a complete combustion. If the exhaust gas is recirculated back into the cylinder, this may have positive or negative effects on the ignition and combustion process. The quantity of recirculated exhaust gas should therefore be carefully determined in dependence of the operating conditions and mass of intake air and fuel. It is usually expressed by ψ_{EGR} [%] which is defined as the mass percent of the recirculated exhaust mass m_{EGR} in the total intake mixture mass m_i (Agarwal et al. 2011):

$$\psi_{\mathrm{EGR}}\,[\%] = \frac{m_{\mathrm{EGR}}}{m_i} \times 100 \tag{3.7}$$

An appropriate amount of EGR can improve cold startability of a diesel engine and promote combustion and emission performance during a cold start. For some operating conditions, EGR may offer a way to drastically reduce NO_x emissions without a significant penalty for the specific fuel consumption and soot emissions. However, under some operating conditions, the opposite effects have also been observed (Maiboom et al. 2008).

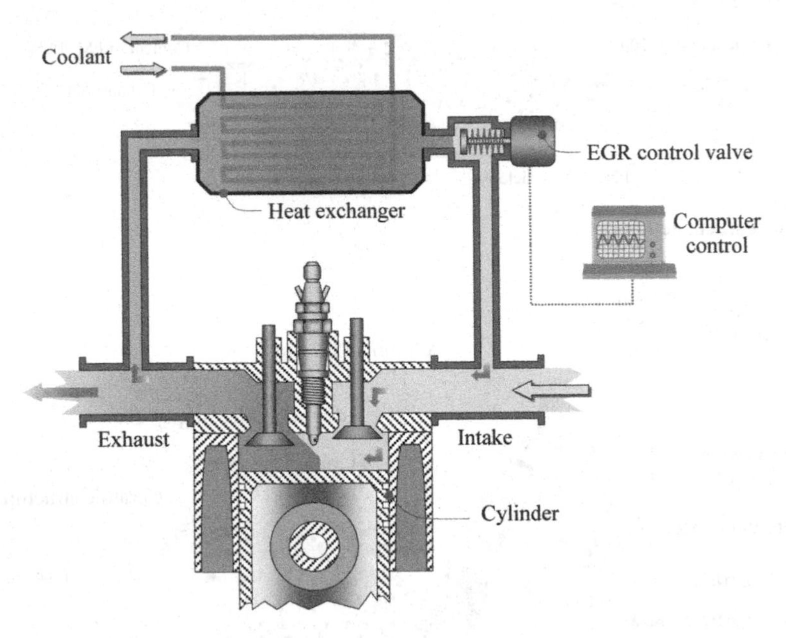

Fig. 3.8 Exhaust gas recirculation scheme

High EGR rates are known to cause large increases in PM emissions. In this view, employing a diesel particulate filter system can be considered to be a favorite solution. This is because a particulate filter provides about 90 % efficiency on reducing PM emissions and thus gives increased flexibility for NO_x control by the EGR.

Two various EGR configurations

- Low Pressure Loop (LPL)
- High Pressure Loop (HPL)

are presented in Figs. 3.9 and 3.10 (Mustel 1997; Zheng et al. 2004).

The LPL EGR (Fig. 3.9) utilize the positive difference between the turbine outlet pressure p_4 and compressor inlet pressure p_1, $p_4 > p_1$. If necessary, the outlet pressure can be elevated by partial throttling to ensure sufficient driving pressure for the EGR flow.

Exhaust is usually taken directly from the filter outlet and lead to the turbo inlet. This is considered to be a clean EGR because the recirculated flow is passing through the particulate filter. It should be noted, however, that conventional compressors and intercoolers are not designed to endure the temperature and fouling of diesel exhausts.

In general, the LPL approach of EGR is not applicable, if the compressor is not designed for exhaust gases. Efforts have also been made to route the exhaust from the turbine outlet to the intercooler outlet directly, bypassing the compressor. Although this circumvents the exhaust fouling problem, an independent EGR pump becomes necessary to counteract the boost pressure. Special EGR pumps

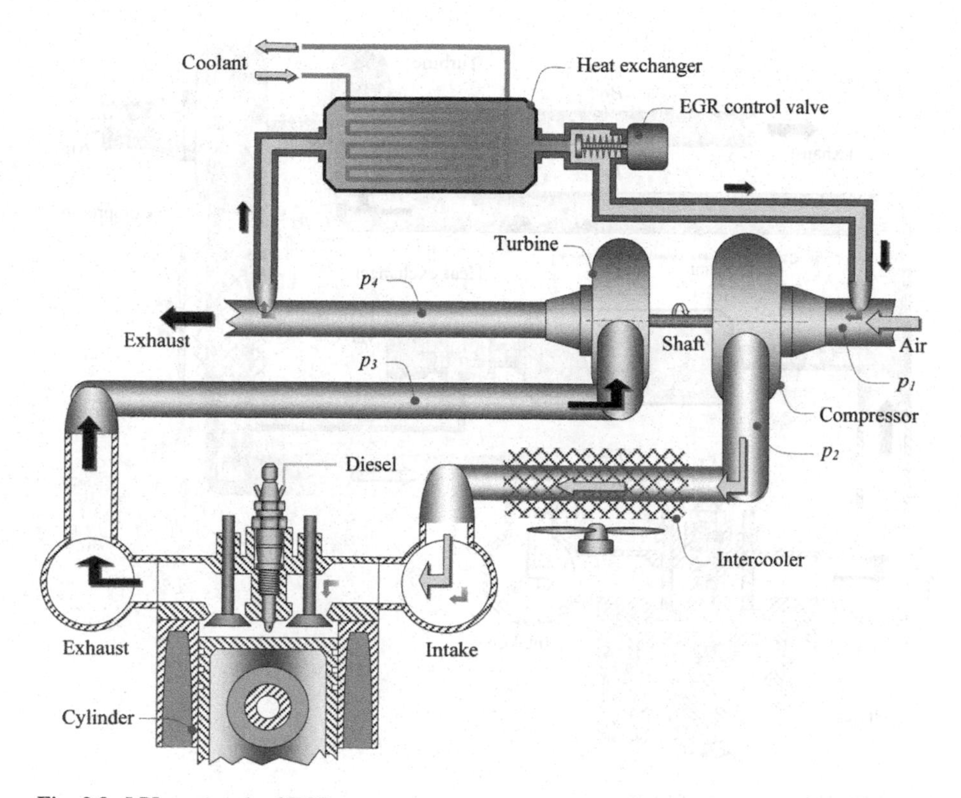

Fig. 3.9 LPL approach of EGR

are needed to withstand the exhaust heat and fouling, in addition to the substantial pumping power requirements.

For a HPL system (Fig. 3.10) it is typical to take the exhaust gases before they enter the turbocharger. After going through the EGR valves, the recirculated exhaust gases are reintroduced in the intake manifold, downstream of the compressor. The compressor and intercooler are, therefore, not exposed to the exhaust. However, such HPL EGR is only applicable when the turbine upstream pressure p_3 is sufficiently higher than the boost pressure p_2, i.e., if $p_3 > p_2$ prevails. In the case that the required pressure difference cannot be met with the original matching between the turbocharger and the engine, remedies must be made by either increasing the turbine upstream pressure or reducing the boost pressure.

Besides of the emissions, EGR may also influence engine performance, carbon deposits, and wear of various parts of a diesel engine (Al-Qurashi et al. 2011; Agarwal et al. 2011; Ghazikhani et al. 2010). EGR displaces oxygen in the intake air by exhaust gas recirculated to the combustion chamber. Exhaust gases lower the oxygen concentration in the combustion chamber and increase the specific heat of the intake air mixture, which results in lower flame temperatures. Reduced oxygen and lower flame temperatures affect performance and emissions of a diesel engine in several ways. Thermal efficiency is slightly increased and specific fuel

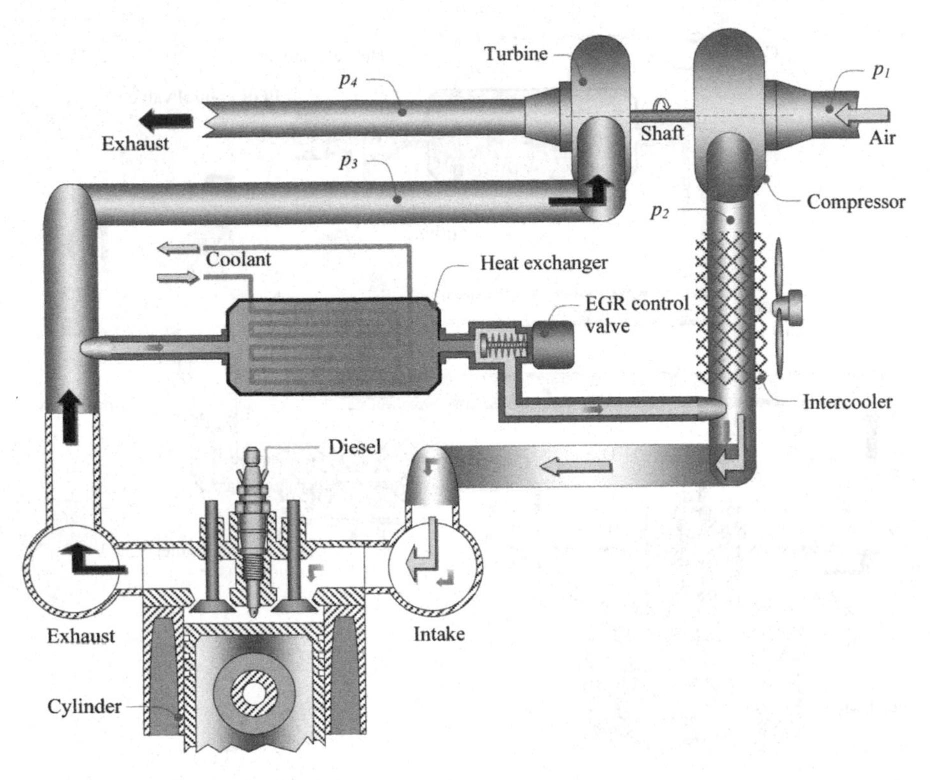

Fig. 3.10 HPL approach of EGR

consumption is decreased at lower loads. But at higher loads, thermal efficiency and specific fuel consumption are almost the same either with or without EGR.

The exhaust gas temperature is decreased with EGR. Hydrocarbons, carbon monoxide, and smoke opacity increase by EGR usage, but NO_x emissions decreases significantly. A 15 % EGR rate is found to be effective to reduce NO_x emission substantially without deteriorating significantly engine performance in terms of thermal efficiency, specific fuel consumption, and emissions. At lower loads, EGR also reduces NO_x without deteriorating performance and emissions. A higher rate of EGR, however, reduces NO_x to a great extent but deteriorates performance and emissions at higher engine loads. Thus, it follows that higher rates of EGR can be applied only at lower loads.

Related to EGR, there is another important point that needs attention. When an engine is equipped with an EGR, increased soot deposits can be expected on the cylinder head, injector tip, and piston crown. Increased wear of the piston rings was also observed. The top compression ring is not influenced significantly, but the wear of the second and the third compression rings may increase notably. The worst wear increase was observed for the oil piston ring. Unfortunately, the EGR may cause a severe wear problem on various engine components such as piston rings, cylinder liner, etc. (Suzuki 1997).

3.3 Alternative Fuels

The usage of conventional mineral diesel fuel is already quite well investigated and the problems related mainly to NO_x and PM emissions are very well known. In the quest to reduce these problems, the usage of alternative fuels may present an interesting option. The evaluation of alternative fuels utilization should include both, emissions and energy considerations. From the emissions point of view, there are of course again trade-offs to be taken into account. For example, fuels with lower PM levels can have higher NO_x or HC emission levels.

In order to evaluate their potential usage in diesel engines, alternative fuels have to be analyzed at least in relation to *cetane number*, *density*, *viscosity*, *cloud point*, *flash point*, *cold filter plugging point*, *fuel corrosiveness*, *lubricity*, *water content*, and *fuel stability*.

The *cetane number* is a measure of the ignition quality of the fuel and affects the combustion process. Up to a certain limit, an increase of the cetane level reduces the NO_x and PM emissions. However, by increasing the cetane number above the level, required for a given engine, may not improve engine performance. The minimum cetane number for mineral diesel fuel is 40.

Fuel density influences directly fuel spray development, specific fuel consumption, engine power, wear, deposits, and exhaust emissions. An increase of fuel density leads to longer spray tip penetration with a narrower spray angle.

The *kinematic viscosity* of fuel is also an important property which impacts the performance of the injection system. Some injection pumps can experience excessive wear and power loss due to injector or pump leakage, if viscosity is too low. If fuel viscosity is too high, it may cause too much pump resistance, filter damage, and adversely affect fuel spray patterns. In general, fuels with low viscosity have poorer lubrication properties. The kinematic viscosity is measured according to the corresponding standard, which defines this property as the resistance to flow of a fluid under gravity at some predefined temperature.

The *cloud point* of fuel is the temperature at which the amount of precipitated wax crystals becomes large enough to make the fuel appear cloudy or hazy. Wax may form because normal paraffins often occur naturally in a fuel. As the temperature of the fuel is lowered, these paraffins become less soluble in the fuel and precipitate out as wax crystals. In some fuel systems, cloud point can indicate the onset of fuel-filter plugging.

The *pour point* is the lowest temperature at which the fuel will flow and is used to predict the lowest temperature at which the fuel can be pumped.

The *cold filter plugging point* (CFPP) is the lowest temperature at which the fuel flow through the filter becomes problematic. This parameter is used to predict the fuel's low temperature operability properties.

The *flash point* is not related directly to engine performance. It has to be verified to meet the safety requirements for fuel handling and storage. The flash point is the lowest fuel temperature at which the vapor above a fuel sample will momentarily ignite under the prescribed test conditions.

The *fuel corrosiveness* properties indicate possible problems with copper, brass, or bronze fuel system components.

The *fuel lubricity* is a very important property, since the diesel fuel injection system relies on the fuel to lubricate the moving parts. If the fuel's lubricating properties are inadequate, this will lead to increased wear on injectors and pumps.

The *water content* in fuel may lead to increased corrosion. During fuel transport and storage, water and sediments can contaminate the fuel. This contamination can contribute to filter plugging and fuel injection system wear.

In order to improve *fuel stability*, adequate additives, fuel mixing, or fuel heating (in case that low temperatures promote gel formation) may be necessary. Unstable fuels can form soluble gums or insoluble organic particulates. Both gums and particulates may contribute to injector deposits, and particulates can clog fuel filters. The formation of gums and particulates may occur gradually during long-term storage or quickly during fuel system recirculation caused by fuel heating.

A number of fuels have been investigated as possible alternatives for diesel engines. Today's most frequently investigated and used alternative fuels are:

- *Water in diesel emulsion*
- *Natural gas* and *liquefied petroleum gas*
- *Methane* and *propane*
- *Dimethyl ether* and *dimethyl carbonate*
- *Fischer–Tropsch diesel*
- *Hydrogen*
- *Alcohols*
- *Vegetable oils*, *bioethanol*, and *biodiesel*

Some alternative fuels can be used pure, while others have to be mixed with mineral diesel. In any case, the diesel engine has to be modified to some extent.

The engines, which use mineral diesel and some gaseous fuel, are referred to as *dual fuel engines*. Natural gas and bio-derived gas appear to be quite attractive alternative fuels for dual fuel engines in view of their environment-friendly nature. In dual fuel gas diesel engine a mixture of air and gaseous fuel is prepared in an external mixing device and compressed in the cylinder. The compressed mixture is then ignited by energy from the combustion of the diesel fuel spray, which is called the *pilot fuel*. The amount of pilot fuel needed for this ignition is between 10 and 20 % of the amount needed for diesel-only operation at normal working loads. This amount varies with the point of engine operation and engine design parameters.

3.3.1 Water in Diesel Emulsion

In the last decade, water addition to mineral diesel before or after injection has been quite intensively investigated (Tesfa et al. 2012; Subramanian 2011; Maiboom and Tauzia 2011; Tauzia et al. 2010; Lif and Homlberg 2006; Armas et al. 2005; Samec et al. 2002; Kadota and Yamasaki 2002; Kegl and Pehan 2001). Namely, it can be

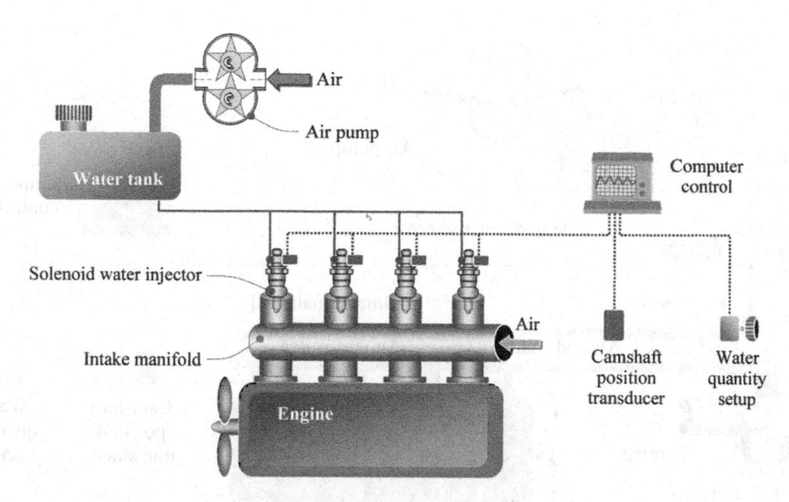

Fig. 3.11 Scheme of a multi-point water injection into the intake manifold

shown that by adding water to mineral diesel, the peak combustion temperatures are reduced which leads to a high reduction of NO_x. Till today, various methods of water addition have been developed, such as:

- *Water injection into the intake manifold*
- *Water injection directly into the combustion chamber*
- *Water injection into emulsion with the fuel*

Water Injection into the Intake Manifold. In order to inject water into the intake manifold, several systems have been investigated till now. The first one is the *multi-point water injection system* (Brusca and Lanzafame 2001; Imahashi et al. 1995; Odaka et al. 1991), which basically consists of the following parts: injectors, electronic control unit, pump, water tank, and pipes (Fig. 3.11).

The water tank is loaded by air pressure, which is controlled by a pressure valve. The injectors are all the time connected with the water tank and directly loaded by water pressure. The electronic control unit (ECU) manages the injectors according to the signals from impulse sensors related to piston position. The injectors have integrated electric coil valves that are activated by an electric signal from the ECU. In the absence of an electric signal from the ECU, the injector valve is closed. The ratio between air and water is adjusted by a potentiometer. The amount of water injected into the air stream depends on pressure that is adjusted by the air pressure in the water tank. The air pressure in the water tank is controlled by an electronic regulator. The injection procedure starts immediately after the intake valve is open and finishes just before the intake valve is closed.

A very cost-effective solution to reduce NO_x emission and thermal loading of the engine is the *mono-point water injection*. In this system, water is added to the air stream with one injector, located just *before* the *turbocharger*. The main advantage of this system is the simplicity of its construction (Fig. 3.12).

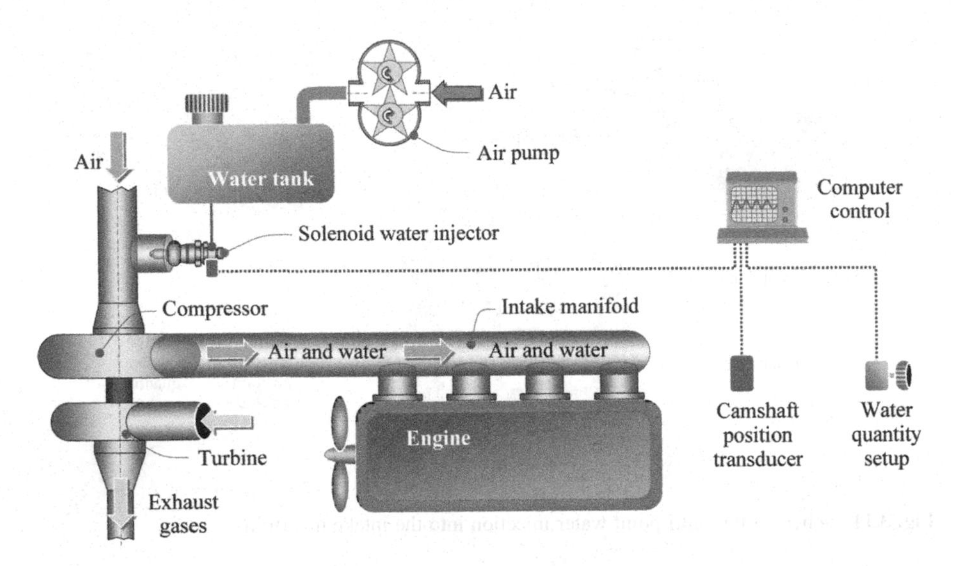

Fig. 3.12 Scheme of mono-point water injection before turbocharger

There are no significant differences in measurable engine characteristics between the mono-point and multi-point water injection systems. Both systems reduce NO_x emission effectively. Some authors also report a positive influence of water injection on PM emission, but this topic is not well investigated.

Water Injection Directly into the Combustion Chamber. Various strategies have been proposed till now to inject water directly into the combustion chamber, with the aim of reducing NO_x emissions and keeping the water quantity small in comparison to water injection into the intake manifold. A good example is the *stratified fuel/water injection system* (Miyano et al. 1995). By this method, water is fed into the injector's dead volume when fuel injection does not take place. When injection begins, fuel and water are injected into the combustion chamber in a stratified condition.

At the end of the previous injection process, the high pressure tube and injector dead volumes contain only pure fuel. At an appropriate moment, determined by the ECU, water is pumped into the injector's dead volume at some pressure, that is higher than the opening pressure of the non-return valve and lower than the needle nozzle opening pressure. The delivered water pushes the fuel in the high pressure system toward the fuel pump. The water quantity is defined by the water supply pressure and by the duration of the solenoid valve opening. At the moment of the solenoid valve closing, water delivery is finished and the non-return valve closes. At this moment, the middle part of the injector's dead volume is filled with water, while the rest is filled with fuel. This means that the injection process starts with a small amount of fuel. This is followed by the injection of water and finally again fuel.

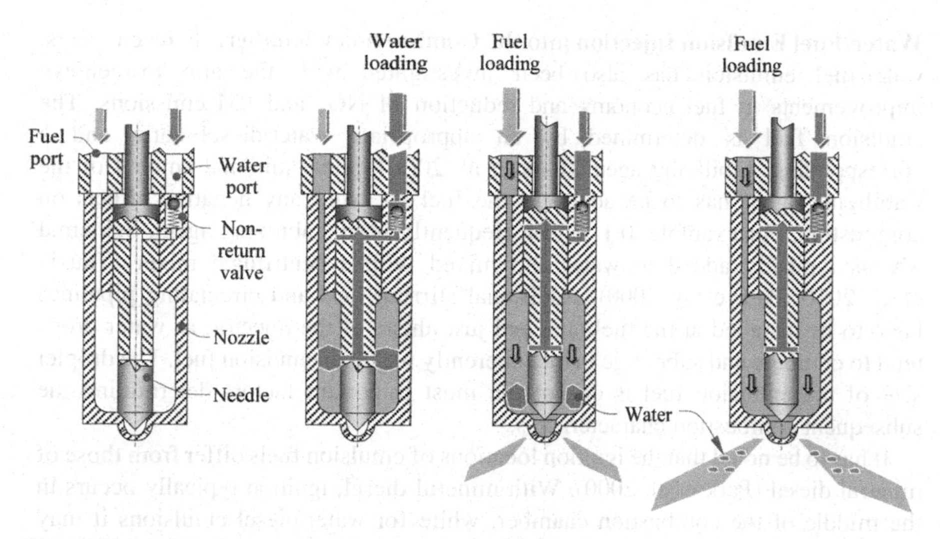

Fig. 3.13 Scheme of the stratified fuel/water injection system

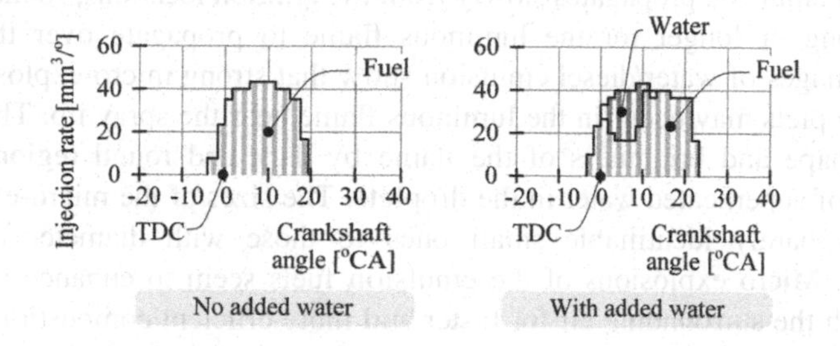

Fig. 3.14 Stratified fuel/water injection process

The stratified fuel/water injection system is shown schematically in Fig. 3.13. The injection rate history ($dq/d\alpha$) of fuel/water injection in dependence on crankshaft angle after top dead center (ATDC) is presented in Fig. 3.14. The graph looks like a three-phase injection; in both graphs the total quantity of injected fuel is the same. This means that water addition leads to increased injection duration.

Injecting water in this way delays slightly the ignition but this delay can be kept at a low level, even if a large quantity of water is injected and significant NO_x reduction is achieved. An important advantage of direct water injection is the possibility to change the water to fuel ratio in dependence on engine parameters (speed and load) or during engine warm-up (cold start) (Stanglmaier et al. 2008). This system allows for varying the percentage of water in the mixture on a cycle-resolved basis. Consequently, a considerable improvement in NO_x and PM emissions can be obtained, both, under steady-state and transient conditions. The control of water percentage on a cycle-resolved basis has also been shown to be an effective method for mitigating NO_x and smoke emissions over step-load transients.

Water/Fuel Emulsion Injection into the Combustion Chamber. In recent years, water/fuel emulsion has also been investigated with the aim to achieve improvements in fuel economy and reduction of NO_x and PM emissions. The emulsion fuel is determined by an appropriate water/diesel ratio and a corresponding stabilizing agent (Park et al. 2000). Some minimal amount of the stabilizing agent has to be added to the fuel to avoid any negative effects on combustion. For example, 0.1 % of a frequently used stabilizing agent, Sekiemal SA, is typically added to water and mixed with a centrifugal mixer (Tauzia et al. 2010; Park et al. 2000). Additional stirring fans and circulating pipelines have to be installed at the fuel tank and just ahead of the injector, as water drops tend to combine and submerge in the inherently unstable emulsion fuel. The droplet size of the emulsion fuel is one of the most important factors determining the subsequent combustion characteristics.

It has to be noted that the ignition locations of emulsion fuels differ from those of mineral diesel (Park et al. 2000). With mineral diesel, ignition typically occurs in the middle of the combustion chamber, while for water/diesel emulsions it may occur in the bottom region or at multiple points simultaneously. The flame of a water/diesel emulsion propagates slowly from the ignition locations, so that it takes twice as long or longer for the luminous flame to propagate over the whole chamber. Images of water/diesel emulsion show that strong micro-explosions of a group of droplets may occur in the luminous flame near the spray tip. They affect the local shape and brightness of the flame by dark and round regions due to explosions of superheated water in the droplets. The sizes of the micro-explosions range from barely identifiable small ones to those with diameter of a few millimeters. Micro explosions of the emulsion fuels seem to enhance mixing of the fuel with the surrounding air for faster and more efficient combustion.

The influence of water injection on combustion (pilot and main injections) has also been studied and shows similar effects as EGR. However, these effects are much more significant for water injection than for EGR for a given dilution ratio. To keep the cycle efficiency, the readjustment of fuel injection parameters, when increasing the dilution ratio, is thus higher in the case of water injection than in the case of EGR (Park et al. 2000).

In general, the results reported suggest that the water emulsification has a potential to slightly improve the brake efficiency and to significantly reduce the NO_x, soot, unburned HC and PM emissions of a diesel engine. However, in order to optimize the emulsion formulation in terms of water content and internal structure, experimental work is practically unavoidable (Armas et al. 2005). By using a water/diesel emulsion in a diesel engine, the cold-start may become impossible. Therefore, the injection system should be purged before engine stop in order to use pure diesel fuel for the engine start (Maiboom and Tauzia 2011). Furthermore, reliability problems may occur at some injection components, in particular in high pressure pumps that might need some specific modifications. Long-term emulsion stability could also turn out to be problematic in a water/diesel emulsion distribution network. Maybe, a possible alternative could be an onboard emulsion fabrication, which could allow the water/fuel ratio to vary.

3.3.2 Natural Gas and Liquefied Petroleum Gas

Natural gas (*NG*) is one of the most important energy carriers today, because it is available in large quantities and its reserves are of the same magnitude as the crude oil reserves. Typical compositions of NG are approximately 92.7 % methane, 3.3 % ethane, 2.2 % nitrogen, 0.5 % carbon dioxide, 0.7 % propane, 0.1 % isobutane, 0.2 % N-butane, 0.1 % pentane, and 0.1 % hexane (Nwafor 2000). NG offers several advantages such as clean combustion, high availability, and an attractive price (Poompipatpong and Cheenkachorn 2011). Additionally, its relatively high auto-ignition temperature is suitable for higher compression engines (Selim 2001). Due to a low cetane number, an engine using natural gas requires injection of mineral diesel fuel as the pilot ignition fuel. Such engines are *dual fuel engines*, which need two fuel systems (Kowalewicz and Wojtyniak 2005). Previous studies investigated the characteristics of *dual fuel* operation in unmodified or slightly modified diesel engines (Selim 2001). For dual fuelling, the in-cylinder pressure and heat release rate are lower than for neat diesel fuelling. By increasing the pilot diesel fuel injection quantity, the pressure in the cylinder and the rate of its rise also increase. Furthermore, an increase of pilot fuel quantity extends the lean burning limit and decreases HC and CO emissions, which are generally higher than for diesel fuelling. The ignition delay is longer for dual fuelling (Cordiner et al. 2008; Kowalewicz and Wojtyniak 2005). The combustion characteristics of a diesel engine with natural gas/diesel fuels show positive effects on thermal efficiency, total specific fuel consumption, soot, and NO_x emissions (Selim 2001). However, the reports indicate that such dual operation system cannot reach high speed operation that is obtainable by a diesel-only engine. Therefore, adequate modifications might be necessary to mitigate this negative effect.

Liquefied petroleum gas (*LPG*) is a mixture of propane and butane and comes with significant variations of its composition in various countries. An engine can be fuelled with a lean homogenous mixture of LPG and air and pilot diesel fuel injection can be used for ignition. In general, LPG usage delivers lower smoke, lower CO at high load, and lower NO_x at low and middle loads (Qi et al. 2007; Kowalewicz and Wojtyniak 2005). The usage of LPG requires only a minor modification of the diesel engine.

3.3.3 Methane and Propane

Compared to mineral diesel, methane has a higher heating value and lower adiabatic flame temperature. Many existing diesel engines can be relatively easily converted to dual fuel operation with methane, keeping the same compression ratios and a diesel-like efficiency, but with much lower emissions of NO_x and PM. On the other hand, direct methane injection is characterized by some practical complications, due to the necessity of a high pressure supply and to the problem of

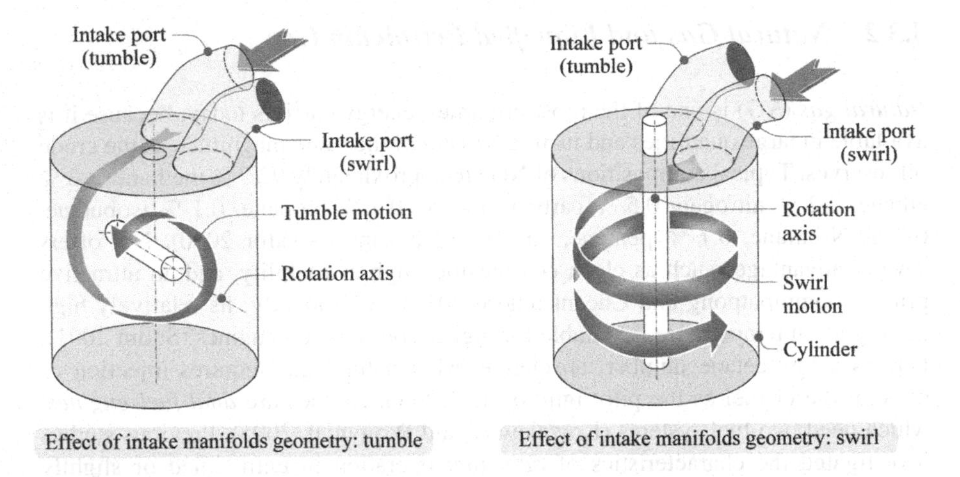

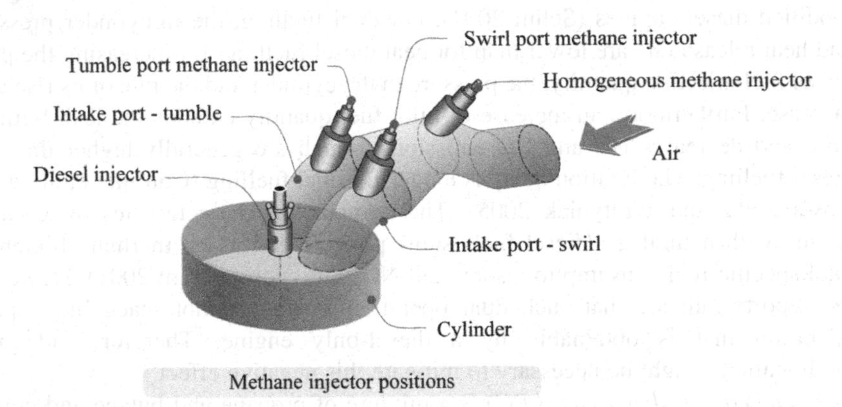

Fig. 3.15 Intake manifold geometries and methane injection positions

positioning another injector into the cylinder head. One possibility to overcome these technical difficulties in *dual fuel* mode is the injection of methane and preparation of *air–methane mixture* in the intake manifold (Carlucci et al. 2011; Sahoo et al. 2009; Abd-Alla et al. 2001).

Several options for methane injection are possible as illustrated in Fig. 3.15. In the case of *homogeneous injection*, a *homogeneous-like air–methane mixture* is produced by injecting methane along the intake air flow at a distance of about 40 cm from the cylinder axis. On the contrary, in order to obtain a certain degree of *mixture stratification* up to the end of the compression stroke, a *port injection* is proposed, in which methane is injected near the inlet valves. In the case, when engine has two intake pipes, two different injection positions are possible (Fig. 3.15). In the case of *swirl port injection*, methane is injected near the swirl inlet valve, while, in the case of *tumble port injection*, the injector is placed very close to the tumble inlet valve.

In Carlucci et al. (2011) the interactions of the methane injection strategies with the various in-cylinder charge motions have been investigated. For this purpose, the

Fig. 3.16 Activation/deactivation of the engine intake ports

swirl helicoidal port and the tumble direct port of the engine intake were activated and deactivated in order to obtain three different in-cylinder bulk motions, denominated as the *swirl*, *tumble*, and *swirl/tumble* ports (Fig. 3.16).

At *low load*, no relevant effects of the methane supply method on the heat release histories were observed (Abd-Alla et al. 2001; Carlucci et al. 2011). On the contrary, the analysis of the combustion images and the luminance curves demonstrated that, for several combinations of the engine-operating parameters, port injection can probably induce some stratification effects on the in-cylinder charge, modifying the intensity and the spreading of the combustion flame. This happened especially when a suitable in-cylinder bulk motion was produced. Concerning pollutant emissions at low load, it was revealed that pilot diesel fuel injection pressure is a very important factor. At high pilot injection pressure, the *swirl port* delivers the lowest levels of HC and NO_x.

At *high load*, the swirl port injection engine shows higher peaks on the heat release curves and a generally more rapid development of the combustion with respect to the swirl/tumble port, while the tumble port shows an intermediate trend (Carlucci et al. 2011). At high loads, the improving effect of swirl port bulk motion affects the combustion, starting from the first steps of the oxidation process. No relevant effects of the methane supply method on the heat release histories were observed. Concerning the pollutant emissions at high load, it was revealed that, with respect to low load tests, the variation of pilot diesel fuel quantity affects less seriously the NO_x production. Moreover, the swirl inlet configuration showed the highest NO_x levels, the lowest ones were obtained with the swirl/tumble ports (especially with methane port injection), while the tumble port showed intermediate values. Regarding the swirl/tumble ports inlet configuration, by changing from homogeneous to port methane injection, a decrease of the NO_x levels was observed, especially when the port injection was performed into the swirl intake duct.

By comparing the low and high load emission results, it is important to note that, for several combinations of the operating parameters, methane port injection was always associated with the lowest values of pollutants emissions. This demonstrates that the port injection, as a methane supply method for dual fuel engines, is a very effective strategy to reduce unburned HC and NO_x concentrations, especially when implemented with variable intake geometry systems to produce suitable in-cylinder bulk motion and turbulence intensity for various engine operating parameters.

The dual fuel engines can also use *mineral diesel fuel* and *propane*. The results show that under the same operating condition, the effective thermal efficiency increases with the increase of propane proportion in the blends (Sahoo et al. 2009; Ma et al. 2008). Increased propane part also leads to decreased ignition delay and combustion duration. Furthermore, the maximum in-cylinder pressure, maximum rate of pressure rise, maximum rate of heat release, and maximum mean combustion temperature increase. The exhaust HC and smoke emissions decrease, while only minor variations of CO levels have been observed, except at high engine load, where CO levels also decreased. The NO_x emission increases with the increase of propane proportion in the blend.

3.3.4 Dimethyl Ether and Dimethyl Carbonate

Dimethyl ether (*DME*) is a liquefied gas that can be produced from a variety of feed stock. It is nontoxic and environmentally benign. DME has a high cetane number, low density, low lubricity, and it is more corrosive than mineral diesel (Arcoumanis et al. 2008). Fast evaporation of DME allows for good mixing with air in the engine cylinder and its high oxygen content can assure smokeless combustion through low formation and high oxidation rates of particulates. DME requires a high injected volume to deliver the same amount of energy as the diesel fuel, due to its lower density and combustion enthalpy. Furthermore, DME-fueled systems also need lubricity enhancing additives and anticorrosive sealing materials to assure leakage-free operation.

The DME spray characteristics and the corresponding combustion process have been investigated for various fuel injection systems, ranging from mechanically controlled injection systems to advanced common rail systems, in a variety of prototype engines with minor modifications (Arcoumanis et al. 2008; Kim et al. 2011; Youn et al. 2011; Zhu et al. 2012).

The DME engine is smoke free and this offers an opportunity to reduce NO_x emission without the trade-off, related to increased PM. By retarding the injection timing of a *mechanically controlled injection system*, a significant decrease of NO_x emission can be achieved, while the CO and formaldehyde emission increase to some extent (Zhu et al. 2012). In *common rail systems*, the injection quantity of DME fuel at the same injection conditions (injection pressure, energizing duration) is larger than that of mineral diesel due to the high return fuel pressure. Although the DME fuel has low fuel density, the high return fuel pressure prolongs the real injection duration of DME. Also, the spray tip penetration and spray cone angle of DME are longer and wider than those of mineral diesel. Among the combustion characteristics, the peak combustion pressure and the ignition delay of DME fuel are higher and shorter than those of mineral diesel, respectively. The DME fuel causes fast burning and reduces the ignition delay because it has a more oxygenated chemical structure and higher evaporating characteristics than diesel. The NO_x emission of DME is slightly higher than that of mineral diesel at the same engine load condition due to the high peak combustion pressure, the higher oxygen content, and the active combustion of the DME. The oxygenated component and volatility of DME result in HC and CO emissions being lower than those of mineral diesel (Kim et al. 2011; Youn et al. 2011).

Overall, DME has been found to be a very promising alternative fuel for diesel engines, capable of providing high thermal efficiency, low combustion noise, and soot free combustion; it thus merits further research and development to evaluate better its potential as a mass production fuel for the automotive market (Arcoumanis et al. 2008).

Dimethyl carbonate (*DMC*) is an oxygenated renewable fuel, which is usually used as an oxygenated additive to blend with mineral diesel in order to improve the combustion and reduce emissions (Cheung et al. 2011; Xiaolu et al. 2006). DMC reduces smoke almost linearly with its concentration, which is directly related to the oxygen content of the fuel. With 10 % of DMC content in mineral diesel, a smoke reduction of 35–50 % is attainable. Furthermore, reductions of HC and CO levels can be expected, but NO_x emissions may increase slightly (Xiaolu et al. 2006).

Attempts have also been made to use DMC as a primary fuel. In this context it should be noted that it is difficult to fuel diesel engines with DMC only, due to its low cetane number and high latent heat of vaporization. Some investigations of DMC combustion in diesel engines show the advantages of combining EGR with a small injection of mineral diesel by an auxiliary pump in order to ignite the DMC. Experiments on a single cylinder two-stroke diesel engine show that DMC also has a positive influence on the spray development (Xiaolu et al. 2006). DMC sprays have smaller atomization particles and more uniform distribution with respect to mineral diesel sprays. A DMC-fueled engine may have a 2–3 % higher effective thermal efficiency than a mineral diesel-fueled engine at moderate and high load conditions, partially due to its lower exhaust gas temperature. At low load operation, the brake thermal efficiency of the DMC fueled engine is slightly lower than that of the mineral diesel fueled engine, owing to higher HC emissions. A DMC-fueled engine with EGR and a small amount of mineral diesel for the ignition has a potential for simultaneous reduction of NO_x and smoke emissions by keeping CO emission almost the same. Moreover, the smoke density can be brought almost to the zero level. It has to be kept in mind that the mineral diesel injection timing influences the NO_x, CO, and HC emissions of the DMC-fueled engine. With the advanced diesel injection timing the NO_x emissions increase, while the CO and HC emissions decrease.

3.3.5 Fischer–Tropsch Diesel

Fischer–Tropsch diesel (*FTD*) is called the fuel obtained from synthetic gas in the presence of a catalyst, at high pressure and temperature. The basic mechanism of the *Fischer–Tropsch process* is $2nH_2 + nCO = n(-CH_2-) + nH_2O$.

In order to reduce NO_x emissions in diesel engines, a typical strategy is to retard the injection at the cost of penalizing the fuel consumption. Because FTD has a higher cetane number than mineral diesel, FTD usage leads to shorter ignition delays (Bermúdez et al. 2011; Gill et al. 2011). This results in an improved specific fuel consumption and thermal efficiency, while maintaining NO_x at an acceptable level. However, the advantages associated with a high cetane number diminish at higher compression ratios, where the soot levels increase due to the reduced premixed combustion. The overall combustion parameters of FTD fuels are very sensitive to high levels of EGR, particularly soot emission which increases dramatically as the level of EGR increases. Since FTD has lower distillation characteristics, the improvement in atomization and dispersion of fuel spray enables faster evaporation, accelerating the fuel mixing with air. In addition, the rates of heat release for such fuels are lower and therefore improve the combustion noise. However, using pilot injection in conjunction with a common rail injection system can also assist the reduction of combustion noise.

3.3.6 Hydrogen

Because of the high self-ignition temperature, pure hydrogen can't be used directly in a diesel engine. Some alternative methods are the *hydrogen enrichment technique*, *hydrogen port/manifold injection*, and *hydrogen into the cylinder injection*, which use mineral diesel as a pilot fuel for the purpose of ignition (Saravanan and Nagarajan 2010).

By the *hydrogen enrichment technique* air is enriched with hydrogen by using a venturi/gas carburetor in the intake manifold. A pilot quantity of diesel is used as an ignition source. The benefit of hydrogen enrichment is that the brake thermal efficiency increases, but the power output of the engine drops due to the partial replacement of air by gaseous hydrogen.

By *hydrogen port/manifold injection*, hydrogen is injected in the intake port by using a mechanically or electronically operated injector. The position of the injector on the manifold determines whether the system is a port fuel injection system or a manifold injection system. For both systems, mineral diesel is usually taken as the ignition source. The advantage of hydrogen injection over the carbureted system is that with proper injection timing some problems, related to backfire and preignition, can be eliminated.

By *hydrogen into the cylinder injection*, hydrogen is injected directly into the combustion chamber at the end of the compression stroke. During idling or part load conditions, the efficiency of the engine may be reduced slightly. In spite of that, compared to other methods of hydrogen usage, this method is the most efficient. Two types of injectors are available for the use in direct injection systems.

The *low pressure direct injector* injects the fuel as soon as the intake valve closes when the pressure is low inside the cylinder. The *high pressure direct injector* injects the fuel at the end of the compression stroke.

The major problem associated with the use of direct injection is that it must be able to withstand higher combustion temperature in addition to preventing the injector from corrosion due to exhaust gases. Lubrication between the injector's moving parts also makes the design of the direct injector more complicated (Saravanan and Nagarajan 2010).

By using hydrogen as a supplementary fuel in a diesel engine, the brake thermal efficiency increases due to a better combustion process (Saravanan et al. 2008) and the specific energy consumption decreases due to the operation of hydrogen fueled engine under lean burn conditions. The effect of hydrogen addition on the combustion process of heavy duty diesel engines depends on the load and the amount of hydrogen added. Addition of a small amount of hydrogen has only a minor or negligible effect on the cylinder pressure and combustion process. To change this situation, a relative large amount of hydrogen is needed (Liew et al. 2010). Hydrogen addition yields modest emission reductions with a limited penalty on engine performance (Lilik et al. 2010).

3.3.7 Alcohols

Alcohols, mainly *methanol* and *ethanol*, have been widely investigated in combination with mineral diesel, with the aim to reduce NO_x and particulate emissions (Agarwal 2007; Kowalewicz and Wojtyniak 2005; Abu-Qudais et al. 2000; Surawski et al. 2010). Methanol is biodegradable, toxic, and corrosive, while ethanol is not considered toxic and is also biodegradable. Methanol is produced mostly from coal and natural gas, but meanwhile it can also be produced from renewable sources, such as wood or waste paper. Ethanol is produced from biomass, such as potatoes, beets, sugar cane, wood, brewery waste and many other agricultural products, and food wastes in the process of fermentation. Additionally, it can also be produced from natural gas and crude oil.

Alcohols and mineral diesel are mostly applied together, either in the *blended mode* or in the *fumigation mode*. In comparison with the blended mode, the fumigation approach seems to be more flexible, despite of the extra fuel injection system required. This is because, firstly, it allows the amount of injected alcohol to vary in dependence on actual requirements. Secondly, since the alcohol is not premixed with mineral diesel, an emulsion additive, to ensure proper mixing of alcohol and diesel, is not required.

A number of studies have demonstrated that NO_x and particulate matter can both be reduced with *methanol* or *ethanol fumigation* (Cheng et al. 2008; Zhang et al. 2009). In general, *methanol fumigation* (Fig. 3.17) was found to decrease NO_x emission and smoke opacity and to have a beneficial effect on fuel efficiency at high engine loads. Some investigation results show a reduction of NO_x and particulate emissions, but an increase of HC and CO emissions (Zhang et al. 2010; Wang et al. 2008; Yao et al. 2008). Furthermore, *ethanol fumigation* influences gaseous emissions and the in-cylinder pressure shows that HC and CO emissions may increase severely,

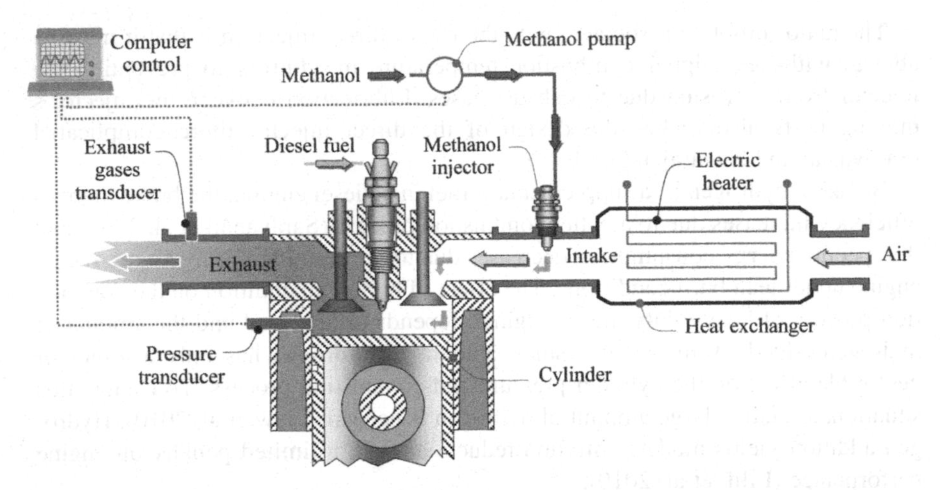

Fig. 3.17 Methanol and diesel fuel injection systems

compared with diesel fuel, but the NO_x emissions usually decrease (Hayes et al. 1988). The maximum rate of pressure rise and peak pressure is also significantly higher (Jiang et al. 1990). Ethanol fumigation could be a viable method to offset mineral diesel while lowering diesel soot emissions and avoiding lubricity problems, potentially associated with ethanol–diesel-blended fuels (Leahey et al. 2007).

3.3.8 Vegetable Oils, Bioethanol, and Biodiesel

Vegetable Oils. Vegetable oils also offer a potential environmental advantage. The most frequently investigated raw *vegetable* oils for diesel engine usage are among edible oils: *corn*, *palm*, *peanut*, *rapeseed*, *sesame seed*, *soybean*, and *sunflower* oil, and among nonedible oils: *cottonseed*, *jatropha*, and *rubberseed* oil (No 2011; Hossain and Davies 2010; Demirbas 2007; Kalam et al. 2003). The chemical and physical properties of these vegetable oils may differ from mineral diesel significantly, as shown in Fig. 3.18.

On average, the *density* of vegetable oils is about 12 % higher than that of mineral diesel, while their *energy content* is around 10 % lower as a result of lower hydrogen content (Hossain and Davies 2010). For most vegetable oils the *cetane number* is around 10–20 % lower than that of mineral diesel. The cetane number depends on the locations and number of double bonds in the molecular structure of the oils. Lower cetane number implies a larger ignition delay and tends to result in lower efficiency. The *viscosity* affects the flow of fuel and spray characteristics. Due to the large molecular size of the triglycerides, making up about 98 % of vegetable oils, viscosity is higher and volatility lower than for mineral diesel. Higher viscosity of vegetable oils leads to poorer combustion. The *flash point temperature* indicates the overall flammability hazard in the presence of air and

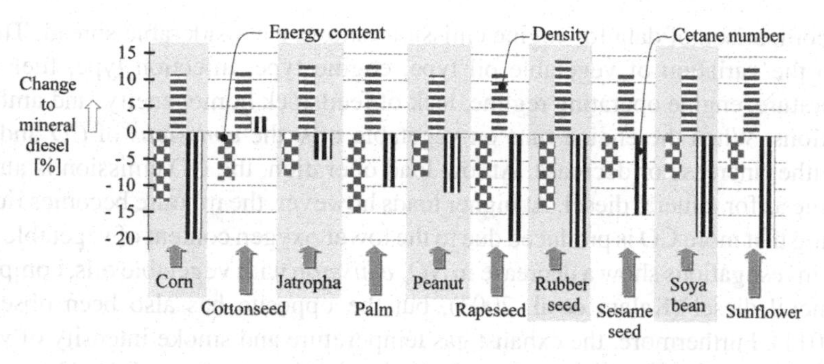

Fig. 3.18 Comparison of vegetable oils properties with respect to mineral diesel

the *pour point temperature* is a measure of fuel performance under cold temperature conditions. In the case of the edible vegetable oils, the flash point temperatures are higher, whereas the pour point temperatures are lower than for mineral diesel. On the other hand, for nonedible vegetable oils both of these temperatures are much higher than for mineral diesel. The *iodine value* is a measure of the number of double bonds and indicates oxidation stability. Iodine value increases with the level of unsaturation. Higher iodine value means lower oxidation stability. Oxidation can lead to polymerization. Iodine values vary significantly among the vegetable oils considered. The *carbon residue value* correlates with the carbonaceous deposits inside the combustion chamber and injector systems. For vegetable oils it is considerably higher than for mineral diesel.

Various properties of vegetable oils lead to various injection, atomization, and combustion characteristics of a diesel engine. Several investigations show significant performance variations obtained among various types of these oils (Ramadhas et al. 2004; Nwafor 2003; Bhattacharyya and Reddy 1994). Of course, those effects vary in dependence of engine load, engine speed, feedstock homogeneity, ambient conditions, engine type, and injection type. However, some general influence of vegetable oils usage on *effective power*, *effective specific fuel consumption*, and *thermal efficiency*, can be roughly estimated as follows.

By using vegetable oils, the *effective engine power* decreases by 2–18 %. The main reasons for this lie in lower energy content of vegetable oils and in higher viscosity, which leads to poor atomization and inefficient mixing of air and fuel. It has to be noted that in some cases (e.g., when using rapeseed and coconut oils), the peak in-cylinder pressure is lower than the one obtained with mineral diesel (Hossain and Davies 2010; Kalam et al. 2003; Nwafor and Rice 1996); meanwhile in some other cases (e.g., using orange oil), the peak in-cylinder pressure and heat release are higher (Hossain and Davies 2010; Purushothaman and Nagarajan 2009).

Effective specific fuel consumption of vegetable oils is the same or higher than that of mineral diesel. Most investigations show an increase in the range of 2–15 %. *Thermal efficiency* of vegetable oils is in the range from +3 to −10 % compared to mineral diesel. High viscosity and low volatility are again the reasons for this efficiency loss. Eventual efficiency gains may be explained by the lower energy content of vegetable oils.

A comparison of data for engine emissions shows a considerable spread. This is due to the variation of vegetable oil type, engine type, injection type, fuel inlet temperature, engine operating regime, lack of feedstock homogeneity, and ambient conditions. When the engine runs on vegetable oils, the *emissions* of *CO* and *HC* may either increase or decrease. At low load operation, the CO emission is almost the same as for mineral diesel. At higher loads however, the mixture becomes richer, meaning that more CO is produced due to the lower oxygen content of vegetable oils. Some investigations show a decrease in NO_x *emission* with vegetable oils, compared to mineral diesel (Kalam et al. 2003), but the opposite has also been observed (No 2011). Furthermore, the exhaust gas temperature and smoke intensity of vegetable oils may either increase or decrease in comparison to mineral diesel.

Vegetable oils can be successfully used in a diesel engine by making adequate engine and fuel modifications. The *engine modifications* include preheating/heated fuel line, dual fueling, injection system modifications, etc. Heating and blending of vegetable oils reduces the viscosity but its molecular structure remains unchanged and hence the polyunsaturated character and low volatility problems still exist. The *fuel modifications* include pyrolysis (thermal cracking), micro-emulsion, transesterification, and hydrodeoxygenation to reduce polymerization and viscosity. It may be worth noting that transesterification, where the kinematic viscosity can be remarkably reduced, is an effective process to overcome many problems associated with vegetable oils (No 2011).

Bioethanol. Bioethanol is produced from carbohydrates (hemicelluloses and cellulose) in lignocellulosic materials (Demirbas 2007). It can be derived from wheat, sugar beet, corn, straw, and wood. Bioethanol can be used as an additive to mineral diesel. The chemical and physical properties of bioethanol differ from those of mineral diesel. Consequently, the presence of bioethanol causes various physical and chemical property variations of the blended mineral diesel, mostly related to lower cetane number, lower energy content, viscosity, flashpoint, pour point, etc. (Torres et al. 2011a; Li et al. 2005). Lower cetane number worsens cold starting ability, increases noise, and shortens engine life. Density and viscosity decrease by addition of bioethanol, thus promoting retarded injection timing in mechanically controlled injection systems (Torres et al. 2011a). High bioethanol volatility may influence significantly injection timing and ignition and combustion characteristics. Furthermore, it can cause some problems related to storage. While the pour point decreases significantly with respect to that of pure diesel, the cloud point increases. Concentrations of bioethanol above 5 % lead to unwanted behavior in the cloud point test. The addition of bioethanol, however, does not modify substantially the cold filter plugging point (CFPP).

Another point worth notion is the *ethanol–diesel mixture stability*. In this context experimental tests are necessary in order to determine whether additional mixing, additives, or heating (in case that low temperatures produce gel formation) are necessary. In one of such experiments, (Torres et al. 2011a), samples of 5, 10, and 15 % of bioethanol (obtained by fermentation of sugars) in mineral diesel were tested at various temperatures: +25 °C (normal ambient temperature), +30 °C (summer), +8 °C (not critical winter), and −18 °C (critical winter in central Europe). Each sample was checked once a week during a 5-week period; it was assumed that this is enough time to estimate the blend behavior during long periods of storage. Figure 3.19 shows some photographs of the 15 % bioethanol and 85 % diesel mixture.

At +30 °C and after 4 weeks, no changes in stability, color, and aggregation state of the blends could be observed in any of the samples. However, already at +25 °C and after 4 weeks, the 10 % sample separated into bioethanol and diesel. Bioethanol bubbles, traveling to the surface, could be observed. At +8 °C in the 15 % sample bioethanol separated from diesel after only 1 week; two layers have been observed. At −18 °C in the 15 % sample three layers have been observed after 5 weeks; two layers were observed within the diesel fuel.

Obviously bioethanol/diesel separation depends on temperature, bioethanol concentration, and water content (Hansen et al. 2005; Lapuerta et al. 2007). The situation can be improved by using various kinds of additives (Weber et al. 2006; Reyes et al. 2009; Ribeiro et al. 2007). Nowadays, the addition of biodiesel, acting as a blend stabilizer, is becoming the preferred method (Lapuerta et al. 2009; Chotwichien et al. 2009; Lebedevas et al. 2009; Kwanchareon et al. 2007) because it has the advantage of increasing the biofuel concentration in the fuel, which is one of the targets proposed by the European Community in the Directive 2003/30/EC to promote the use of biofuels for transport.

Bioethanol addition to mineral diesel influences the injection and spray characteristics, combustion performance, and engine emissions (Torres et al. 2011b, 2010; Huang et al. 2009; Rakopolulus et al. 2008a, b; Rakopoulus et al. 2007; He et al. 2004). Investigations show that bioethanol addition to mineral diesel up to 30 % reduces smoke emissions at high loads, decreases NO_x emissions at low loads, but can somewhat increase NO_x emissions at medium and high loads. Bioethanol increases HC and CO emissions with the exception of CO emission at high loads. The combustion duration shortens and ignition retards by increasing the bioethanol content in mineral diesel. Bioethanol also improves the effective specific fuel consumption at full load.

Biodiesel. Biodiesel is a transesterified vegetable oil. Biodiesel is technically competitive with mineral diesel or even offers some technical advantages (Demirbas 2007). This can be even further improved by adding bioethanol. In this context the addition of bioethanol up to 15 % to biodiesel has been investigated. The *bioethanol–biodiesel blends* have some specific properties. In general, however, investigations show that bioethanol addition improves the most important fuel properties, related to the injection process and engine characteristics. It was also observed that *filter plugging tendency* improves proportionally to the bioethanol concentration. Owing to the hygroscopic nature of bioethanol and biodiesel, special attention must be paid during storage and transportation to avoid the absorption of water from the ambient humidity. Properties related to cold climates, such as cloud point, pour point, and CFPP, improve by bioethanol addition because phase separation does not occur and bioethanol shows good behavior at low temperatures. When testing fuel blends, it is important to note that the flash point will show the value of the more volatile of the components, i.e., bioethanol in this case. Therefore, additives are typically necessary to raise the blend flash points within the limits required by the standards.

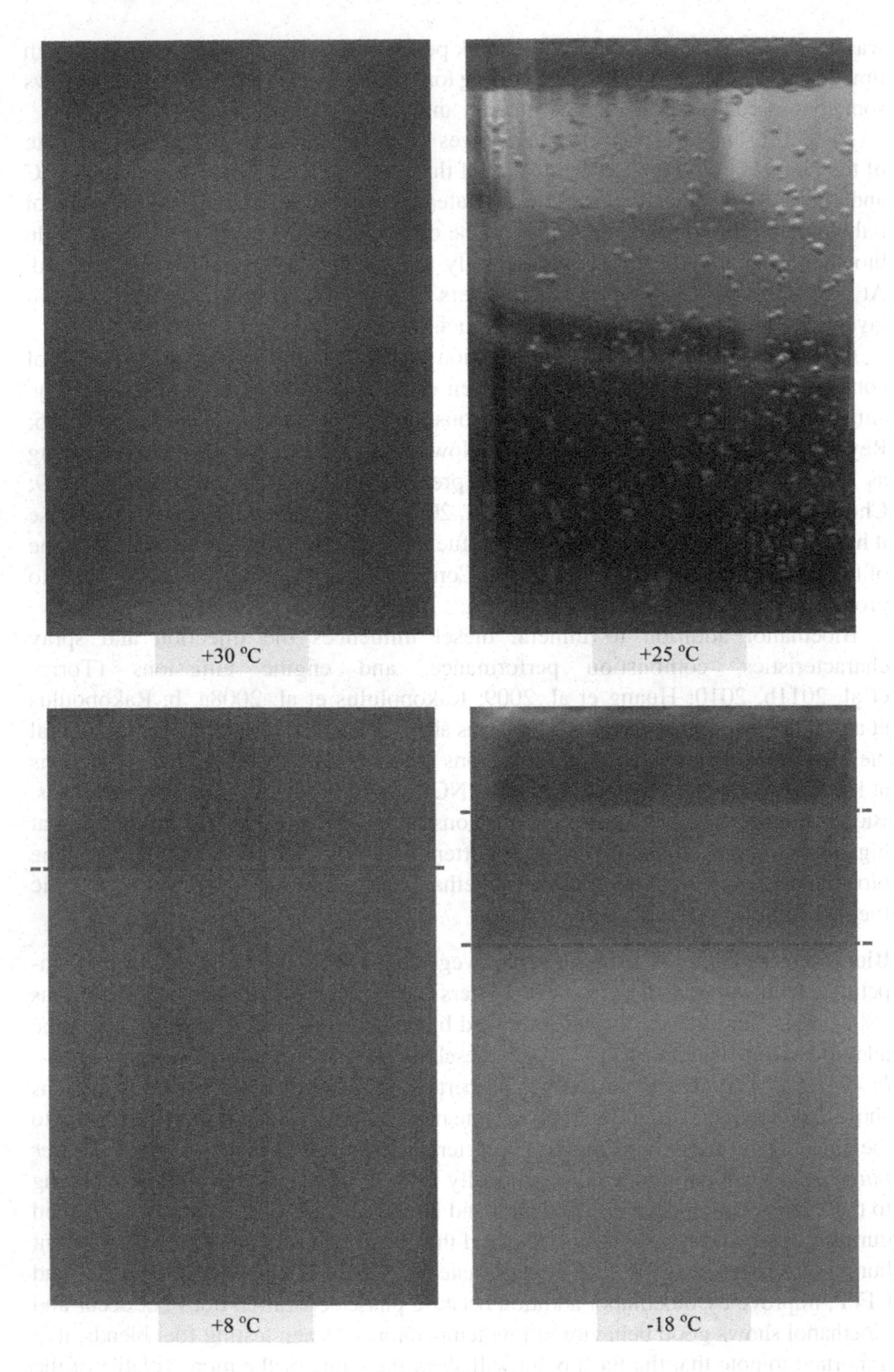

Fig. 3.19 Stability of bioethanol–diesel blend D85E15

Fig. 3.20 Stability of bioethanol–biodiesel blend B85E15

In order to evaluate the bioethanol–biodiesel blend stability, samples with up to 15 % of bioethanol were tested at +25 °C, +30 °C, +8 °C, and −18 °C (Torres et al. 2011b). Each sample was checked once a week during a 5-week period. When neat biodiesel warmed up from the solid state and became liquid, some white particles could easily be observed (Fig. 3.20). These particles are paraffins. In case of bioethanol–biodiesel blends, bioethanol changes the melting process in such a way that paraffins do not become visible. Invisibility of the paraffins is an improvement from the point of winter properties, such as the cloud point and cold filter plugging point. Namely, these properties are linked to the moment at which paraffins appear during the solidification process. This means that bioethanol also acts as a winter additive. Furthermore, bioethanol–biodiesel blends need less time to become liquid than neat biodiesel.

3.4 Discussion

To improve the worldwide acceptance of diesel engines, it is necessary to increase engine effective power and torque and reduce fuel consumption and harmful emissions, especially NO_x and smoke emissions.

The NO_x concentration can be reduced by a reduction of the flame temperature, which can be achieved, for example, by lower oxygen content or increased content of inert gas in the intake air (as with EGR) or by better premixing in order to achieve partly homogeneous lean mixture before the start of combustion. Reduction of smoke concentration in the exhaust gas can be achieved either by decreasing smoke formation or increasing smoke oxidation. Smoke formation can be mitigated by more thorough mixing that prevents the appearance of rich flames with local relative air/fuel ratios below 0.6. Oxidation can be improved by sustaining a high temperature level for a sufficiently long time period in the presence of oxygen. Unfortunately, this is counterproductive with regard to the NO_x problem. Anyhow, because there are many trade-offs in the quest for a better diesel engine, the development process is quite tedious. In spite of this, some general guidelines may still be given and the measures that can be implemented can be outlined as follows:

- Improved in-cylinder gas flow such as controlled swirl and turbulence through a variable intake geometry, multivalve cylinder head, and optimized bowl shape
- Improved injection system with a vertical nozzle, injection rate shaping such as pre-injection and split injection or a shorter deactivation period, and finer sprays, achieved by an electronically controlled injection system with injection pressures up to 2,000 bar (e.g., by unit-injector or common-rail systems)
- Improved and controlled combustion procedure through reduced flame temperature (higher EGR level, cooled EGR, better premixing), avoidance of over-rich zones, and faster start of oxidation
- Improved fuel quality and usage of alternative fuels with lower sulfur and aromatic content and increased cetane number

The influence of exhaust gas after treatment, engine management, and alternative fuels on engine characteristics was and still is a lively investigated topic. The interdependent relationships between the involved quantities are quite sophisticated. In spite of that, much helpful insight can be gained by a careful study of many thorough investigations, for example, like the one provided by (Armas et al. 2010). In this investigation the effects of *injection control* and *alternative fuels* on engine characteristics have been investigated by using the DDC/VM Motori 2.5 L turbocharged diesel engine with a *common rail injection system* and an ultra low sulfur mineral diesel fuel (D100), *soybean biodiesel* (SoBIO) and *Fisher–Tropsch* (FT) fuel. Test were run at 2,400 rpm and 64 Nm torque with *single* and *split* (pilot and main) *injection strategies* and without exhaust gas recirculation (EGR) In the case of single injection, the experimental results have been obtained by using advanced (5.83°CA BTC), baseline (3.83°CA BTC), and retarded injection timing (1.83°CA BTC). In the case of split injection strategy, advanced (31.4°CA BTC), baseline (29.4°CA BTC) and retarded (27.4°CA BTC) injection start was used for the pilot injection. The main injection started always at the same angle after top center (1.4°CA ATC). The results are summarized in the following.

Figure 3.21 shows that effective specific fuel consumption increases with SoBIO and decreases with FT fuelling when the injection timing and other parameters are

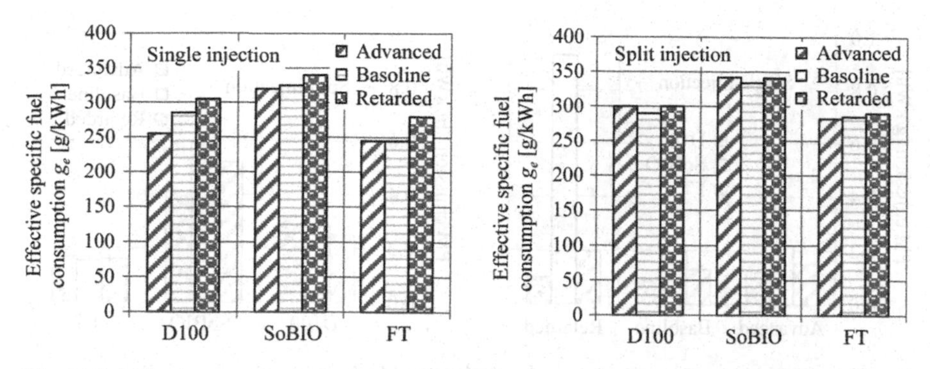

Fig. 3.21 Effective specific fuel consumption for various injection timings and fuels

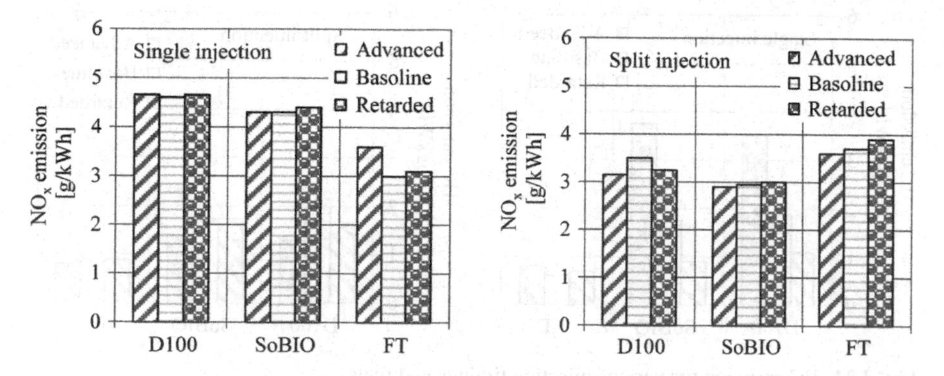

Fig. 3.22 NO_x emission for various injection timings and fuels

constant at single and split injection. Compared to D100, the effective specific fuel consumption of SoBIO fuel increased approximately by 12 % for single and 15 % for split injection. This fact is consistent with the lower energy content of SoBIO, which is about 12.9 % lower than the energy content of D100. Compared to D100, a decrease of g_e for the FT fuel (2 % for the advanced injection timing, 8 % for the delayed injection timing) was observed. This is consistent with the energy content of FT fuel, being 2.3 % higher than that of D100. The results show that in the case of split injection various injection timings have no significant effect on g_e for any of the tested fuels.

Figure 3.22 shows that NO_x emission is lower when SoBIO and FT fuels are used instead of D100. Independent of the injection timing, the reduction of NO_x emissions was approximately 7 % when SoBIO is used. Compared to mineral diesel, the use of FT led to a reduction of NO_x emissions approximately by 22 % and 33 %, when operating with advanced and retarded injection timing, respectively. The influence of the start of pilot injection on NO_x emission is less significant than the influence of fuel composition.

Figure 3.23 shows that PM emission of SoBIO is high with the retarded injection timing in the case of single injection strategy. This is because the delayed

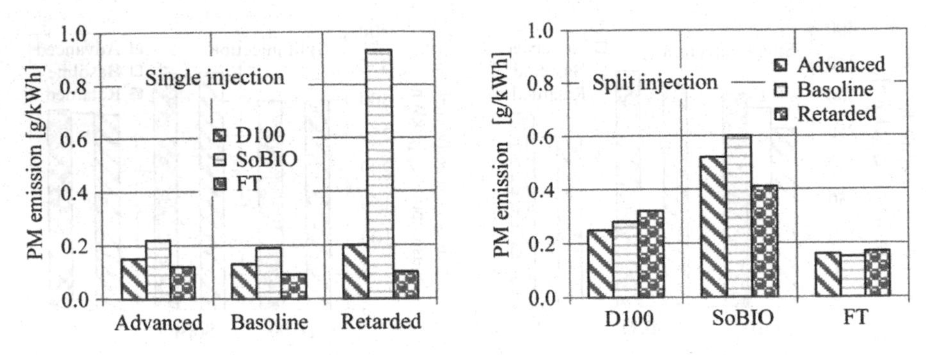

Fig. 3.23 PM emission for various injection timings and fuels

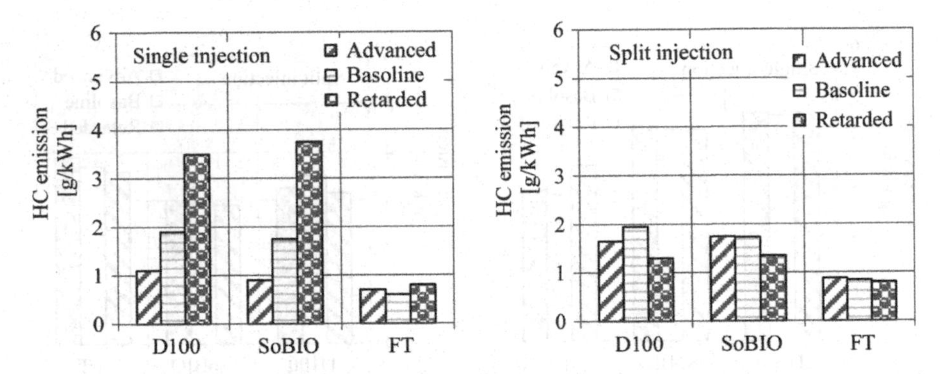

Fig. 3.24 HC emission for various injection timings and fuels

combustion process (being practically without premixed combustion phase) is combined with a flat high temperature distillation curve of SoBIO.

Considering the NO_x–PM trade-off observed for D100, SoBIO, and FT, when using the single injection strategy, the emissions results show that biodiesel usage can lead to a slight decrease in NO_x emissions when the engine is operated within the range of the considered injection timing. However, this operating condition can produce an increase in PM emissions, which increase as the injection timing is retarded. This trend can be explained by the longer injection duration as a consequence of the lower energy content of biodiesel. The FT fuel produced a significant decrease of both NO_x and PM emissions, independently of the tested injection timing. In the case of split injection strategy, fuel composition has a greater impact on the emissions than the pilot injection timing. As with the single injection strategy, the FT fuel produces the lowest PM emissions without a significant increase of NO_x emissions.

With single injection strategy, the HC emission increased by retarding the injection timing for D100 and SoBIO (Fig. 3.24). When FT fuel is used, the level of HC emissions is practically independent of the injection timing, being lower than for D100 by 38, 67, and 78 % for advanced, baseline, and retarded injection timing, respectively. With the split injection strategy, the HC emission, produced by D100

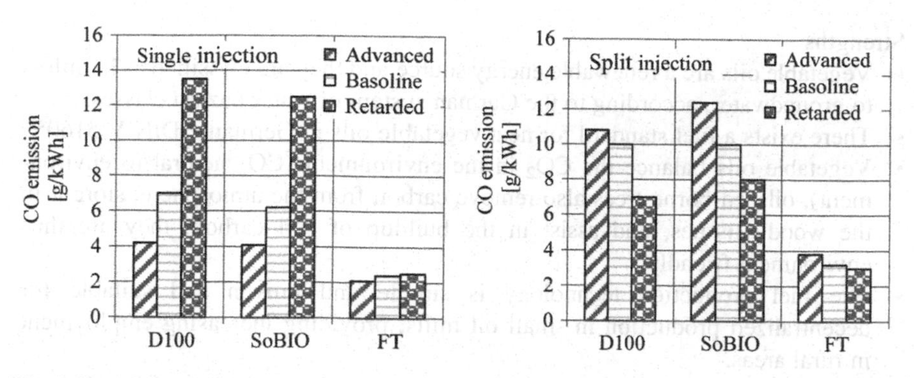

Fig. 3.25 CO emission for various injection timings and fuels

and SoBIO, slightly decreases when the pilot injection is retarded (Fig. 3.24). In all tests, the FT fuel has the lowest HC emission compared to D100 and SoBIO. SoBIO produced the highest PM emission in all the tested starts of pilot injection.

In the case of single injection, the CO emissions follow the same trend as HC emissions, being lower than those produced by D100 by 56, 70, and 81 % for advanced, baseline, and retarded injection timing, respectively (Fig. 3.25). In the case of split injection, the CO emissions produced by D100 and SoBIO show a slight decrease when the start of pilot injection is retarded. In all tests, the FT fuel has the lowest CO emission compared to D100 and SoBIO.

The tests described in the forgoing text were performed without making use of the exhaust gas recirculation (EGR). It should be noted, however, that EGR usage may, in general, reduce the total engine NO_x emissions significantly.

Diesel exhaust also contains sulfuric salts and other abrasive and corrosive substances (Zheng et al. 2004). It has been argued whether EGR should be applied to diesel engines because of the increased piston–cylinder wear. Heavy use of EGR can also deteriorate the energy efficiency, operational stability, and PM generation of the engine. In spite of that, the concerns about increased wear and deteriorated performance have been somewhat pushed into background, because of ever-stringent emission regulations. Therefore, the current concern is mostly about how aggressively EGR should be applied at various speeds and loads. It should be noted, however, that increased wear, related to EGR, continues to be a problem, affecting engine durability and performance.

Currently, EGR is still the most viable technique to reduce NO_x substantially. Energy efficient after treatment systems, dealing with NO_x and PM simultaneously, are still in the early development stages. The inability of available catalytic after treatment technologies further encourages the use of EGR.

Another topic worth of attention are *vegetable oils*. Some general advantages and disadvantages of their *usage* in *diesel engines* have been summarized in the *strength–weakness–opportunities–threat* (SWOT) analysis by (Misra and Murthy 2010; Russo et al. 2012).

Strengths

- Vegetable oils are a renewable energy source and they are classified as harmless to groundwater, according to the German system of water hazard classes.
- There exists a fuel standard for neat vegetable oils in Germany, DIN V 51605.
- Vegetable oils balance the CO_2 in the environment (CO_2 neutral to environment), oilseed-borne trees also remove carbon from the atmosphere, store it in the woody tissues, and assist in the buildup of soil carbon; they are thus environment friendly.
- The fuel production technology is simple and proven and suitable for decentralized production in small oil mills, providing increasing employment in rural areas.
- A number of nonedible oil crops is perennial and is not affected by climatic changes.
- The cetane number is similar or close to that of mineral diesel.
- The energy contents of various vegetable oils are nearly 90 % of that of mineral diesel.
- A higher flash point of vegetable oils allows them be stored at high temperatures without any fire hazard; their flashpoint makes them easy to handle and store.
- By mixing a vegetable oil into the mineral diesel at low ratios, the engine performance and exhaust emissions are, in general, improved with respect to neat mineral diesel.
- The additional oxygen molecule in its chemical structure is beneficial for the combustion process.
- Vegetable fuels seldom contain sulfur, which is not the case for mineral diesel.
- Compared to mineral diesel, NO_x emissions are drastically lower, up to around 30 % for 100 % straight vegetable oil.
- Vegetable oils are easily available in rural areas where their usage is advantageous, especially for smaller engines in agriculture.

Weaknesses

- Non-modified diesel engines can't run on neat vegetable oils.
- Vegetable oils in their natural form have higher viscosity compared to mineral diesel.
- Variable output, variable oil content, long gestation period of the crops.
- At present their availability is scattered.
- There are no large-scale production plants.
- The economic viability depends on seed yields and the income from by-products (press cake).
- In Europe there exists no common standard for neat vegetable oils.
- The presence of chemically bound oxygen in vegetable oil lowers their energy content.
- Very high viscosity and low volatility of vegetable oil lead to poor fuel atomization and lower engine performance.
- Lube oil dilution, high carbon deposits, ring sticking, scuffing of the engine liner, and injection nozzle failure are the major problems, associated with direct use of straight vegetable oils.

- The high flash point attributes to lower volatility characteristics.
- Both, cloud and pour points are significantly higher than that of mineral diesel; these high values may cause problems during cold weather.
- Long storage problems related to viscosity; long-term storage of vegetable oils may cause degradations of certain fuel properties; additives are needed.
- No commercial output is available without ample farming inputs.

Opportunities

- There are a number of nonedible oils which are good fuels; about 300 varieties of tree born oilseeds have been identified; selected crops can be grown on arid and semiarid lands which are presently not cultivable.
- Vegetative propagation is possible in many varieties.
- They have carbon credit value (Kyoto protocol).
- There is a relatively large demand worldwide owing to environmental problems.
- Mineral diesel fuel dissolves quite well with vegetable oils.
- Preheating, blending of vegetable oils with diesel fuel, and blending vegetable oils with solvents would greatly remove many problems associated with diesel engine operation with neat vegetable oil.
- A great opportunity for reduction of NO_x emissions.
- Energy policies that provide tax reductions/exemptions and biofuel obligations could largely increase use of vegetable oils,
- Their use and development can enhance energy supply security.
- Waste oil can be used as cheap feedstock for vegetable oil.
- Other oil plants than rapeseed can be cultivated in Europe (e.g., sun flower, soya).
- There is a significant potential of employment generation capacity in rural areas.

Threats

- The behavior of the plants can vary in dependence of agro climatic zones and regions.
- It is still to be established how these plants would behave once removed from its original habitat and put under high density and intensive cropping system.
- Costly input materials at present.
- As of now these oils are difficult to sustain economically without subsidies.
- No sustainable procurement mechanism available in the market.
- Requirement of seeds in large quantity, even for the modest scenario of 5 % blending with mineral diesel.
- In Europe, mainly rapeseed is used as feedstock source; rapeseed can be cultivated only every 4 years on the same field.
- No standard for vegetable oils feedstocks exists, which means the variation in feedstock properties could vary for many reasons.

Direct use of vegetable oils is impractical and unsatisfactory either for direct or indirect diesel engines (Bozbas 2008). The most important short-term and long-term problems, their probable reasons, and the potential solutions are known to a great extent. Briefly, they can be summarized as follows.

Short-term problem:	*cold weather starting*
Probable cause:	high viscosity, low cetane number, low flash point of vegetable oils
Potential solution:	heat fuel prior to injection process chemically alter the fuel to ester

Short-term problem:	*plugging and gumming of filters, lines, and injectors*
Probable cause:	natural gums in vegetable oils
Potential solution:	partially refine the oil to remove gums

Short-term problem:	*engine knocking*
Probable cause:	very low cetane number of some oils, unsuitable injection timing
Potential solution:	adjust injection timing, heat fuel prior to injection, and chemically alter the fuel to ester

Long-term problem:	*coking of injectors on the piston and head of the engine, carbon deposit on the piston and head of the engine*
Probable cause:	high viscosity of vegetable oils, poor combustion at partial load
Potential solution:	heat fuel prior to injection process chemically alter the fuel to ester

Long-term problem:	*excessive engine wear*
Probable cause:	high viscosity of vegetable oils, poor combustion at partial load, possible free fatty acids in vegetable oils, and dilution of engine lubricating oil due to blow-by of vegetable oils
Potential solution:	heat fuel prior to injection process chemically alter the fuel to ester, increase motor oil changes and motor oil additives to inhibit oxidation

Long-term problem:	*failure of engine lubricating oil due to polymerization*
Probable cause:	collection of polyunsaturated vegetable oil blow-by in crankcase to the point where polymerization occurs
Potential solution:	heat fuel prior to injection process, chemically alter fuel to ester, increase motor oil changes and motor oil additives to inhibit oxidation

References

Abd-Alla, G. H., Soliman, H. A., Badr, O. A., & Abd-Rabbo, M. F. (2001). Effects of diluent admissions and intake air temperature in exhaust gas recirculation on the emissions of an indirect injection dual fuel engine. *Energy Conversion and Management, 42*(8), 1033–1045.

Abu-Qudais, M., Haddad, O., & Qudaisat, M. (2000). The effect of alcohol fumigation on diesel performance and emissions. *Energy Conversion & Management, 41*, 389–399.

Agarwal, A. K. (2007). Biofuels (alcohols and biodiesel) applications as fuels for internal combustion engines. *Progress in Energy and Combustion Science, 33*, 233–271.

Agarwal, D., Singh, S. K., & Agarwal, A. K. (2011). Effect of exhaust gas recirculation (EGR) on performance, emissions, deposits and durability of a constant speed compression ignition engine. *Applied Energy, 88*, 2900–2907.

Alkemade, U. G., & Schumann, B. (2006). Engines and exhaust after treatment systems for future automotive applications. *Solid State Ionics, 177*, 2291–2296.

Al-Qurashi, K., Lueking, A. D., & Boehman, A. L. (2011). The deconvolution of the thermal, dilution, and chemical effects of exhaust gas recirculation (EGR) on the reactivity of engine and flame soot. *Combustion and Flame, 158*(9), 1696–1704.

Arcoumanis, C., Bae, C., Crookes, R., & Kinoshita, E. (2008). The potential of di-methyl ether (DME) as an alternative fuel for compression-ignition engines: A review. *Fuel, 87*, 1014–1030.

Armas, O., Ballesteros, R., Martosb, F. J., & Agudeloc, J. R. (2005). Characterization of light duty Diesel engine pollutant emissions using water-emulsified fuel. *Fuel, 84*, 1011–1018.

Armas, O., Yehliu, K., & Boehman, A. L. (2010). Effect of alternative fuels on exhaust emissions during diesel engine operation with matched combustion phasing. *Fuel, 89*, 438–456.

Bermúdez, V., Lujan, J. M., Pla, B., & Linares, W. G. (2011). Comparative study of regulated and unregulated gaseous emissions during NEDC in a light-duty diesel engine fuelled with Fischer Tropsch and biodiesel fuels. *Biomass and Bioenergy, 35*(2), 789–798.

Bhattacharyya, S., & Reddy, C. S. (1994). Vegetable oils as fuels for internal combustion engines: A review. *Journal of Agricultural Engineering Research, 57*, 157–166.

Bozbas, K. (2008). Biodiesel as an alternative motor fuel: Production and policies in the European Union. *Renewable and Sustainable Energy Reviews, 12*, 542–552.

Brusca, S., Lanzafame, R. (2001). Evaluation of the effects of water injection in a single cylinder CFR cetane engine. *SAE paper* 2001-01-2012

Carlucci, A. P., Laforgia, D., Saracino, R., & Toto, G. (2011). Combustion and emissions control in diesel–methane dual fuel engines: The effects of methane supply method combined with variable in-cylinder charge bulk motion. *Energy Conversion and Management, 52*(8–9), 3004–3017.

Cheng, C. H., Cheung, C. S., Chan, T. L., Lee, S. C., & Yao, C. D. (2008). Experimental investigation on the performance, gaseous and particulate emissions of a methanol fumigated diesel engine. *Science of the Total Environment, 389*, 115–124.

Cheung, C. S., Zhu, R., & Huang, Z. (2011). Investigation on the gaseous and particulate emissions of a compression ignition engine fueled with diesel–dimethyl carbonate blends. *Science of the Total Environment, 409*(3), 523–529.

Chotwichien, A., Luengnaruemitchai, A., & Jai-In, S. (2009). Utilization of palm oil alkyl esters as an additive in ethanol–diesel and butanol–diesel blends. *Fuel, 88*(9), 1618–1624.

Cordiner, S., Gambino, M., Iannaccone, S., Rocco, V., & Scarcelli, R. (2008). Numerical and experimental analysis of combustion and exhaust emissions in a dual-fuel diesel/natural gas engine. *Energy and Fuels, 22*(3), 1418–1424.

Demirbas, A. (2007). Progress and recent trends in biofuels. *Progress in Energy and Combustion Science, 33*, 1–18.

Fino, D., Fino, P., Saracco, G., & Specchia, V. (2003). Innovative means for the catalytic regeneration of particulate traps for diesel exhaust cleaning. *Chemical Engineering Science, 58*, 951–958.

Forzatti, P., Lietti, L., Nova, I., & Tronconi, E. (2010). Diesel NO_x aftertreatment catalytic technologies: Analogies in LNT and SCR catalytic chemistry. *Catalysis Today, 151*, 202–211.

Ghazikhani, M., Feyz, M. E., & Joharchi, A. (2010). Experimental investigation of the exhaust gas recirculation effects on irreversibility and brake specific fuel consumption of indirect injection diesel engines. *Applied Thermal Engineering, 30*, 1711–1718.

Gill, S. S., Tsolakis, A., Dearn, K. D., & Rodríguez-Fernández, J. (2011). Combustion characteristics and emissions of Fischer-Tropsch diesel fuels in IC engines. *Progress in Energy and Combustion Science, 37*(4), 503–523.

Gray, P. G., & Frost, J. C. (1998). Impact of catalyst on clean energy in road transportation. *Energy & Fuels, 12*, 1121–1129.

Hansen, A. C., Zhang, Q., & Lyne, P. W. L. (2005). Ethanol–diesel fuel blends—a review. *Bioresource Technology, 96*, 277–285.

Hayes, T.K., Savage, L.D., White, R.A., Sorenson, S.C. (1988). The effect of fumigation of different ethanol proofs on a turbocharged diesel engine. *SAE Paper* 880497

He, B.Q., Wang, J.X., Shuai, S.J., Yan, X.G. (2004). Homogenous charge combustion and emissions of ethanol ignited by pilot diesel on diesel engines. *SAE Paper* 2004-01-0094

Hossain, A. K., & Davies, P. A. (2010). Plant oils as fuels for compression ignition engines: A technical review and life-cycle analysis. *Renewable Energy, 35*, 1–13.

Huang, J., Wang, Y., Li, S., Roskilly, A. P., Yu, H., & Li, H. (2009). Experimental investigation on the performance and emissions of a diesel engine fuelled with ethanol-diesel blends. *Applied Thermal Engineering, 29*, 2484–2490.

Imahashi, T., Hashimoto, K., Hayashi, J. I., Yamada, T. (1995). Research on NO_x reduction for large marine diesel engines. ISME Yokohama

Jiang, Q. Q., Ottikkutti, P., Vangerpen, J. (1990). The effect of alcohol fumigation on diesel flame temperature and emissions. *SAE Paper* 900386

Kadota, T., & Yamasaki, H. (2002). Recent advances in the combustion of water fuel emulsion. *Progress in Energy and Combustion Science, 28*, 385–404.

Kalam, M. A., Husnawan, M., & Masjuki, H. H. (2003). Exhaust emission and combustion evaluation of coconut oil-powered indirect injection diesel engine. *Renewable Energy, 28*, 2405–2415.

Kamasamudram, K., Currier, N. W., Chen, X., & Yezerets, A. (2010). Overview of the practically important behaviors of zeolite-based urea-SCR catalysts, using compact experimental protocol. *Catalysis Today, 151*, 212–222.

Kegl, B., Pehan, S. (2001). Reduction of diesel engine emissions by water injection. *SAE paper* 2001-01-3259

Kim, H., Kim, Y., & Lee, K. (2008). A study of the characteristics of mixture formation and combustion in a PCCI engine using an early multiple injection strategy. *Energy & Fuels, 22*, 1542–1548.

Kim, H. J., Park, S. H., Lee, K. S., & Lee, C. S. (2011). A study of spray strategies on improvement of engine performance and emissions reduction characteristics in a DME fueled diesel engine. *Energy, 36*(3), 1802–1813.

Kook, S., Park, S., & Bae, C. (2008). Influence of early fuel injection timings on premixing and combustion in a diesel engine. *Energy & Fuels, 22*, 331–337.

Kowalewicz, A., & Wojtyniak, M. (2005). Alternative fuels and their application to combustion engines. *Proceedings of the Institution of Mechanical Engineers, Part D: Journal of Automobile Engineering, 219*, 103–125.

Kwanchareon, P., Luengnaruemitchai, A., & Jai-In, S. (2007). Solubility of a diesel–biodiesel–ethanol blend, its fuel properties, and its emission characteristics from diesel engine. *Fuel, 86* (7–8), 1053–1061.

Lapuerta, M., Armas, O., & García-Contreras, R. (2007). Stability of diesel–bioethanol blends for use in diesel engines. *Fuel, 86*(10–11), 1351–1357.

Lapuerta, M., Armas, O., & García-Contreras, R. (2009). Effect of ethanol on blending stability and diesel engine emissions. *Energy & Fuels, 23*(9), 4343–4354.

Leahey, D. M., Jones, B. C., Gilligan, J. W., Brown, L. P., Hamilton, L. J., Gutteridge, C. E., Cowart, J. S., Caton, P. A. (2007). Combustion of biodiesel- and ethanol–diesel mixtures with intake injection. *SAE Paper* 2007-01-4011

Lebedevas, S., Lebedeva, G., Makareviciene, V., Janulis, P., & Sendzikiene, E. (2009). Usage of fuel mixtures containing ethanol and rapeseed oil methyl esters in a diesel engine. *Energy & Fuels, 23*(1), 217–223.

Li, D., Zhen, H., Xingcai, L., Wu-gao, Z., & Jian-guang, Y. (2005). Physico-chemical properties of ethanol-diesel blend fuel and its effect on performance and emissions of diesel engines. *Renewable Energy, 30*, 967–976.

Liew, C., Li, H., Nuszkowski, J., Liu, S., Gatts, T., Atkinson, R., & Clark, N. (2010). An experimental investigation of the combustion process of a heavy-duty diesel engine enriched with H_2. *International Journal of Hydrogen Energy, 35*, 11357–11365.

Lif, A., & Homlberg, K. (2006). Water-in-diesel emulsions and related systems. *Advances in Colloid and Interface Science, 123–126*, 231–239.

Lilik, G. K., Zhang, H., Herreros, J. M., Haworth, D. C., & Boehman, A. L. (2010). Hydrogen assisted diesel combustion. *International Journal of Hydrogen Energy, 35*, 4382–4398.

Ma, Z., Huang, Z., Li, C., Wang, X., & Miao, H. (2008). Combustion and emission characteristics of a diesel engine fuelled with diesel–propane blends. *Fuel, 87*(8–9), 1711–1717.

Maiboom, A., & Tauzia, X. (2011). NO_x and PM emissions reduction on an automotive HSDI Diesel engine with water-in-diesel emulsion and EGR: An experimental study. *Fuel, 90*, 3179–3192.

Maiboom, A., Tauzia, X., & Hetet, J. F. (2008). Experimental study of various effects of exhaust gas recirculation (EGR) on combustion and emissions of an automotive direct injection diesel engine. *Energy, 33*, 22–34.

Misra, R. D., & Murthy, M. S. (2010). Straight vegetable oils usage in a compression ignition engine—A review. *Renewable and Sustainable Energy Reviews, 14*, 3005–3013.

Miyano, H., Yoshida, N., Nakai, T., Nagae, Y., Yasueda, S. (1995). Stratified fuel-water injection system for NO_x reduction of diesel engine. ISME Yokohama

Mustel, W. (1997). *Achieving the 2004 heavy-duty diesel emissions using combination between electronic EGR and particulate trap regenerated by a cerium based fuel borne catalyst, 2.* Dresden: Dresdner Motorenkolloquium.

No, S. Y. (2011). Inedible vegetable oils and their derivatives for alternative diesel fuels in CI engines: A review. *Renewable and Sustainable Energy Reviews, 15*, 131–149.

Nwafor, O. M. I. (2000). Effect of choice of pilot fuel on the performance of natural gas in diesel engines. *Renewable Energy, 21*(3–4), 495–504.

Nwafor, O. M. I. (2003). The effect of elevated fuel inlet temperature on performance of diesel engine running on neat vegetable oil at constant speed conditions. *Renewable Energy, 28*, 171–181.

Nwafor, O. M. I., & Rice, G. (1996). Performance of rapeseed oil blends in a diesel engine. *Applied Energy, 4*(4), 345–354.

Odaka, M., Koike, N., Tsokamoto, Y., Narusawa, K., Yoshida, K. (1991). Effects of EGR with a supplemental manifold water injection to control exhaust emissions from heavy-duty diesel powered vehicles. *SAE paper* 910739

Park, J. W., Huh, K. Y., & Park, K. H. (2000). Experimental study on the combustion characteristics of emulsified diesel in a rapid compression and expansion machine. *Proceedings of the Institution of Mechanical Engineers, Part D: Journal of Automobile Engineering, 214*, 579–586.

Peng, H., Cui, Y., Shi, L., & Deng, K. (2008). Effects of exhaust gas recirculation (EGR) on combustion and emissions during cold start of direct injection (DI) diesel engine. *Energy, 33*, 471–479.

Poompipatpong, C., & Cheenkachorn, K. (2011). A modified diesel engine for natural gas operation: Performance and emission tests. *Energy, 36*(12), 6862–6866.

Purushothaman, K., & Nagarajan, G. (2009). Performance, emission and combustion characteristics of a compression ignition engine operating on neat orange oil. *Renewable Energy, 34*, 242–245.

Qi, D. H., Bian, Y. Z. H., Ma, Z. H. Y., Zhang, C. H. H., & Liu, S. H. Q. (2007). Combustion and exhaust emission characteristics of a compression ignition engine using liquefied petroleum gas–diesel blended fuel. *Energy Conversion and Management, 48*, 500–509.

Rakopolulus, C. D., Antopoulos, K. A., Rakopoulos, D. C., & Hountalas, D. T. (2008a). Multi-zone modeling of combustion and emissions formation in DI diesel engine operating on ethanol-diesel fuel blends. *Energy Conversion & Management, 49*, 625–643.

Rakopolulus, D. C., Rakopoulus, C. D., Giakoumis, E. G., Papagiannakis, R. G., & Kyritsis, D. C. (2008b). Experimental-stochastic investigation of the combustion cyclic variability in HSDI diesel engine using ethanol-diesel fuel blends. *Fuel, 87*, 1478–1491.

Rakopoulus, C. D., Antonopoulus, K. A., & Rakopoulus, D. C. (2007). Experimental heat release analysis and emissions of a HSDI diesel engine fueled with ethanol-diesel fuel blends. *Energy, 32*, 1791–1808.

Ramadhas, A. S., Jayaraj, S., & Muraleedharan, C. (2004). Use of vegetable oils as I.C. engine fuels—a review. *Renewable Energy, 29*, 727–742.

Reyes, Y., Aranda, D. A. G., Santander, L. A. M., Cavado, A., & Belchior, C. R. P. (2009). Action principles of cosolvent additives in ethanol diesel blends: stability studies. *Energy & Fuels, 23* (5), 2731–2735.

Ribeiro, N. M., Pinto, A. C., Quintella, C. M., da Rocha, G. O., Teixeira, L. S. G., & Guarieiro, L. L. N. (2007). The role of additives for diesel and diesel blended (ethanol or biodiesel) fuels: a review. *Energy & Fuels, 21*(4), 2433–2445.

Russo, D., Dassisti, M., Lawlor, V., & Olabi, A. G. (2012). State of the art of biofuels from pure plant oil. *Renewable and Sustainable Energy Reviews, 16*, 4056–4070.

Sahoo, B. B., Sahoo, N., & Saha, U. K. (2009). Effect of engine parameters and type of gaseous fuel on the performance of dual-fuel gas diesel engines—A critical review. *Renewable and Sustainable Energy Reviews, 13*(6–7), 1151–1184.

Samec, N., Kegl, B., & Dibble, R. W. (2002). Numerical and experimental study of water/oil emulsified fuel combustion in a diesel engine. *Fuel, 81*, 2035–2044.

Sampara, C. S. (2008). *Global reaction kinetics for oxidation and storage in diesel oxidation catalysts*. Dissertation. University of Michigan

Saravanan, N., & Nagarajan, G. (2010). Performance and emission studies on port injection of hydrogen with varied flow rates with Diesel as an ignition source. *Applied Energy, 87*, 2218–2229.

Saravanan, N., Nagarajan, G., Kalaiselvan, K. M., & Dhanasekaran, C. (2008). An experimental investigation on hydrogen as a dual fuel for diesel engine system with exhaust gas recirculation technique. *Renewable Energy, 33*, 422–427.

Selim, M. Y. E. (2001). Pressure–time characteristics in diesel engine fueled with natural gas. *Renewable Energy, 22*, 473–489.

Sitshebo, S., Tsolakis, A., & Theinnoi, K. (2009). Promoting hydrocarbon-SCR of NO_x in diesel engine exhaust by hydrogen and fuel reforming. *International Journal of Hydrogen Energy, 34*, 7842–7850.

Stanglmaier, R. H., Dingle, P. J., & Stewart, D. W. (2008). Cycle-controlled water injection for steady-state and transient emissions reduction from a heavy-duty diesel engine. *Journal of Engineering for Gas Turbine and Power, 130*, 103–111.

Stanislaus, A., Marafi, A., & Rana, M. S. (2010). Recent advances in the science and technology of ultra low sulfur diesel (ULSD) production. *Catalyst Today, 153*, 1–68.

Subramanian, K. A. (2011). A comparison of water-diesel emulsion and timed injection of water into the intake manifold of a diesel engine for simultaneous control of NO and smoke emissions. *Energy Conversion and Management, 52*, 849–857.

Surawski, N., Miljevic, B., Roberts, B. A., Modini, R., Situ, R., Brown, R. J., Bottle, S. E., & Ristovski, Z. D. (2010). Particle emissions, volatility, and toxicity from an ethanol fumigated compression ignition engine. *Environmental Science and Technology, 44*, 229–235.

Suzuki, T., (1997). Development and perspective of the diesel combustion system for commercial vehicles. *IMechE Combustion Engine Group Prestige Lecture*, London.

Tauzia, X., Maiboom, A., & Shah, S. R. (2010). Experimental study of inlet manifold water injection on combustion and emissions of an automotive direct injection Diesel engine. *Energy, 35*, 3628–3639.

Tesfa, B., Mishra, R., Gu, F., & Ball, A. D. (2012). Water injection effects on the performance and emission characteristics of a CI engine operating with biodiesel. *Renewable Energy, 37*, 333–344.

Torres-Jimenez, E., Dorado, M. P., & Kegl, B. (2011a). Experimental investigation on injection characteristics of bioethanol–diesel fuel and bioethanol–biodiesel blends. *Fuel, 90*, 1968–1979.

Torres-Jimenez, E., Svoljšak-Jerman, M., Gregorc, A., Lisec, I., Dorado, M. P., & Kegl, B. (2010). Physical and chemical properties of ethanol–biodiesel blends for diesel engines. *Energy & Fuels, 24*, 2002–2009.

Torres-Jimenez, E., Svoljšak-Jerman, M., Gregorc, A., Lisec, I., Dorado, M. P., & Kegl, B. (2011b). Physical and chemical properties of ethanol–diesel fuel blends. *Fuel, 90*, 795–802.

Twigg, M. V. (2007). Progress and future challenges in controlling automotive exhaust gas emissions. *Applied Catalyst B: Environmental, 70*, 2–15.

Wang, J., Mao, X., Zhu, K., Song, J., & Zhuo, B. (2009). An intelligent diagnostic tool for electronically controlled diesel engine. *Mechatronics, 19*, 859–867.

Wang, L. J., Song, R. Z., Zou, H. B., Liu, S. H., & Zhou, L. B. (2008). Study on combustion characteristics of a methanol diesel dual fuel compression ignition engine. *Proceedings of the Institution of Mechanical Engineers, Part D: Journal of Automobile Engineering, 222*, 619–627.

Weber de Menezes, E., da Silva, R., Cataluña, R., & Ortega, R. J. C. (2006). Effect of ethers and ether/ethanol additives on the physicochemical properties of diesel fuel and on engine tests. *Fuel, 85*(5–6), 815–822.

Xiaolu, L., Hongyan, C., Zhiyong, Z., & Zhen, H. (2006). Study of combustion and emission characteristics of a diesel engine operated with dimethyl carbonate. *Energy Conversion and Management, 47*(11–12), 1438–1448.

Yao, C., Cheung, C. S., Cheng, C., Wang, Y., Chan, T. L., & Lee, S. C. (2008). Effect of diesel/methanol compound combustion on diesel engine combustion and emissions. *Energy Conversion and Management, 49*(6), 1696–1704.

Youn, I. M., Park, S. H., Roh, H. G., & Lee, C. S. (2011). Investigation on the fuel spray and emission reduction characteristics for dimethyl ether (DME) fueled multi-cylinder diesel engine with common-rail injection system. *Fuel Processing Technology, 92*(7), 1280–1287.

Zhang, Z. H., Cheung, C. S., Chan, T. L., & Yao, C. D. (2009). Emission reduction from diesel engine using fumigation methanol and diesel oxidation catalyst. *Science of the Total Environment, 407*, 4497–4504.

Zhang, Z. H., Cheung, C. S., Chan, T. L., & Yao, C. D. (2010). Experimental investigation on regulated and unregulated emissions of a diesel/methanol compound combustion engine with and without diesel oxidation catalyst. *Science of the Total Environment, 408*(4), 865–872.

Zheng, M., Reader, G. T., & Hawley, J. G. (2004). Diesel engine exhaust gas recirculation—A review on advanced and novel concepts. *Energy Conversion and Management, 45*(6), 883–900.

Zhu, Z., Li, D. K., Liu, J., Wei, Y. J., & Liu, S. H. (2012). Investigation on the regulated and unregulated emissions of a DME engine under different injection timing. *Applied Thermal Engineering, 35*, 9–14.

Chapter 4
Biodiesel as Diesel Engine Fuel

In recent years, the interest to use biodiesel as a substitute for mineral diesel has been increasing steadily. Biodiesel is a renewable fuel, consisting of various fatty acid methyl esters with the exact composition depending on the feedstock. This is a distinctly different composition than the hydrocarbon content of mineral diesel. In spite of that, biodiesel has many properties very close to those of mineral diesel. Consequently, the required biodiesel-related modifications of the diesel engine are typically rather minor. On the other hand, because of its different chemical character, biodiesel has several properties, which differ from those of mineral diesel just enough to offer an opportunity to reduce harmful emissions without worsening other economy and engine performances. It should be noted, however, that biodiesel properties may depend heavily on its raw materials.

4.1 Biodiesel Sources

Biodiesel is derived from renewable resources, such as vegetable oils, animal fats, and waste restaurant greases. Biodiesel is an environmentally friendly alternative fuel, but typically comes at a higher price than mineral diesel (Demirbas 2008a, 2009). The cost of biodiesel varies in dependence on the base stock, geographic area, variability in crop production from season to season, the price of the crude petroleum, and other factors.

The typical raw materials for biodiesel are oils from rapeseed, canola, soybean, sunflower, palm, beef and sheep tallow, poultry, fish, jatropha, almond, barley, camelina, coconut, copra, groundnut, karanja, laurel, oat, coffee beans, poppy seed, okra seed, rice bran, sesame, sorghum, wheat, and microalgae (Altiparmak et al. 2007; Arkoudeas et al. 2003; Balat and Balat 2008; Çetinkaya et al. 2005; Chen and Chen 2011; Chisti 2007; Demirbas 2006, 2008a; Giannelos et al. 2005; Gomez et al. 2002; Goodrum and Eitchman 1996; Hu et al. 2005; Jain and Sharma 2010; Jha et al. 2008; Kalam and Masjuki 2002; Lapinskiene et al. 2006; Lebedevas et al. 2006; Misra and Murthy 2011; Murugesan et al. 2009; Oliveira et al. 2008;

B. Kegl et al., *Green Diesel Engines*, Lecture Notes in Energy 12,
DOI 10.1007/978-1-4471-5325-2_4, © Springer-Verlag London 2013

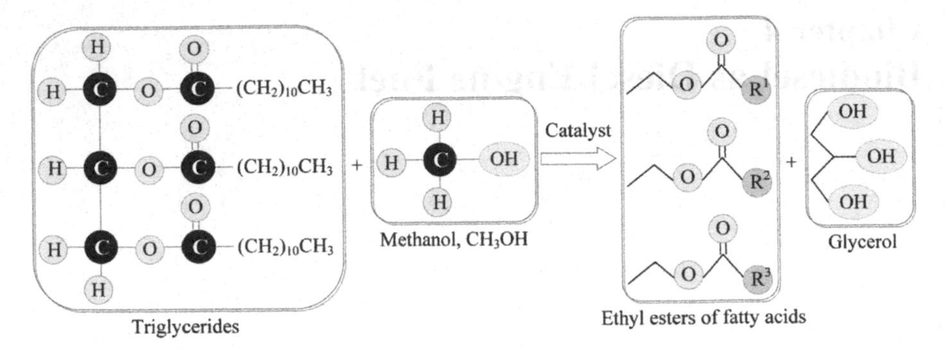

Fig. 4.1 Transesterification process

Park et al. 2008; Peng et al. 2006; Peterson and Hustrulid 1998; Senzikiene et al. 2006; Sharma et al. 2008; Shumaker et al. 2008; Smith et al. 2010; Tang et al. 2008; Yuan et al. 2005).

Biodiesel is generally produced by transesterification of vegetable oils and animal fats as shown in Fig. 4.1. Transesterification is a chemical reaction between triglycerides and a short-chain alcohol in the presence of a catalyst to produce monoesters (Demirbas 2009; West et al. 2008). The commonly used catalysts in the ester reaction are lipase catalyst, acid catalyst, and alkali catalyst. The most of the commercial biodiesel is produced from plant oils, by using very effective alkali catalysts such as sodium or potassium hydroxides, carbonates, or alkoxides.

The long- and branched-chain triglyceride molecules are transformed to monoesters and glycerin. Commonly used short-chain alcohols are methanol, ethanol, propanol, and butanol. Methanol is the most used commercially because of its low price (Lin et al. 2011; Sinha et al. 2008). The by-product glycerin is used in pharmaceuticals.

Transesterification of vegetable oils and animal fats is a costly and time-consuming process that needs expensive reactants and laboratory equipment. In this context, numerical simulation, for example, by using an artificial neural network model, can be useful to simulate biodiesel production through the transesterification process (Yuste and Dorado 2006) in order to reduce cost and long time-consuming laboratory tests.

Biodiesel produced from food-grade oils is called *biodiesel first generation*, while biodiesel produced from microalgae is called *biodiesel new generation*. Like plants, microalgae use sunlight, CO_2, and water to produce oil but they do so more efficiently than crop plants. Oil content in microalgae can exceed 80 % by weight of dry biomass (Vijayaraghavan and Hemanathan 2009). Algae can grow extremely rapidly by doubling their biomass within 24 h. The biodiesel from microalgae is produced through growth, harvest, extraction, and conversion phase (Fig. 4.2) (Amaro et al. 2011; Lin et al. 2011). In the conversion phase, the transesterification process can be performed by treating the triglycerides extracted in hexane with ethanol in the presence of potassium hydroxide as a catalyst (Singh

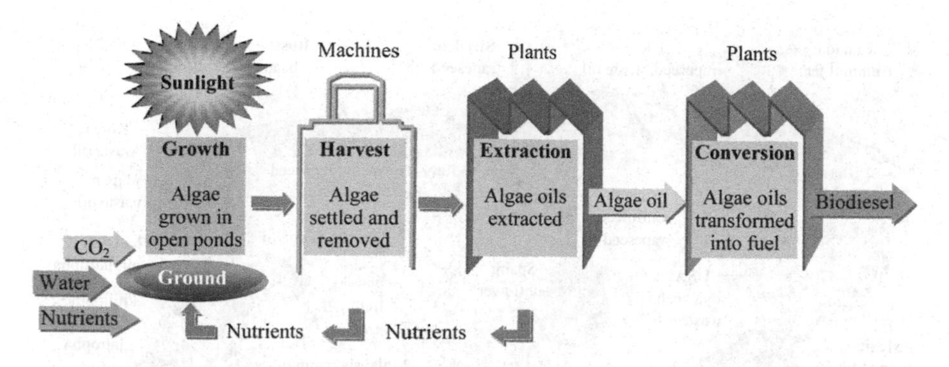

Fig. 4.2 Processing of microalgae to biodiesel

and Olsen 2011; Singh et al. 2011; Singh and Singh 2010; Vijayaraghavan and Hemanathan 2009).

The raw materials for biodiesel production are quite different in various parts of the world and they depend greatly on climate, local soil conditions, and availability (Fig. 4.3) (Lin et al. 2011).

In the USA, soybean oil is the most commonly biodiesel feedstock, whereas the rapeseed (canola) oil and palm oil are the most common source for biodiesel in Europe and in tropical countries, respectively (Singh and Singh 2010). World biodiesel production has risen from 1.8 billion liters in 2003 (Bozbas 2008) to 19 billion liters in 2010. The largest producer of biodiesel is the European Union, which generated 53 % of all biodiesel in 2010. The top three biodiesel producing nations in Europe are Germany, France, and Italy.

The biodiesel source material should fulfill at least the following two requirements: low production costs and large production scale. On one hand, refined oils have high production costs and low production scale. On the other side, nonedible seeds, algae, and sewerage have low production costs and are more easily available than refined or recycled oils. Additionally, the percentage of oil and the yield per hectare are also important parameters to be considered when evaluating a biodiesel source. It is worth noting that the risks of handling, transporting, and storing biodiesel are much lower than those associated with mineral diesel (Demirbas 2009).

4.2 Biodiesel Properties

Biodiesel is an oxygenated, sulfur-free, biodegradable, nontoxic, and environmentally friendly alternative diesel fuel. The most important physical and chemical properties of biodiesel that influence diesel engine performance, ecology, and economy characteristics are long-time chemical and temperature *stability*, *density*, *viscosity*, *sound velocity*, *bulk modulus*, *cetane number*, *cloud point* (CP), *pour*

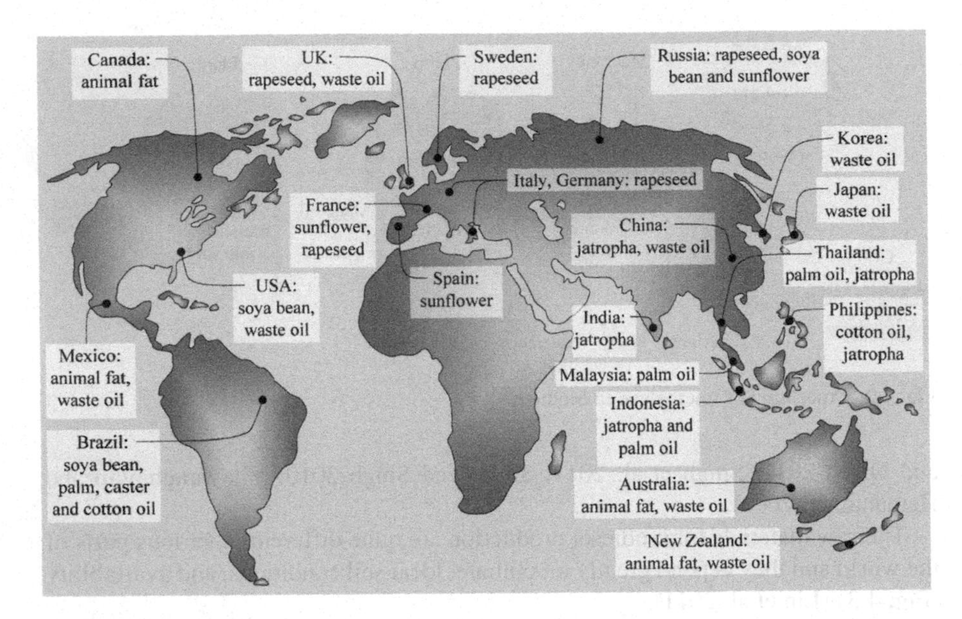

Fig. 4.3 Biodiesel sources around the world

point (PP), *cold filter plugging point* (CFPP), *flash point*, *filter plugging tendency* (FPT), *corrosiveness*, *lubricity*, *chemical composition*, *contaminants content*, and *water content* (Hoekman et al. 2012; Schleicher et al. 2009; Torres et al. 2010). Furthermore, quality control of fuel-related properties of biodiesel, such as volatility, is also needed in order to assure the expected engine performance. In this context, the vapor pressures and boiling points that can be determined by the thermogravimetric analysis method can be also be taken as quality control metrics for biodiesel (Conceição et al. 2007; Goodrum 2002).

Fuel Stability. Fuel stability is related to many factors like fuel degradation, phase separation, oxidation, polymerization, and so on. In the scope of this book only fuel *degradation* and *phase separation* will be addressed briefly since biodiesel usage, either as an additive or pure, influences these properties significantly.

Biodiesel exhibits, from the ecological point of view, a rather good and welcome biodegradability. On the other hand, this property implicates a bunch of problems when storing biodiesel (Atadashi et al. 2010). The reasons for enhanced biodegradability are mostly unknown (Schleicher et al. 2009). It is assumed however, that one of the reasons could be the additionally offered carbon source that can influence the *growth of microorganisms*. Furthermore, the excretion of biodiesel metabolizing microorganisms may positively affect the growth of microorganisms metabolizing mineral oil and be the source of enzymes needed for degrading mineral oil. Anyhow, microbiological growth and the resulting fuel degradation is simply one of the consequences of the biodegradability of biodiesel.

Phase separation is another fuel stability problem since biodiesel is often blended with other fuels. This must be prevented even during a long time periods and various temperatures in order to enable appropriate fuel injection and combustion processes (Torres et al. 2010). If a blend of various fuels separates after some period of time, this means that variable fuel is injected into the cylinder. Because the engine management is optimized for a specific fuel, the engine characteristics will vary with respect to time. In this case, some additional mixing of blends will be necessary. Besides of this, in the case of solidification of some fuel components at lower temperatures, fuel delivery to some cylinders may become seriously hindered, which is of course completely unacceptable.

Finally, it may be noted that the *oxidative stability* may also be problematic under some circumstances. It mostly depends on biodiesel sources, the age of biodiesel, and the conditions under which the fuel has been stored (Hoekman et al. 2012).

Density, Viscosity, Bulk Modulus, and Sound Velocity. Fuel density, viscosity, bulk modulus of elasticity, and sound velocity influence the injection characteristics significantly. At low temperature *fuel density* and *viscosity* rise, which increases both the pressure drop through the filter and the flow resistance through the low-pressure pump gallery (Allen et al. 1999; Kegl 2008; Torres et al. 2010). Viscosity is a measure of resistance to flow of a liquid due to internal friction. Higher viscosity affects the atomization of the fuel after injection and, thereby, ultimately increases the tendency of the formation of engine deposits. Especially in mechanical injection systems, higher viscosity leads to reduced fuel losses during the injection process, to faster evolution of pressure, and thus to advanced injection timing. The temperature dependency of viscosity of biodiesels must be known in order to design properly eventual heat exchangers in the fuel supply system (Kerschbaum and Rinke 2004). *Bulk modulus* is defined as a measure of compressibility of a fluid under hydrostatic pressure. As the pressure is increased, less space between molecules is available, and the resistance of compression increases. Thus, the bulk modulus increases with the increase of injection pressure (Yoon et al. 2009). A higher bulk modulus leads to a more rapid pressure wave propagation from the pump to the needle nozzle and an earlier needle lift. Finally, it should be noted that viscosity and bulk modulus define the *sound velocity* in the fuel which is a very important property for mechanically controlled injection systems.

Cetane Number. *Cetane number* is one of the most significant fuel properties that *specify the ignition quality* of a fuel in a diesel engine. The cetane number of biodiesel depends upon various parameters, such as biodiesel raw materials, climate conditions where the raw material is collected, and the biodiesel production technology. For these reasons, the cetane number of biodiesel fuels varies in a wide range of 48–67 (Ramadhas et al. 2006). The cetane number can impact startability, noise level, and exhaust emissions of a diesel engine.

Cloud Point, Pour Point, and Cold Filter Plugging Point. Low temperature performance is one of the most important considerations for diesel fuels. *Cloud* *point* (CP), *pour point* (PP), and *cold filter plugging point* (CFPP) provide important information on cold weather behavior of the fuel. The CP usually occurs at a higher temperature than the PP. This is because the CP is the temperature below which wax (paraffin) in diesel fuel or biowax in biodiesel form a cloudy appearance and the PP is the lowest temperature at which the fluid still flows. If CP, PP, and CFPP are too high, this means that solids and crystals grow rapidly, agglomerate, clog fuel lines and filters, and cause major operability problems (Bhale et al. 2009; Hoekman et al. 2012; Joshi and Pegg 2007; Torres et al. 2010). To avoid possible problems when fuel passes through the filtration system, a low CFPP is recommended. This is important in low-temperate countries since a high CFPP will clog vehicle engines more easily.

Flash Point. *Flash point* is the lowest temperature at which the fuel can form an ignitable mixture with air. This property affects the *shipping* and *storage classification*. A lower flash point means higher precautions in handling and transporting the fuel. The flash point is inversely related to fuel volatility (Hoekman et al. 2012; Torres et al. 2010).

Filter Plugging Tendency. *Filter plugging tendency* (FPT) is related to the quality of fuels and predicts the *applicability of these fuels after long storage periods* (Torres et al. 2010).

Lubricity. *Lubricity* refers to the reduction of friction between solid surfaces in relative motion (Hoekman et al. 2012). Fuel lubricity characteristics play a significant role in the engine lubrication systems. In diesel engines, the fuel is part of the engine lubrication process and gains special interest when oxygenated blended fuels are used (Torres et al. 2010). In many cases, the fuel itself is the only lubricant within a fuel injector. With increasing fuel injection pressure, injector rate shaping, and multiple injections per cycle, maintaining an adequate lubricity may become critical (Hoekman et al. 2012). For lubricity, the presence of a polarity-imparting heteroatom, preferable oxygen, with the nature and number of the oxygen moiety, and a carbon chain of sufficient length, has a significant role. If sufficient oxygenated moieties of lubricity imparting capability are present, then even a short-chain compound will possess excellent lubricity in the neat form, although a lack of miscibility with hydrocarbons such as mineral diesel will not impart lubricity when using this material as an additive (Knothe and Steidley 2005a, b). Lubricity strongly depends on the nature in which oxygen atoms are bound in the molecule. Lubricity properties of a fuel are important factors influencing *friction wear* in the engine (Demirbas 2009).

Chemical Composition. The main chemical elements in biodiesel and mineral diesel fuels are *carbon* and *hydrogen*, while the rest is mainly oxygen and nitrogen (Balat and Balat 2008). An excess of *oxygen* in biodiesel fuels is expected to have a beneficial effect on the emissions profile, compared to mineral diesel, because it improves the combustion process which leads to lower soot emission. Mineral diesel can comprise maximal 0.05 wt.% of *sulfur*, while biodiesel contains virtually no sulfur. This offers an extra advantage in potential reduction of acid rain production.

Contaminants Content. Contaminants content determines whether a fuel can be commercialized at all. This is because the presence of *contaminants* can lead to severe operational problems, such as *engine deposits*. Fourier transformation infrared (FT-IR) analysis can be used to determine if any undesirable component (contaminant) is present in the fuel. Using FT-IR analysis, it is rarely, if ever, possible to identify an unknown compound, which is in contrast to the gas chromatography (GC) technique. However, FT-IR analysis is becoming a preferred method for quality control of biofuels, because of its operational ease, rapidity of measurement, and nondestructiveness (Conceição et al. 2005; Lira et al. 2010; Torres et al. 2010). The IR spectrum of a given compound is unique and can therefore serve as a fingerprint for this compound. Consequently, by referring to known spectra, this analysis makes possible to identify any known compound. On the other hand, GC tests need to be used when it is necessary to obtain the fatty acid profile of any source, and these tests also deliver information on the raw material used in biodiesel production.

Water Content and Corrosiveness. When the *water content* within biodiesel exceeds some limits, aqueous microorganisms can appear. Water is part of the respiration system of most microbes. Biodiesel is a great food for microbes and water is necessary for microbe respiration. Therefore, the presence of water accelerates the growth of microbe colonies, which can seriously plug up a fuel injection system. Thus, water content must be checked and controlled (Torres et al. 2010). The presence of water has also negative effects on the yields of methyl esters in the catalyzed method (Demirbas 2009). Furthermore, water content in biodiesel reduces the heat of combustion. This leads to more smoke, harder starting, and reduced power. Below 273 K water begins to form ice crystals in biodiesel. These crystals provide sites of nucleation and accelerate the gelling of the residual fuel. Water can cause *corrosion* of several system components. Consequently, parts of the fuel injection system and fuel tank can potentially be damaged by the use of new fuels, if their water content is not negligible.

At the bottom line, the biggest advantage of biodiesel against mineral diesel is its environmental friendliness. Further *advantages of biodiesel* as a diesel fuel are its portability, availability, renewability, higher combustion efficiency, lower sulfur and aromatic content, higher cetane number, higher biodegradability, domestic origin, its potential for reducing a given economy's dependency on imported petroleum, high flash point, and inherent lubricity in the neat form. The main *disadvantages of biodiesel* are its higher viscosity, lower energy content, higher cloud point and pour point, injector cooking, and high price (Demirbas 2009; Kegl 2006, 2008; Torres et al. 2010).

4.2.1 *Measurement Methods for Biodiesel Properties*

Physical and chemical properties of biodiesel have to be investigated carefully by using various measurement methods. Some of the techniques published recently (Gomez et al. 2002; Hoekman et al. 2012; Kerschbaum and Rinke 2004; Knothe 2005; Lapuerta et al. 2009; Schleicher et al. 2009; Tate et al. 2006a, b; Torres et al. 2010; Wain et al. 2005) are briefly outlined in the following.

Fuel Stability. In order to estimate fuel stability in the context of biodegradation, the corresponding methods focus on fuel behavior in aqueous milieu and in anhydrous milieu as occurring during storage in vehicle tanks. In either case, the fuel is stored under controlled conditions and observed over an adequate period of time. In anhydrous milieu the fuel samples can be incubated over a time period of more than 100 days at various temperatures (Schleicher et al. 2009). In this case, sterile fuel or various fuel blends are inoculated with blends of microorganisms and analyzed over a given time period.

Fuel stability is also related to possible phase separation. In order to estimate possible problems, the fuel has to be observed for a period of several weeks at various temperatures. The results may be used in order to determine if additional mixing, additives, or heating (in the case that low temperatures produce solid aggregation of fuel) are necessary (Torres et al. 2010). The samples typically have to be tested at ambient, representative summer, representative winter, and representative critical winter temperatures. Usually, each sample is observed optically at regular time intervals during the test. A test period of 4–6 weeks is estimated to be long enough to know the fuel behavior during long periods of storage.

Density. One of the latest density measurement devices is the *APAAR density meter DMA 48*. It enables to measure the density accurately at some reference temperature (for example, at 15 °C) and to calculate the density at other temperatures (Torres et al. 2010). To calculate the density value, a small portion (less than 1 ml) of test fuel is introduced into a temperature-controlled sample cell. The oscillation frequency f is noted and the density of the test sample is calculated. Some measuring techniques such as the one by using a simple density meter, the technique based on sound velocity, and the technique based on the natural frequency response (*DMA 48 density meter* is a good example), are presented in Fig. 4.4. In the case of DMA 48 density meter, a U-shaped glass tube of known volume and mass is filled with the liquid sample and excited electronically by a piezo element. The U-tube is kept oscillating continuously at some characteristic frequency f. This frequency is inversely proportional to the density ρ of the filled-in sample. The optical pick-ups record the oscillation period P as $P = 1/f$. The density ρ is calculated from the period P and the coefficients A and B as

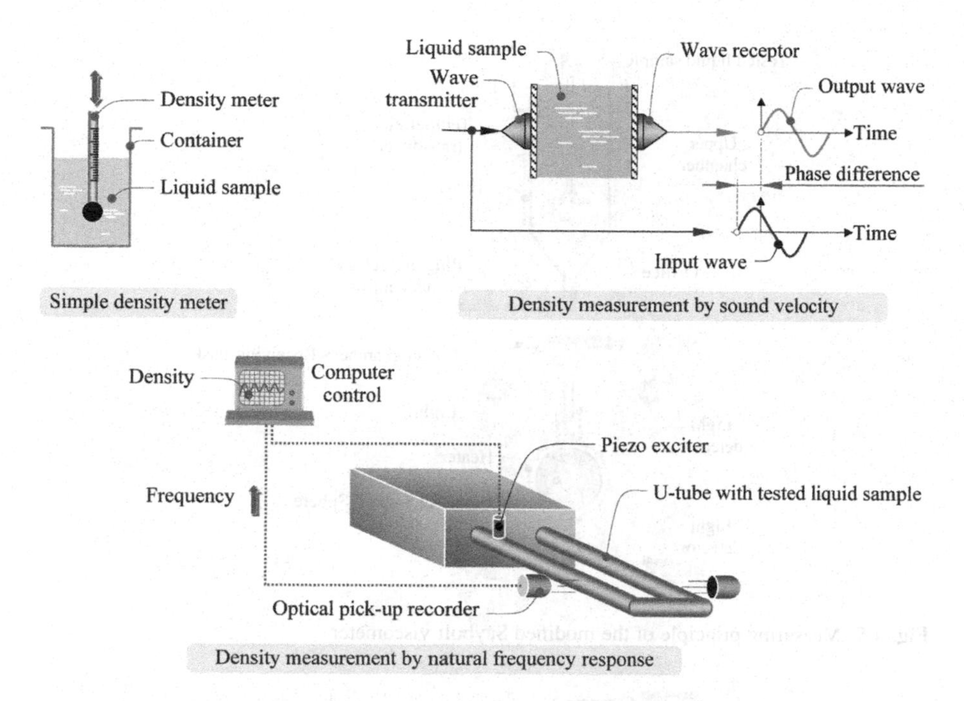

Fig. 4.4 Density measuring techniques

$$\rho = A \times P^2 - B, \tag{4.1}$$

where ρ is the density [kg/m^3], P is the time period [s], and A and B are the corresponding cell constants, determined by measuring the oscillation frequencies when the cell is filled with a calibration fluid of a known density.

Viscosity. *Kinematic viscosity* can be measured by the resistance to flow of a fluid under gravity. It can be determined, for example, by a *Walter Herzog GmbH MP-480* device (Torres et al. 2010). The test consists of measuring the time for a fixed volume of liquid to flow under gravity through a capillary at a known and well-controlled temperature. To determine the temperature dependency of viscosity, the *Thermo–Haake viscometer* (Kerschbaum and Rinke 2004) can be used. In this case, the measurement starts at somewhat higher temperature which is then gradually decreased during the test. The end temperature of the measurement is the one at which the samples become solid.

In order to perform measurements with special accuracy and repeatability requirements, special devices may be needed. For example, to measure the kinematic viscosity within 0.056 mm^2/s with 2 % repeatability, a *modified Saybolt viscometer* (Tate et al. 2006a, b) with three part chambers can be used (Fig. 4.5). The detection system at the lower chamber contains two lights and two photodiodes. When the fuel level reaches the detector, the fuel impedes the amount of light transmitted to the diode and a distinct decrease in the signal voltage from the diode is observed.

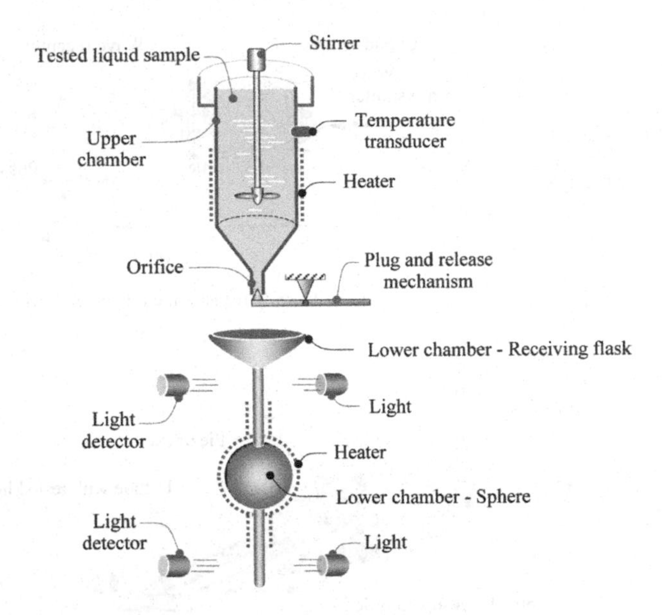

Fig. 4.5 Measuring principle of the modified Saybolt viscometer

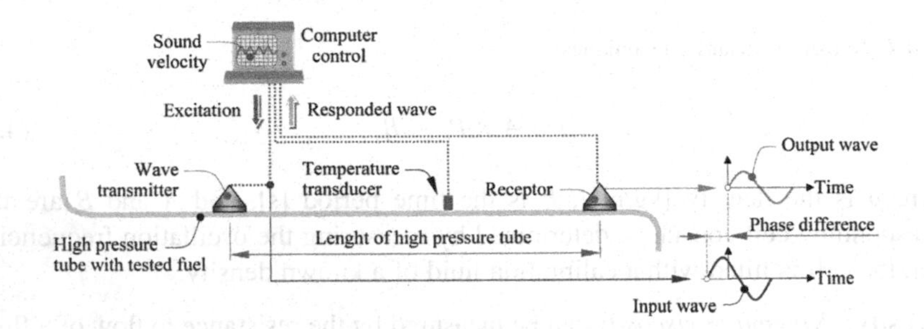

Fig. 4.6 Principle of sound velocity measurement

Sound Velocity. The measurement of sound velocity in fuel is usually based on the principle of pressure wave propagation. The time difference between two pressure waves is typically measured on some specified length of the diesel engine high pressure tube (Fig. 4.6).

Cetane Number. The cetane number can be determined by using the Co-operative Fuel Research (*CFR*) *engine*, which is a one-cylinder, four-stroke cycle engine with a cylindrical precombustion chamber and a compression ratio capability ranging from 6:1 to 28:1 (Fig. 4.7). The standard cetane number measuring method, *ASTM 613*, consists of running the engine with the test fuel at the specified conditions of speed, load, and intake temperature (Knothe 2005; Lapuerta et al. 2009). The injection timing is adjusted so that the start of injection is 13° before top dead center. The compression ratio is adjusted until ignition occurs at top dead center.

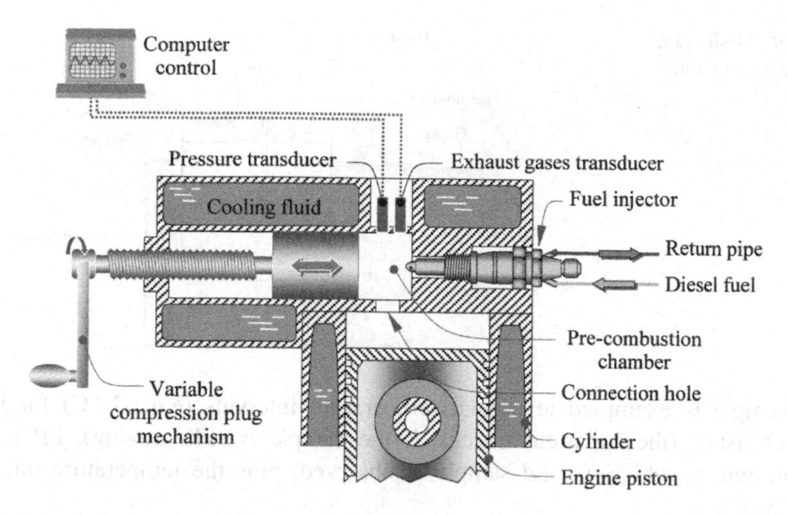

Fig. 4.7 Combustion chamber of the cetane rating engine

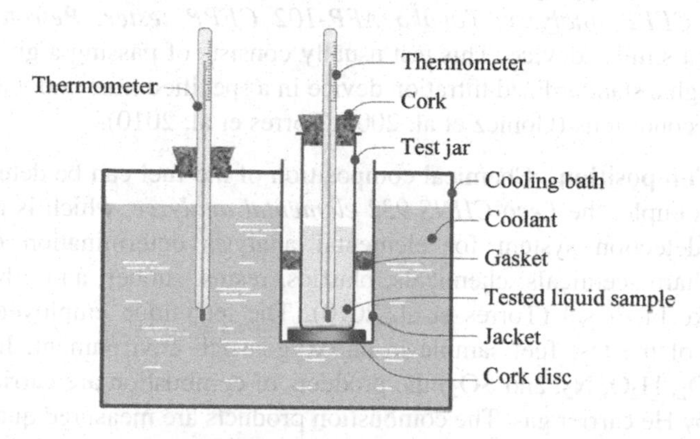

Fig. 4.8 Scheme of the Herzog CPP 97-2 device

The test fuel is then replaced with blends of the reference fuels (blends of hexadecane and hepamethylnonane) until one finds a blend with a slightly higher compression ratio and a blend with a slightly lower compression ratio than the test fuel. The cetane number of the test fuel is determined by linear interpolation of the cetane numbers of the reference fuels.

Cloud Point, Pour Point, and Cold Filter Plugging Point. The cloud point (CP) test can be performed in a *Herzog CPP 97-2* device (Fig. 4.8) by using a prescribed fuel sample which is cooled at some specified rate and examined periodically (Torres et al. 2010). The CP is recorded as the temperature at which a cloud is first observed at the bottom of the test jar.

The pour point (PP) test can be performed in a *Herzog HCP 852 Combi* by using a prescribed fuel sample. At the test start, the fuel has a somewhat higher temperature (e.g., 45 °C) and is then cooled down at some specified rate (Torres et al. 2010).

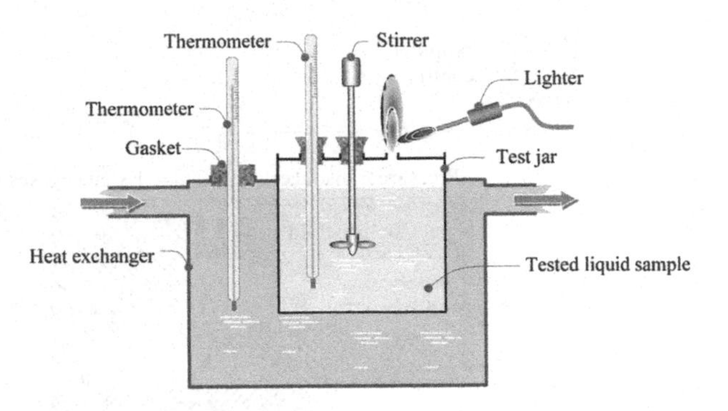

Fig. 4.9 Flash point measurement device

The sample is examined at regular temperature intervals (e.g., 3 °C) for flow characteristics (the apparatus checks, if the sample is still moving). PP is the temperature at which a solid sample is observed, plus the temperature interval (e.g., 3 °C).

The cold filter plugging point (CFPP) can be determined by using, for example, the *Herzog CFPP analyzer, Tanaka AFP-102 CFPP tester*, *Petrotech CFPP analyzer*, or a similar device. This test usually consists of passing a given volume of fuel through a standardized filtration device in a specified time and under certain temperature conditions (Gomez et al. 2002; Torres et al. 2010).

Chemical Composition. Chemical composition of the fuel can be determined by using, for example, the *Leco CHNS 932 elemental analyzer*, which is an infrared absorption detection system for elemental analysis determination of various materials: pharmaceuticals, chemicals, plastics, resins, rubber, and other organic matrices, like biodiesel (Torres et al. 2010). The technique employed relies on combustion of the test fuel sample in an oxygen-rich environment. In a CHNS analysis (CO_2, H_2O, N_2, and SO_2) the products of combustion are carried through the system by He carrier gas. The combustion products are measured quantitatively by means of a nondispersive IR absorption detection system, except for the N_2 which is determined via a thermal conductivity detector. Oxygen has to be measured in a separate furnace. The sample is combusted at high temperature in a carbon-rich environment. The resulting CO_2 is then measured by using the CO_2 IR cell so that the percentage of oxygen can be determined.

Flash Point. The fuel flash point can be determined, for example, by the *Walter Herzog MP-329* automatic flash point analyzer (Fig. 4.9) which utilizes the *Pensky–Martens method* as one of the closed cup methods. It is suitable for flash point measurements in the interval from 10 °C up to 370 °C. To determine the flash point, a prescribed volume of fuel sample is placed into the test cup of Pensky–Martens apparatus and heated at the specified rate with continuous stirring (Torres et al. 2010). An ignition source is directed into the cup at regular intervals with simultaneous interruption of stirring until a flash that spreads throughout the inside of the cup is observed. The corresponding temperature is the flash point of the fuel.

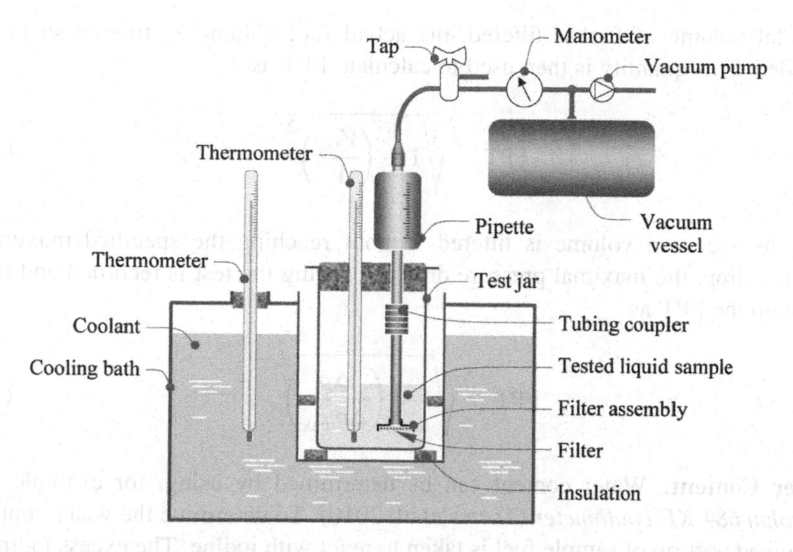

Fig. 4.10 FPT determination device

Corrosiveness. Many parts of the fuel injection system and fuel tank may be built of copper or copper alloys. To evaluate the potential corrosion problems, the *copper/silver corrosion tests* can be used (Torres et al. 2010). In this test a polished copper strip is immersed in a specific volume of the sample being tested and heated under the conditions of temperature and time that are specific to the class of material being tested. At the end of the heating period, the copper strip is removed, washed, and the color and tarnish level assessed against the ASTM Copper Strip Corrosion Standard.

Contaminants Content. Contaminant content in fuel can be determined from the IR spectrum, obtained by using, for example, the *Perkin Elmer FT-IR spectroscopy device* (Model MSP 9250 PE) (Torres et al. 2010). To perform the test, a prescribed volume of fuel sample is sandwiched between two plates to produce a thin capillary film and placed into the testing device. Then, the sample is lightened (beamed) through with IR light. When the sample is hit by the light having the wavelength, corresponding to the energy needed to stimulate the oscillations of inter atomic bonds, the light is absorbed. The absorption of the light in the IR spectrum is observed as an absorption zone, characteristic for specific molecule in the sample. A certain vibration zone at a specific wavelength corresponds to a specific functional group of molecules.

Filter Plugging Tendency. The filter plugging tendency (FPT) of biodiesel fuels can be determined with a *Filter plugging tendency analyzer* (Fig. 4.10) (Torres et al. 2010). To determine the FPT, a specified fuel volume V_{fix} is passed at a constant rate through a glass fiber filter medium and the pressure drop across the filter is monitored. If the specified maximum pressure drop $\Delta p_{\max}$ is reached before the total volume of fuel is filtered, the actual fuel volume V, filtered so far, is recorded. This quantity is then used to calculate FPT as

$$\mathrm{FPT} = \sqrt{1 + \left(\frac{V_{\mathrm{fix}}}{V}\right)^2} \tag{4.2}$$

If the specified volume is filtered without reaching the specified maximum pressure drop, the maximal pressure drop Δp during the test is recorded and used to obtain the FPT as

$$\mathrm{FPT} = \sqrt{1 + \left(\frac{\Delta p}{\Delta p_{\mathrm{max}}}\right)^2} \tag{4.3}$$

Water Content. Water content can be determined by using, for example, the *Metrohm 684 KF coulometer* (Torres et al. 2010). To determine the water content, a weighed portion of sample fuel is taken to react with iodine. The excess iodine is detected and, based on the stoichiometry of the reaction (1 mol of iodine reacts with 1 mol of water), the mass of water can be obtained.

Lubricity. Lubricity is the measure of the reduction of friction, caused by a lubricant. One possible way to determine the lubricity is to use a *test ball* that is oscillated and pushed against a stationary steel plate. The interface with the plate has to be fully immersed in a fuel sample. The wear scar diameter observed on the ball is a measure of the fluid's lubricity. The mean wear scar diameter is determined as $d_{\mathrm{MWSD}} = x + y/2$, where x is the scar dimension, perpendicular to oscillation direction [μm] and y is the scar dimension parallel to oscillation direction [μm]. Lower d_{MWSD} means better lubricity (Torres et al. 2010). The fuels can also be evaluated by the *four-ball test* (Fig. 4.11) (Wain et al. 2005). This test is also used to compare wear characteristics of new and used engine oils, as well as to simulate wear in hydraulic pump components.

4.2.2 Requirements for Biodiesel Properties

The most important biodiesel properties, along with the EN 14214 limits, units, and corresponding test methods, are presented in Table 4.1 (Torres et al. 2010).

4.2.3 Properties of Biodiesels from Various Sources

Even though approximately 350 oil-bearing crops are currently identified, only few of them are practically used for biodiesel production. Most notable, these crops are rapeseed, soybean, palm, canola, and sunflower (Demirbas 2008b; Hoekman

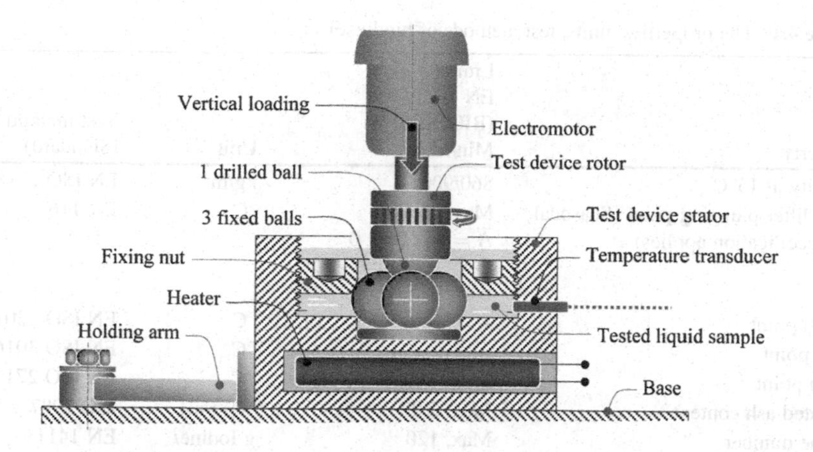

Fig. 4.11 Four-ball test method

et al. 2012; Koçak et al. 2007). The most frequently investigated biodiesel fuels are those made from *rapeseed* (RaBIO), *soybean* (SoBIO), *palm* (PaBIO), *canola* (CaBIO), *sunflower* (SuBIO), *coconut* (CoBIO), *hazelnut* (HaBIO), *jatropha* (JaBIO), *karanja* (KaBIO), *olive* (OlBIO), and *waste cooking oil* (WcBIO).

Stability. Stability is related to *biodegradability*, *phase separation*, and so on. In general, it seems that biodiesels from vegetable oils tend to stronger biodegradability than biodiesels from used edible oils (Basha et al. 2009; Schleicher et al. 2009; Sinha et al. 2008). In any case, biodegradability depends on biodiesel type. For neat RaBIO and its 5 and 20 % blends with mineral diesel, the microbiological stability under aerobic and anaerobic conditions was investigated in (Schleicher et al. 2009). The obtained results show that the most rapid microbial growth happened in the blend of 20 % biodiesel and 80 % mineral diesel. In this blend, 84 days after incubation, microbial count was 1,700 times higher than at the beginning. In the blend of 5 % biodiesel and 95 % mineral diesel and in neat biodiesel, the microbial growth was 20 times and 13 times higher than at the test start, respectively. Furthermore, it was observed that the growth conditions also vary in dependence on fuel composition and environmental factors. In neat biodiesel, the microorganisms grew better under anaerobic conditions and temperatures of around 37 °C than under aerobic conditions. Regarding the fuel phase separation, in Torres et al. (2010), the stability of RaBIO was investigated under low temperatures, especially the transition from solid to liquid biodiesel state. When RaBIO warms up from −18 °C, it becomes liquid. During this process white particles in the fuel can be easily observed. These particles are paraffins (Fig. 4.12).

Density, Viscosity, Sound Velocity, and Bulk Modulus. The *density* and *kinematic viscosity* of biodiesel fuels at 40 °C are compared to those of mineral diesel in Table 4.2 (Agarwal and Chaudhury 2012; Alptekin and Canakci 2009; Cecrle et al. 2010; Tate et al. 2006a). It can be seen that the densities of all biodiesels are higher than that of mineral diesel. A similar observation can be made for the

Table 4.1 The properties, units, test methods of biodiesel

Property	Limits EN 14214 (BIODIESEL) Min/Max	Unit	Test method (Standard)
Density at 15°C	860/900	kg/m^3	EN ISO 12185
Cold filter plugging point (Seasonal specification applies)	Max. Grade: A = +5, B = 0 C = −5, D = −10 E = −15, F = −20	°C	EN 116
Cloud point	–	°C	EN ISO 23015
Pour point	–	°C	EN ISO 3016
Flash point	Min. 120	°C	EN ISO 2719
Sulfated ash content	Max. 0.02	%[m/m]	ISO 3987
Iodine number	Max. 120	g Iodine/100 g	EN 14111
Acidity number	Max. 0.5	mg KOH/g	EN 14104
Ester content	Min. 96.5	% [m/m]	EN 14103
Linolenic acid methyl ester	Max. 12.0	% [m/m]	EN 14103
Methanol content	Max. 0.20	% [m/m]	EN 14110
Phosphorus content	Max. 10.0	mg/kg	PML.07.30
Oxidation stability, 110 °C	Min. 6	h	EN 14112
Monoglycerides content	Max. 0.8	% [m/m]	EN 14105
Diglycerides content	Max. 0.20	% [m/m]	EN 14105
Triglycerides content	Max. 0.20	% [m/m]	EN 14105
Free glycerol	Max. 0.02	% [m/m]	EN 14105
Total glycerol	Max. 0.25	% [m/m]	EN 14105
Water content	Max. 500	mg/kg	EN ISO 12937
Solid impurities	Max. 24	mg/kg	EN 12662
Sulfur content, WD-XRF	Max. 10	mg/kg	EN ISO 20884
Kinematic viscosity at 40 °C	3.5/5	mm^2/s	EN ISO 3104
Carbon residue	Max. 0.30	% [m/m]	EN ISO 10370
Corrosiveness to copper, 3 h at 50 °C	1	Classification	EN ISO 2160
Lubricity	–	μm	EN ISO 12156-1
FPT: Pressure/volume	–	kPa/ml	ASTM D 2068
Element composition (CHN)	–		ASTM D 5291

viscosity. Higher viscosity values are a consequence of various chain lengths and position, number, and nature of double bonds, as well as the nature of oxygenated moieties (Knothe and Steidley 2005a, b).

Based on experimental work, several expressions for biodiesel viscosity have been developed for specific temperature intervals. For example, the *kinematic viscosity* ν [mm^2/s] of CaBIO can be expressed in dependence of temperature T [K] as follows (Krisnangkura et al. 2006; Tate et al. 2006a):

Fig. 4.12 Paraffins in rapeseed biodiesel during the melting process

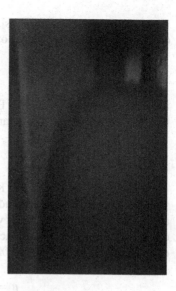

Table 4.2 Density and viscosity of biodiesel and mineral diesel fuels

Fuel	Density [kg/m^3]	Kinematic viscosity [mm^2/s]
Mineral diesel	835	2.8
Canola (CaBIO)	880–886	3.9–4.7
Coconut (CoBIO)	865–875	2.7–3.0
Hazelnut (HaBIO)	863–885	3.1–5.4
Jatropha (JaBIO)	873–875	4.2–4.3
Karanja (KaBIO)	871–883	4.2–4.3
Olive (OlBIO)	870	4.6
Palm (PaBIO)	870–880	4.3–5.7
Rapeseed (RaBIO)	877–893	4.2–5.1
Soybean (SoBIO)	880–885	3.9–4.6
Sunflower (SuBIO)	880–884	4.0–4.7
Waste cooking oil (WcBIO)	880–888	4.9

$$\nu = e^{\left(-0.341-\frac{498}{T}+\frac{338,301}{T^2}\right)} \tag{4.4}$$

and for SoBIO as (Krisnangkura et al. 2006; Tate et al. 2006a):

$$\nu = e^{\left(0.067-\frac{1,078}{T}+\frac{469,741}{T^2}\right)} \tag{4.5}$$

The differences between the calculated and experimental values of kinematic viscosity are lower than 5 % within the temperature interval of 20–80 °C (Krisnangkura et al. 2006; Tate et al. 2006a). At other temperatures the accuracy may be lower.

Dynamic viscosities μ [mPas] of various biodiesel fuels, like WcBIO and RaBIO can be expressed in dependence of temperature T [K] (Kerschbaum and Rinke 2004) as

$$\mu = e^{\left(\frac{2,359}{T} - 6.1\right)} \tag{4.6}$$

for temperatures above 273 K. Below 273 K, the dynamic viscosity for RaBIO can be expressed as (Kerschbaum and Rinke 2004) as

$$\mu = e^{\left(\frac{7,139.2}{T} - 23.283\right)} \tag{4.7}$$

At temperatures below 273 K, the viscosities rise steeply and can reach values, which are one order of magnitude higher than at ambient temperature (Kerschbaum and Rinke 2004). The dynamic viscosity of blends from diesel and fish oil biodiesel can be determined in dependence of fuel temperature T [K] and volume fraction ψ_{B} of biodiesel (Joshi and Pegg 2007) as

$$\mu = e^{\left(-5.98 + \frac{2,151}{T} + 0.00374\psi_{\mathrm{B}}\right)} \tag{4.8}$$

For a given fuel, the *density* ρ varies in dependence of temperature and pressure. The density of RaBIO and mineral diesel blends can be expressed in dependence on temperature T [°C], pressure p [bar] and volume fraction ψ_{B} of biodiesel by the following empirical expression (Kegl 2006):

$$\rho = \psi_{\mathrm{B}}(0.0543\,p + 877) + (1 - \psi_{\mathrm{B}})(0.0611\,p + 826) - 0.88\,(T - 20) \tag{4.9}$$

The comparison of fuel densities of neat mineral diesel (D100), neat RaBIO, and their blends, obtained experimentally and numerically at ambient pressure, is presented in Fig. 4.13. The density increases by increasing the content of RaBIO and by decreasing temperature.

For a given fuel, the *sound velocity* a[m/s] depends on pressure and temperature. On the basis of experimentally obtained results, the following empirical expression for the sound velocity a of RaBIO and D100 blends, in dependence of pressure p [bar], volume biodiesel fraction ψ_{B}, and temperature T [°C] have been derived (Kegl 2006):

$$\begin{aligned}
a &= (a_0 + a_1 p + a_2 p^2)(C_1 + C_2 + C_3) \\
C_1 &= 0.975 \\
C_2 &= 0.03\ \psi_{\mathrm{B}} \\
C_3 &= 4 \times 10^{-5}\ p\,\psi_{\mathrm{B}} \\
a_0 &= 1376.2 - 1.9676T - 9.625 \times 10^{-3}T^2 \\
a_1 &= 0.5327 - 1.644 \times 10^{-3}T + 3.394 \times 10^{-5}T^2 \\
a_2 &= -2.637 \times 10^{-4} + 4.43 \times 10^{-6}T - 5.358 \times 10^{-8}T^2
\end{aligned} \tag{4.10}$$

The calculated sound velocity in RaBIO and its blends with D100 is compared with the experimentally obtained data in Fig. 4.14 (Kegl 2006).

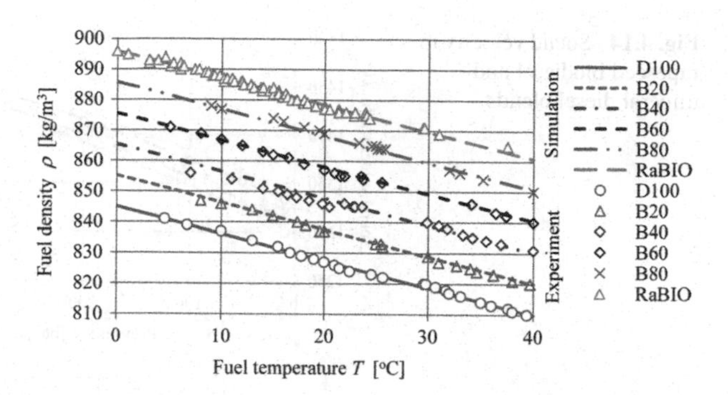

Fig. 4.13 Density of rapeseed biodiesel and mineral diesel blends

Based on the experimental results and the empirical formulas for sound velocity and density, the *bulk modulus of elasticity* E can be calculated as $E = a^2 \times \rho$. Figure 4.15 shows the calculated bulk modulus in dependence on pressure and fuel temperature (Kegl 2006). It can be seen that the curve slopes of all considered fuels are practically the same.

Cetane Number. The *cetane number* of several biodiesels is given in Table 4.3 (Agarwal 2007; Demirbas 2008b; Hoekman et al. 2012; Koçak et al. 2007; Singh and Singh 2010). For practically all biodiesels, the average cetane number is larger than 50.

Cloud Point, Pour Point, Cold Filter Plugging Point, and Flash Points. For blends of biodiesel and mineral diesel, the cloud point (CP) temperature T_{CP} can be expressed as a function of biodiesel volume fraction ψ_{B}. For example, the T_{CP} [K] of fish oil biodiesel can be expressed as (Joshi and Pegg 2007; Sinha et al. 2008):

$$T_{\mathrm{CP}} = 256.4 + 0.1991\psi_{\mathrm{B}} + 0.000223\psi_{\mathrm{B}}^2 \tag{4.11}$$

The temperature T_{PP} [K] of pour point (PP) can be expressed as a function of ψ_{B} as (Joshi and Pegg 2007):

$$T_{\mathrm{PP}} = 253.9 + 0.1865\psi_{\mathrm{B}} + 0.000335\psi_{\mathrm{B}}^2 \tag{4.12}$$

The CP, PP, cold filter plugging point (CFPP), and flash points of biodiesel and mineral diesel fuels are compared in Table 4.4 (Alptekin and Canakci 2009; Echim et al. 2012; Gao et al. 2009; Joshi and Pegg 2007; Koçak et al. 2007; Sinha et al. 2008; Smith et al. 2010; Wang et al. 2011).

Lubricity. Experimental lubricity results for biodiesel and mineral diesel, obtained by using industrial test methods, indicate that there is a notable improvement in lubricity when biodiesel is added to mineral diesel. Even biodiesel levels below 1 % can provide up to a 30 % increase in lubricity. The beneficial effects of biodiesel, produced from rapeseed oil, sunflower oil, corn oil, olive oil, and waste oil on *lubricity* of various fuel blends has been confirmed in (Anastopoulus

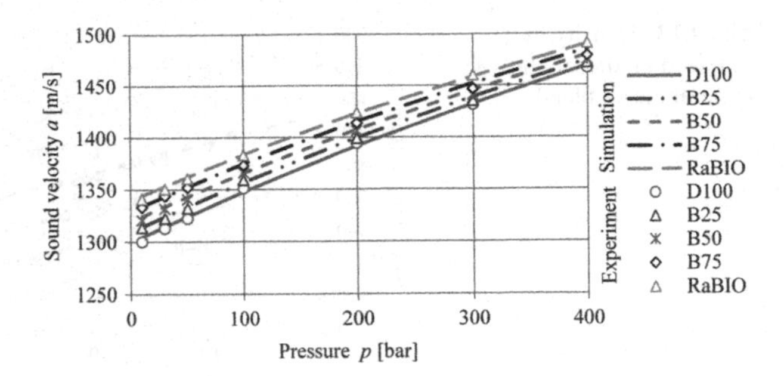

Fig. 4.14 Sound velocity in rapeseed biodiesel and mineral diesel blends

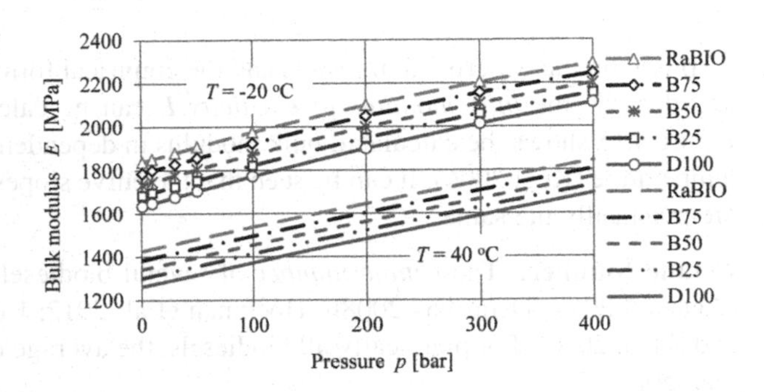

Fig. 4.15 Bulk modulus of RaBIO and D100 blends

Table 4.3 Cetane number of biodiesel and mineral diesel fuels

Fuel	Cetane number
Mineral diesel	48–50
Canola (CaBIO)	52–55
Coconut (CoBIO)	60–65
Hazelnut (HaBIO)	52–55
Jatropha (JaBIO)	48–58
Karanja (KaBIO)	55–58
Olive (OlBIO)	61–62
Palm (PaBIO)	62–70
Rapeseed (RaBIO)	51–64
Soybean (SoBIO)	45–56
Sunflower (SuBIO)	49–52
Waste cooking oil (WcBIO)	51–53

et al. 2005; Demirbas 2009). The addition of these biodiesels to mineral diesel in concentrations of 0.15–0.5 % by volume, decreases wear scar diameter significantly. At biodiesel concentration higher than 1 % by volume, the wear scar diameter asymptotically approaches a constant value.

Chemical Composition and Water Content. Biodiesel fuels contain more oxygen than mineral diesel. For example, RaBIO contains 76.7–79.6 wt.% *carbon*, 10.5–11.1 wt.% *hydrogen*, 8.6–11.0 wt.% *oxygen*, and 0.09–1.3 wt.% *nitrogen* (Balat and Balat 2010; Torres et al. 2010). Due to the presence of electronegative

Table 4.4 CP, PP CFPP, and flash point of biodiesel and mineral diesel fuels

Fuel	CP [°C]	PP [°C]	CFPP [°C]	Flash point [°C]
Mineral diesel	−15 to +6	−35 to −15	−10 to −20	60–80
Canola (CaBIO)	−2 to −1	−9 to −6	−5	140–150
Coconut (CoBIO)	−2 to +5	−4 to −3	−3	107–144
Hazelnut (HaBIO)	−15 to −14	−15 to −14	−10	180–182
Jatropha (JaBIO)	4	−3	0–3	155–165
Karanja (KaBIO)	7–14	4–5	2–3	163–182
Olive (OlBIO)	−2	−6 to −3	−1	168–170
Palm (PaBIO)	8–13	6–16	11–14	144–164
Rapeseed (RaBIO)	−3	−6	−14 to −10	140–192
Soybean (SoBIO)	0–2	−4 to −2	−4	141–178
Sunflower (SuBIO)	−6 to −3	−6 to −1	−3 to 4	164–183
Waste cooking oil (WcBIO)	3	−2.5	−10 to −2	146–156

element oxygen, biodiesel is slightly more polar than mineral diesel; as a result, the viscosity of biodiesel is higher than that of mineral diesel. The presence of elemental oxygen lowers the heating value of biodiesel when compared to mineral diesel. In general, biodiesels are very hygroscopic because of strong polar interactions between esters and water. For example, the *water content* in RaBIO is around 150 mg/kg, while in mineral diesel it is around 50 mg/kg (Torres et al. 2010, 2011).

4.2.4 Improvement of Biodiesel Properties

Biodiesel properties depend on raw material and production technology. Many various methods for biodiesel production from various sources have been developed so far. Thus, nowadays, biodiesel can be produced by using technologies such as ultrasonic cavitation, hydrodynamic cavitation, microwave irradiation, response surface technology, two-step reaction process, etc. Of course, these technologies continue to be developed and improved.

The *response surface methodology*, based on central composite rotatable design, can be used to optimize the three important reaction variables, namely, methanol quantity, acid concentration and reaction time for reduction of free fatty acid content of the oil (Basha et al. 2009). Some investigations show that the preparation of fatty acids methyl esters by using the *microwave irradiation* as a fast method for alcoholysis of triglycerides with methanol may lead to high yields of fatty acid methyl ester (Basha et al. 2009).

Furthermore, another promising approach seems to be the *non-catalytic transesterification reaction by employing supercritical methanol conditions* (Balat and Balat 2010). Transesterification of vegetable oils in supercritical methanol is carried out without using any catalyst. Supercritical methanol has a high potential for transesterification of triglycerides to methyl esters for mineral diesel fuel substitutes. Furthermore, the supercritical transesterification method is more tolerant to the presence of water and free fatty acids than the conventional alkali catalyzed technique, and hence more tolerant to various types of vegetable oils.

To improve the low temperature properties of biodiesel, the *blending with mineral diesel*, the *chemical* or *physical modification* of either the oil feedstock or the biodiesel product, and the use of *additives* can be utilized (Smith et al. 2010).

Blending with mineral diesel is effective at low content (up to 30 vol.%) of biodiesel with cloud points at around –10 °C. Clearly, blending with mineral diesel does not change the chemical nature and therefore the properties of biodiesel. Some properties of biodiesel/mineral diesel blends have been shown in Figs. 4.13, 4.14, and 4.15.

Chemical or *physical modification* of the *oil feedstock* can be done by *winterization*, which is a method for separating those fractions of oils which have a solidification temperature below a specific cut-off (Smith et al. 2010). One of such techniques involves refrigeration of the oils for a prescribed period at a specific temperature, followed by decanting of the remaining liquid. Another, more energy efficient method is to allow tanks of oil to stand outside in cold temperatures for extended periods of time. In either case, the fraction that remains molten is separated from the solid part, producing an oil with improved pour and handling qualities. A more direct method for altering biodiesel feedstock is to genetically modify the fatty acid profile of oilseeds.

Chemical or *physical modification* of biodiesel with the aim to modify its low-temperature properties can be done by *crystallization fractionation*, which is a similar process as that applied in the winterization of oil feedstock (Smith et al. 2010). Options are the dry fractionation and the solvent fractionation. Dry fractionation involves crystallization from the melt without the addition of a solvent and is the simplest and least expensive method. Solvent fractionation has significant advantages over the dry fractionation, including reduced crystallization times and improved yields, but suffers from reduced safety and increased costs. Fractionation of biodiesel via any means results in a reduction in the proportion of saturated esters and therefore in an increase of the fraction of unsaturated esters. This has a significant effect on other biodiesel properties, especially oxidative stability and ignition quality.

Fractionation has also a large impact on the cost of production and therefore price competitiveness of biodiesel, compared to mineral diesel. Fractionation not only introduces a new set of unit operations including solvent addition, crystallization and solvent recovery but also results in significantly reduced yield. The loss of 75 % of the original material would be unacceptable for any commercial process. To mitigate this situation, another approach exploits the *p*-bonds of the unsaturated fatty acids via *electrophilic addition* to produce branched or bulky esters. Isopropyl oleate is first epoxidised via in situ peroxyformic acid method to produce around 95 % epoxidised isopropyl oleate.

The *additives* can be classified into *traditional mineral diesel additives* and additives from emerging *new technologies*, developed specifically for biodiesel fuels. Traditional mineral diesel additives can be described as either pour point (PP) depressants or wax crystalline modifiers (Smith et al. 2010). *Pour point depressants* are developed to improve the pumpability of crude oil and do not affect nucleation habit. Instead, these additives inhibit crystalline growth, thereby eliminating agglomeration. They are typically composed of low molecular weight copolymers, similar in structure to aliphatic alkane molecules. The most widely applied group are copolymers of ethylene vinyl ester. *Wax crystalline modifiers* are copolymers that disrupt part of the crystallization process to produce a larger number of smaller, more compact wax crystals. By using these additives, the PP and cold filter plugging point (CFPP) of neat biodiesel fuels can be decreased by more than 6 °C, but without a noticeable improvement in the cloud point (CP). A potential mechanism to reduce the CP of biodiesels is the use of *bulky moieties* that disrupt the orderly stacking of ester molecules during crystal nucleation (Knothe et al. 2000). Some additives, such as Tween-80, dihydroxy fatty acid, acrylated polyester pre-polymer, palm-based polyol, and castor oil ricinoleate for palm biodiesel, reduce pour point and cloud point values of biodiesel fuels significantly (Ming et al. 2005).

Oxidative stability of biodiesel fuels depends on the type and amount of *antioxidants* (Ryu 2010). Experience shows that the natural α-tocopherol improves oxidative stability of biodiesel fuels less than synthetic antioxidants, like as butylated hydroxyanisole, tert-butylhydroquinone, butylated hydroxytoluene, and propyl gallate. Tert-butylhydroquinone offers the best oxidative stability for all biodiesel fuels. It has been shown that antioxidants have a practically negligible effect on the exhaust emissions of an unmodified four cylinder, four stroke, water-cooled, IDI diesel engine, running on biodiesel (Ryu 2010; Xue et al. 2011). Regarding the specific fuel consumption, it turned out that biodiesel with antioxidants performs better (i.e., the specific fuel consumption is lower) than biodiesel without antioxidants. However, no specific trends have been detected with regard to the type and amount of antioxidants.

The reduction in harmful emissions and improvement in engine efficiency can also be achieved by use of fuel additives. *Metal based additives* are employed as combustion catalyst to promote the combustion and to reduce fuel consumption and emissions for hydrocarbon fuels. These metal-based additives include cerium (Ce), cerium–iron (Ce–Fe), platinum (Pt), platinum–cerium (Pt–Ce), iron (Fe), manganese (Mn), barium, calcium, and copper (Kannan et al. 2011). The reduction of emissions while using a metal-based additive may be either due to the fact that metals react with water vapor to produce hydroxyl radicals or serve as an oxidation catalyst, thereby reducing the oxidation temperature that results in increased particle burnout. Metallic fuel additives can also improve some properties of biodiesel fuels, such as pour point and viscosity.

Biodiesel properties can also be improved by using other fuels as additives. One of such additives is *bioethanol*. In (Torres et al. 2010) some blends of RaBIO and bioethanol have been tested in a mechanically controlled fuel injection system. The investigation focused on various properties of neat RaBIO, 5 % (v/v) bioethanol/biodiesel blend (B95E05), 10 % (v/v) bioethanol/biodiesel blend (B90E10), and 15 % (v/v) bioethanol/biodiesel blend (B85E15). The tested fuel properties were the

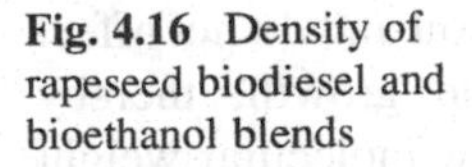
Fig. 4.16 Density of rapeseed biodiesel and bioethanol blends

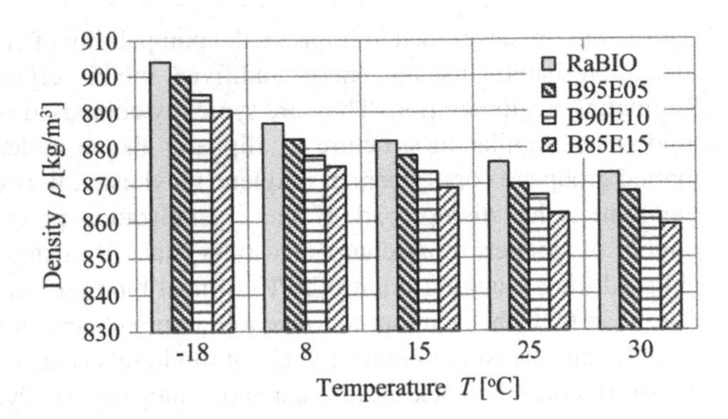

density, sound velocity, viscosity, lubricity, CP, CFPP, PP, and water content. Some results of this investigation are briefly outlined in the following.

The *density* of the blends has been found to decrease almost proportionally to bioethanol concentration and temperature (Fig. 4.16). In any case, for the considered blends the density values remain within the standard range at the reference temperature.

Because of the higher biodiesel density, compared to mineral diesel, it is expected that the *sound velocity* in biodiesel will be higher than in mineral diesel, causing advanced injection timing in mechanically controlled injection systems. However, the lower density of bioethanol compensates for this effect so that in the end effect the influence on injection timing is not so significant.

The average *viscosity* value for neat bioethanol (1.1 mm^2/s) is lower than the viscosity of RaBIO (4.477 mm^2/s). The investigation has shown that when bioethanol concentration increases, the kinematic viscosity of the blend decreases proportionally (Fig. 4.17).

The addition of bioethanol to biodiesel reduces slightly the *lubricity* of the blend, if added in larger quantities (Fig. 4.18). In fact, the measurements of B85E15 have shown only a slightly higher mean wear scar diameter, which means lower lubricity, compared to that of neat biodiesel. In general, bio-based oils appear to be very effective in improving the lubricity of the fuel. Practical experiences support this supposition, for example, the inclusion of 1 % castor oil in a 95 % bioethanol fuel that was used successfully for a fleet of trucks and buses in Brazil.

Regarding the CP, CFPP, and PP the following was observed for the considered blends. All of the samples (B95E05 to B85E15) remained in one phase regardless of the temperature and ethanol concentration. When neat biodiesel warms up from very low temperatures and becomes liquid, emerging paraffins can be easily seen. In the case of bioethanol and biodiesel blends, bioethanol changes the melting process in such a way that paraffins are different in size and shape; effectively, the paraffins are not visible. Invisible paraffins lead to an improvement from the point of view of winter properties, such as CP and CFPP, because these properties are linked to the moment in which paraffins appear during the solidification process. No problems related to phase separation were observed for any blend. Therefore, as

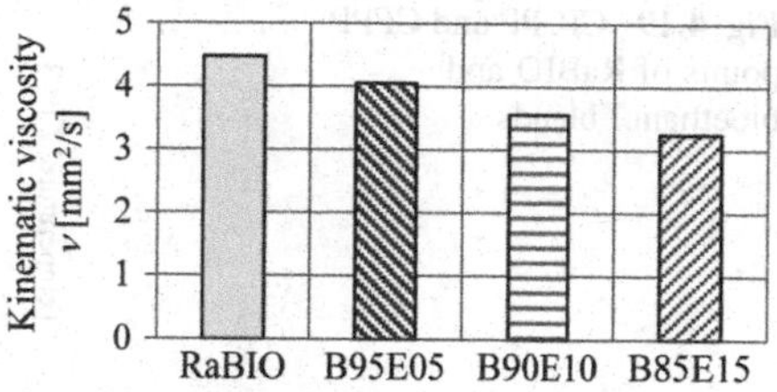

Fig. 4.17 Kinematic viscosity of rapeseed biodiesel and bioethanol blends

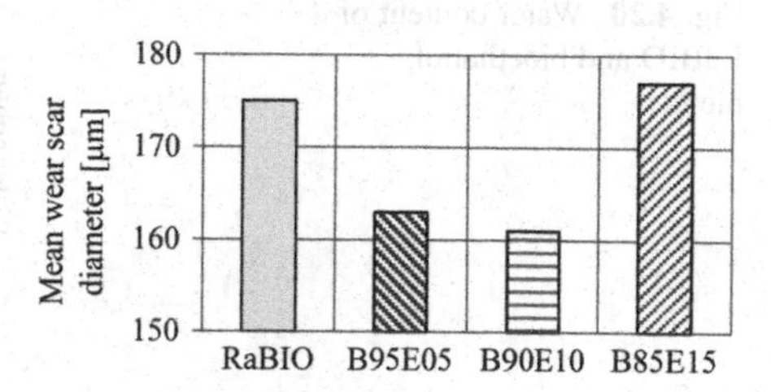

Fig. 4.18 Mean wear scar diameter of RaBIO and bioethanol blends

expected, the CP decreased by increasing bioethanol concentration (Fig. 4.19). The PP also improved when the bioethanol concentration increased, Fig. 4.19. Furthermore, one can say that the influence of bioethanol addition on CFPP is not significant, but a decreasing tendency in CFPP could still be observed (Fig. 4.19). Thus, bioethanol addition improves the fuel filtration process slightly. In general, bioethanol addition improves winter properties. Therefore, it may be used as a winter additive.

As expected, the *water content* increases with the bioethanol concentration (Fig. 4.20). Besides this, bioethanol and biodiesel are both very hygroscopic, compared to mineral diesel. In the case of biodiesel, this fact comes from the strong polar interactions between esters and water. Free glycerol also has a high affinity for water. Therefore, water-related problems may occur during storage, if the tank is not adequately sealed. In that case, the blend will absorb water from ambient humidity.

4.3 Discussion

Nowadays, the most commonly used oils for the production of biodiesel are canola, coconut, hazelnut, jatropha, karanja, olive, palm, rapeseed, soybean, sunflower, and waste cooking oil. The source of biodiesel usually depends on the crops amenable to the regional climate. In Europe, *rapeseed* oil is the most common biodiesel feedstock, whereas the *soybean* oil and *palm* oil are the most common source for biodiesel in the USA and in tropical countries, respectively.

The physical and chemical properties of biodiesel are determined by its chemical composition. Due to its considerable *oxygen content* (typically about 11 %),

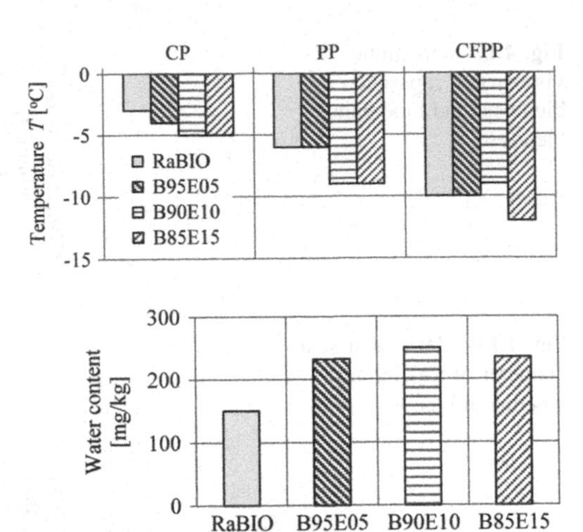

Fig. 4.19 CP, PP and CFPP points of RaBIO and bioethanol blends

Fig. 4.20 Water content of RaBIO and bioethanol blends

biodiesel has lower carbon and hydrogen contents compared to mineral diesel. This results in a reduction in mass energy content by about 10 %.

The two most common sets of *regulatory standards* for *biodiesel fuels* are *ASTM D6751* in the USA and *EN 14214* in Europe. Some of the specifications comprising these standards are directly related to the chemical composition of biodiesels, for example, viscosity, cetane number, cloud point, distillation, etc. Other specifications are related to the purity of biodiesel and address issues pertaining to production processes, transport, and storage, such as flash point, methanol content, metals content, acid level, and filter plugging tendency. Oxidative stability is also an important property of biodiesel that is influenced by both biodiesel chemical composition and by storage and handling conditions.

Regarding the broader usage of *biodiesels* as *diesel engine fuels*, some general advantages and disadvantages can be outlined by the *strength–weakness–opportunities–threat* (SWOT) analysis (Misra and Murthy 2011; Russo et al. 2012).

Strengths

- Biodiesel is environmentally friendly, biodegradable, and sustainable; compared to mineral diesel, it has a lower toxicity; in Europe and USA its quality is regulated by corresponding standards.
- Biodiesel reduces net carbon dioxide emissions up to 78 % on a lifecycle basis when compared to mineral diesel.
- Biodiesels exhibits low sulfur and aromatic content; this results in almost no sulfur dioxides emissions.
- The properties of biodiesel (viscosity, ignition properties) and mineral diesel fuels are quite similar; therefore, biodiesel is an alternative fuel that can be used in existing, virtually unmodified, diesel engines.
- Biodiesels exhibit high combustion efficiency and high cetane numbers, which corresponds to short ignition delays and greater efficiency.
- Biodiesels exhibit good lubrication properties that reduce wear in the engine.
- The raw materials for biodiesel production are renewable and widely available.

- The coproduct glycerine of the transesterification process can be commercially exploited.
- The side products and agricultural residues can be used as fodder, fertilizers, or raw materials for second-generation biofuels.

Weaknesses

- Long-term storage of biodiesel fuels may cause degradations in certain fuel properties; biodiesel is hygroscopic; additives are often necessary.
- There are problems related to fuel solidification at low temperatures.
- There are problems related to situations when biodiesels are first introduced into equipment that has a long history of neat mineral diesel usage; mineral diesel fuel typically forms a layer of deposits on the inside of tanks and fuel tubes; biodiesels loosen these deposits, causing the block of fuel filters; proper filter maintenance during some period, following the introduction of biodiesels, is therefore required.
- Due to high oxygen content, biodiesels produce relatively high NO_x emission levels during combustion.
- The kinematic viscosity is higher than found in mineral diesel fuel; this affects fuel atomization during injection and requires a somewhat modified fuel injection system.
- Biodiesel fuels are relatively aggressive toward fuel pipes, sealings, and filters; these elements need to be adapted adequately.
- Oxidative stability is lower than that of mineral diesel so that, under extended storage conditions, it is possible that oxidation products may cause harm to engine components.
- The production of biodiesel is not sufficiently regulated; biodiesels that do not conform to the European or US standards can cause corrosion, fuel system blockage, seal failures, filter clogging, and deposits in the injection systems.

Opportunities

- Currently, biodiesels are the dominant biofuels in Europe.
- Broader biodiesel usage can improve energy supply and consequently strengthen the economy.
- Energy policy that provides tax reductions/exemptions and biofuel obligations could largely increase the use of pure plant oil and biodiesels.
- Waste oil can be used as cheap feedstock for biodiesel.
- Oil plants other than rapeseed can be cultivated in Europe (e.g., sunflower, soya).

Threats

- Biodiesel fuels may become toxic and polluting in case that they contain harmful solvents and other chemical additives.
- In Europe, rapeseed is mainly used as the feedstock source; rapeseed can be cultivated only every 4 years on the same field.
- A standard for biodiesel feedstock does not exist; variations in feedstock properties may lead to undesirable variations of biodiesels.

References

Agarwal, A. K. (2007). Biofuels (alcohols and biodiesels) applications as fuels for internal combustion engines. *Progress in Energy and Combustion Science, 33*, 233–271.

Agarwal, A. K., & Chaudhury, V. H. (2012). Spray characteristics of biodiesel/blends in a high pressure constant volume spray chamber. *Experimental Thermal and Fluid Science, 42*, 212–218.

Allen, C. A. W., Watts, R. G., Ackman, R. G., & Pegg, M. J. (1999). Predicting the viscosity of biodiesel fuel from their fatty acid ester composition. *Fuel, 78*, 1319–1326.

Alptekin, E., & Canakci, M. (2009). Characterization of the key fuel properties of methyl ester-diesel fuel blends. *Fuel, 88*, 75–80.

Altiparmak, D., Keskin, A., Koca, A., & Gürü, M. (2007). Alternative fuel properties of tall oil fatty acid methyl ester-diesel fuel blends. *Bioresource Technology, 98*, 241–246.

Amaro, H. M., Guedes, A. C., & Malcata, F. X. (2011). Advances and perspectives in using microalgae to produce biodiesel. *Applied Energy, 88*, 3402–3410.

Anastopoulus, G., Lois, E., Karonis, D., Kallingeros, S., & Zannikos, F. (2005). Impact of oxygen and nitrogen compounds on the lubrication properties of low sulfur diesel fuels. *Energy, 30*, 415–426.

Arkoudeas, P., Kallingeros, S., Zannikos, F., Anastopoulos, G., Karonis, D., Korres, D., & Lois, E. (2003). Study of using JP-8 aviation fuel and biodiesel in CI engines. *Energy Conversion & Management, 44*, 1013–1025.

Atadashi, I. M., Aroua, M. K., & Aziz, A. A. (2010). High quality biodiesel engine application: A review. *Renewable and Sustainable Energy Reviews, 14*, 1999–2008.

Balat, M., & Balat, H. (2008). A critical review of bio-diesel as a vehicular fuel. *Energy Conversion and Management, 49*, 2727–2741.

Balat, M., & Balat, H. (2010). Progress in biodiesel processing. *Applied Energy, 87*, 1815–1835.

Basha, S. A., Gopal, K. R., & Jebaraj, S. (2009). A review on biodiesel production, combustion, emission and performance. *Renewable and Sustainable Energy Reviews, 13*, 1628–1634.

Bhale, P. V., Deshpande, N. V., & Thombre, S. B. (2009). Improving the low temperature properties of biodiesel fuel. *Renewable Energy, 34*, 794–800.

Bozbas, K. (2008). Biodiesel as an alternative motor fuel: Production and policies in the European Union. *Renewable and Sustainable Energy Reviews, 12*, 542–552.

Cecrle, E., Depcik, C., Duncan, E., Guo, J., Mangus, M., Peltier, E., Stagg-Williams, S., & Zhong, Y. (2010). Investigation of the effects of biodiesel feedstock on the performance and emissions of a single-cylinder diesel engine. *Energy & Fuels, 26*, 2331–2341.

Çetinkaya, M., Ulusoy, Y., Tekìn, Y., & Karaosmanoğlu, F. (2005). Engine and winter road test performances of used cooking oil originated biodiesel. *Energy Conversion and Management, 46*, 1279–1291.

Chen, H., & Chen, G. Q. (2011). Energy cost of rapeseed-based biodiesel as alternative energy in China. *Renewable Energy, 36*, 1374–1378.

Chisti, Y. (2007). Biodiesel from microalgae. *Biotechnology Advances, 25*, 294–306.

Conceição, M. M., Candeia, R. A., Dantas, H. J., Soledade, L. E. B., Fernandes, J. V. J., & Souza, A. G. (2005). Rhelogical behavior of castor oil biodiesel. *Energy & Fuels, 19*, 2185–2188.

Conceição, M. M., Fernandes, J. V. J., Araújo, A. S., Farias, M. F., Santos, I. M. G., & Souza, A. G. (2007). Thermal and oxidative degradation of castor oil biodiesel. *Energy & Fuels, 21*, 1522–1527.

Demirbas, A. (2006). Biodiesel production via non-catalytic SCF method and biodiesel fuel characteristics. *Energy Conversion & Management, 47*, 2271–2282.

Demirbas, A. (2008a). Biofuels sources, biofuel policy, biofuel economy and global biofuel projections. *Energy Conversion and Management, 49*, 2106–2116.

Demirbas, A. (2008b). Relationships derived from physical properties of vegetable oil and biodiesel fuels. *Fuel, 87*, 1743–1748.

Demirbas, A. (2009). Progress and recent trends in biodiesel fuels. *Energy Conversion and Management, 50*, 14–34.

Echim, C., Maes, J., & Greyt, W. D. (2012). Improvement of cold filter plugging point of biodiesel from alternative feedstocks. *Fuel, 93*, 642–648.

Gao, Y., Deng, J., Li, C., Dang, F., Liao, Z., Wu, Z., & Li, L. (2009). Experimental study of the spray characteristics of biodiesel based on inedible oil. *Biotechnology Advances, 27*, 616–624.

Giannelos, P. N., Sxizas, S., Lois, E., Zannikos, F., & Anastopoulus, G. (2005). Physical, chemical and fuel related properties of tomato seed oil for evaluating its direct use in diesel engines. *Industrial Crops and Products, 22*, 193–199.

Gomez, M. E. G., Howard-Hildige, R., Leahy, J. J., & Rice, B. (2002). Winterization of waste cooking oil methyl ester to improve cold temperature fuel properties. *Fuel, 81*, 33–39.

Goodrum, J. W. (2002). Volatility and boiling points of biodiesel from vegetable oils and tallow. *Biomass & Bioenergy, 22*, 205–211.

Goodrum, J. W., & Eitchman, M. A. (1996). Physical properties of low molecular weight triglycerides for the development of bio-diesel fuels model. *Bioresource Technology, 56*, 55–60.

Hoekman, S. K., Broch, A., Robbins, C., Ceniceros, E., & Natarajan, M. (2012). Review of biodiesel composition, properties, and specifications. *Renewable and Sustainable Energy Reviews, 16*, 143–169.

Hu, J., Du, Z., Li, C., & Min, E. (2005). Study on the lubrication properties of biodiesel as fuel lubricity enhancers. *Fuel, 84*, 1601–1601.

Jain, S., & Sharma, M. P. (2010). Biodiesel production from *Jatropha curcas* oil. *Renewable and Sustainable Energy Reviews, 14*, 3140–3147.

Jha, S. K., Fernando, S., & To, S. D. F. (2008). Flame temperature analysis of biodiesel blends and components. *Fuel, 87*, 1982–1988.

Joshi, R. M., & Pegg, M. J. (2007). Flow properties of biodiesel fuel blends at low temperatures. *Fuel, 86*, 143–151.

Kalam, M. A., & Masjuki, H. H. (2002). Biodiesel from palm oil—an analysis of its properties and potential. *Biomass and Bioenergy, 23*, 471–479.

Kannan, G. R., Karvembu, R., & Anand, R. (2011). Effect of metal based additive on performance emission and combustion characteristics of diesel engine fuelled with biodiesel. *Applied Energy, 88*(11), 3694–3703.

Kegl, B. (2006). Numerical analysis of injection characteristics using biodiesel fuel. *Fuel, 85*, 2377–2387.

Kegl, B. (2008). Biodiesel usage at low temperature. *Fuel, 87*, 1306–1317.

Kerschbaum, S., & Rinke, G. (2004). Measurement of the temperature dependent viscosity of biodiesel fuels. *Fuel, 83*, 287–291.

Knothe, G. (2005). Dependence of biodiesel fuel properties on the structure of fatty acid alkyl esters. *Fuel Processing Technology, 86*, 1059–1070.

Knothe, G., Dunn, R. O., Shockley, M. W., & Bagby, M. O. (2000). Synthesis and characterization of some long-chain diesters with branched or bulky moieties. *Journal of the American Oil Chemists' Society, 77*(8), 865–871.

Knothe, G., & Steidley, R. (2005a). Lubricity of components of biodiesel and petrodiesel. The origin of biodiesel lubricity. *Energy and Fuels, 19*, 1192–1200.

Knothe, G., & Steidley, K. R. (2005b). Kinematic viscosity of biodiesel fuel components and related compounds. Influence of compound structure and comparison to petrodiesel fuel components. *Fuel, 84*, 1059–1065.

Koçak, M. S., Ileri, E., & Utlu, Z. (2007). Experimental study of emission parameters of biodiesel fuels obtained from canola, hazelnut, and waste cooking oils. *Energy & Fuel, 21*, 3622–3626.

Krisnangkura, K., Yimsuwan, T., & Pairintra, R. (2006). An empirical approach in predicting biodiesel viscosity at various temperatures. *Fuel, 85*, 107–113.

Lapinskiene, A., Martinkus, P., & Rebždaite, V. (2006). Eco-toxicological studies of diesel and biodiesel fuels in aerated soil. *Environmental Pollution, 142*, 432–437.

Lapuerta, M., Rodríguez-Fernández, J., & de Mora, E. F. (2009). Correlation for the estimation of the cetane number of biodiesel fuels and implications on the iodine number. *Energy Policy, 37*, 4337–4344.

Lebedevas, S., Vaicekauskas, A., Lebedeva, G., Makareviciene, V., Janulis, P., & Kazancev, K. (2006). Use of waste fats of anima and vegetable origin for the production of biodiesel fuel: Quality, motor properties, and emissions of harmful components. *Energy & Fuels, 20*, 2274–2280.

Lin, L., Cunshan, Z., Vittayapadung, S., Xianquian, S., & Mingdong, D. (2011). Opportunities and challenges for biodiesel fuel. *Applied Energy, 88*, 1021–1031.

Lira, L. F. B., Vasconcelos, F. V. C., Pereira, C. F., Paim, A. P. S., Stragevitz, L., & Pimentel, M. F. (2010). Prediction of properties of diesel/biodiesel blends by infrared spectroscopy and multivariate calibration. *Fuel, 89*, 405–409.

Ming, T. C., Ramli, N., Lye, O. T., Said, M., & Kasim, Z. (2005). Strategies for decreasing the pour point and cloud point of palm oil products. *European Journal of Lipid Science and Technology, 107*, 505–512.

Misra, R. D., & Murthy, M. S. (2011). Jatropa—The future fuel of India. *Renewable and Sustainable Energy Reviews, 15*, 1350–1358.

Murugesan, A., Umarani, C., Subramanian, R., & Nedunchezhian, N. (2009). Biodiesel as an alternative fuel for diesel engines—A review. *Renewable and Sustainable Energy Reviews, 13*, 653–662.

Oliveira, L. S., Franca, A. S., Camaegos, R. R. S., & Ferraz, V. P. (2008). Coffee oil as a potential feedstock for biodiesel production. *Bioresource Technology, 99*, 3244–3250.

Park, J. Y., Kim, D. K., Lee, J. P., Park, S. C., Kim, Y. J., & Lee, J. S. (2008). Blending effects of biodiesels on oxidation stability and low temperature flow properties. *Bioresource Technology, 99*, 1196–1203.

Peng, C. Y., Lan, C. H., & Dai, Y. T. (2006). Speciation and quantification of vapor phases in soy biodiesel and waste cooking oil biodiesel. *Chemosphere, 65*, 2054–2062.

Peterson, C. L., & Hustrulid, T. (1998). Carbon cycle for rapeseed oil biodiesel fuels. *Biomass and Bioenergy, 14*, 91–101.

Ramadhas, A. S., Jayaraj, S., Muraleedharan, C., & Padmakumari, K. (2006). Artifical neural networks used for the prediction of the cetane number of biodiesel. *Renewable Energy, 31*, 2524–2533.

Russo, D., Dassisti, M., Lawlor, V., & Olabi, A. G. (2012). State of the art of biofuels from pure plant oil. *Renewable and Sustainable Energy Reviews, 16*, 4056–4070.

Ryu, K. (2010). The characteristics of performance and exhaust emissions of a diesel engine using a biodiesel with antioxidants. *Bioresource Technology, 101*, 578–582.

Schleicher, T., Werkmeister, R., Russ, W., & Meyer-Pittroff, R. (2009). Microbiological stability of biodiesel-diesel-mixtures. *Bioresource Technology, 100*, 724–730.

Senzikiene, E., Makareviciene, V., & Janulis, P. (2006). Influence of fuel oxygen content on diesel engine exhaust emissions. *Renewable Energy, 31*, 2505–2512.

Sharma, Y. C., Singh, B., & Upadhyay, S. N. (2008). Advancements in development and characterization of biodiesel: A review. *Fuel, 87*, 2355–2373.

Shumaker, J. L., Crofcheck, C., Tackett, S. A., et al. (2008). Biodiesel synthesis using calcined layered double hydroxide catalyst. *Applied Catalyst B: Environmental, 82*, 120–130.

Singh, A., Nigam, P. S., & Murphy, J. D. (2011). Mechanism and challenges in commercialization of algal biofuels. *Bioresource Technology, 102*, 26–34.

Singh, A., & Olsen, S. I. (2011). A critical review of biochemical conversion, sustainability and life cycle assessment of algal biofuels. *Applied Energy, 88*(10), 3548–3555.

Singh, S. P., & Singh, D. (2010). Biodiesel production thorough the use of different sources and characterization of oils and their esters as the substitute of diesel: A review. *Renewable and Sustainable Energy Reviews, 14*, 200–216.

Sinha, S., Agarwal, A. K., & Garg, S. (2008). Biodiesel development from rice bran oil: Transesterification process optimization and fuel characterization. *Energy Conversion and Management, 49*, 1248–1257.

Smith, P. C., Ngothai, Y., Nguyen, Q. D., & O'Neill, B. K. (2010). Improving the low-temperature properties of biodiesel: Methods and consequences. *Renewable Energy, 35*, 1145–1151.

Tang, H., Salley, S. O., & Ng, K. Y. S. (2008). Fuel properties and precipitate formation at low temperature in soy-, cottonseed-, and poultry fat-based biodiesel blends. *Fuel, 87*, 3006–3017.

Tate, R. E., Watts, K. C., Allen, C. A. W., & Wilkie, K. I. (2006a). The viscosities of three biodiesel fuels at temperatures up to 300 °C. *Fuel, 85*, 1010–1015.

Tate, R. E., Watts, K. C., Allen, C. A. W., & Wilkie, K. I. (2006b). The densities of three biodiesel fuels at temperatures up to 300 °C. *Fuel, 85*, 1004–1009.

Torres-Jimenez, E., Svoljšak Jerman, M., Gregorc, A., Lisec, I., Dorado, M. P., & Kegl, B. (2011). Physical and chemical properties of ethanol–diesel fuel blends. *Fuel, 90*, 795–802.

Torres-Jimenez, E., Svoljšak-Jerman, M., Gregorc, A., Lisec, I., Dorado, M. P., & Kegl, B. (2010). Physical and chemical properties of ethanol–biodiesel blends for diesel engines. *Energy & Fuels, 24*, 2002–2009.

Vijayaraghavan, K., & Hemanathan, K. (2009). Biodiesel production from freshwater algae. *Energy & Fuels, 23*, 5448–5453.

Wain, K. S., Perez, J. M., Chapman, E., & Boehman, A. L. (2005). Alternative and low sulfur fuel options: boundary lubrication performance and potential problems. *Tribology International, 38*, 313–319.

Wang, W., Mab, S., Zhao, M., Kuang, L., Nie, J., & Riley, W. W. (2011). Improving the cold flow properties of biodiesel from waste cooking oil by surfactants and detergent fractionation. *Fuel, 90*(3), 1036–1040.

West, A. H., Posarac, D., & Ellis, N. (2008). Assessment of four biodiesel production using HYSYS. *Plant, Bioresource Technology, 99*, 6587–6601.

Xue, J., Grift, T. E., & Hansen, A. C. (2011). Effect of biodiesel on engine performances and emissions. *Renewable and Sustainable Energy Reviews, 15*, 1098–1116.

Yoon, S. H., Suh, H. K., & Lee, C. S. (2009). Effect of spray and EGR rate on the combustion and emission characteristics of biodiesel fuel in a compression ignition engine. *Energy & Fuels, 23*, 1486–1493.

Yuan, W., Hansen, A. C., & Zhang, Q. (2005). Vapor pressure and normal boiling point predictions for pure methyl esters and biodiesel fuels. *Fuel, 84*, 943–950.

Yuste, A. J., & Dorado, M. P. (2006). A neural network approach to simulate biodiesel production from waste olive oil. *Energy & Fuels, 20*, 399–402.

Chapter 5
Effects of Biodiesel Usage on Injection Process Characteristics

A replacement of mineral diesel by biodiesel causes variations in injection process characteristics. Most notably, biodiesel usage influences the *injection pressure*, *injection timing*, and *injection rate*. Clearly, these effects depend to a great extent on *biodiesel properties* (raw material, production technology, ingredients, etc.), *operating regime*, and *ambient temperature*. This means that one has to deal with many highly interdependent influencing parameters. Consequently, the determination of these effects is a sophisticated process that is typically done only partially (e.g., only for one injection system type, for one biodiesel type, etc.) in individual research laboratories. Once such results are known, it is very difficult to estimate how these results can be used in slightly modified circumstances (e.g., when a somewhat different fuel is used). This is why one can often make only some general observations or guidelines. For example, many investigators show that in engines equipped with *mechanically controlled fuel injection systems (MCFIS)* the cyclic fuel delivery, pressure wave propagation time, average injection rate, and maximum pressure during injection are significantly affected when neat biodiesel replaces mineral diesel (Kegl 2006a; Luján et al. 2009a, b). In general, the injection pressure and injection rate are higher and the needle opens earlier for neat biodiesel (Caresana 2011; Kegl 2006a, b). For *electronically controlled fuel injection systems (ECFIS)* the situation is somewhat similar in the sense that various biodiesel types may result in various effects. Although in electronically controlled systems these effects are less exposed than in mechanical systems, numerous tests have to be performed in order to evaluate the influence of biodiesels for various injection pressure levels or injection time strategies (Boudy and Seers 2009; Luján et al. 2009a, b; Ye and Boehman 2010).

B. Kegl et al., *Green Diesel Engines*, Lecture Notes in Energy 12,
DOI 10.1007/978-1-4471-5325-2_5, © Springer-Verlag London 2013

5.1 Injection Pressure

Fuel injection pressure has a significant effect on NO_x and soot emissions. In general, an increase of the injection pressure results in an increase in NO_x emission and a decrease in soot emissions (Song et al. 2012). Biodiesel influences significantly the fuel injection pressure; however, this influence is more exposed in MCFIS and less in ECFIS (Boudy and Seers 2009; Kegl 2006a; Kegl and Hribernik 2006; Luján et al. 2009a, b; Payri et al. 2012; Postrioti et al. 2004; Tziourtzioumis and Stamatelos 2012; Zhang and Boehman 2007; Ye and Boehman 2012; Yoon et al. 2009).

In *MCFIS* (*IDI, DI, and M systems*) fuel is transported from the pump through the high pressure tube to the injector, which is actuated by an adequate fuel pressure rise. The fuel transport path is therefore relatively long and plagued by phenomena like cavitation, pressure wave reflection, hydraulic losses, and so on. It should therefore be obvious that fuel properties like density, bulk modulus, and viscosity influence directly the pressure wave propagation and consequently the injection pressure (Kousoulidou et al. 2012; Kegl 2006a; Luján et al. 2009a, b; Ozsezen and Canakci 2010; Usta 2005). Furthermore, biodiesel effects related to the injection pressure depend also on the engine operating regime and fuel temperature.

In *ECFIS* (*unit injector systems, common rail systems*) the phenomena related to pressure wave propagation and reflection are rather minor since the transport way is rather short and the pressure is quite well controlled at some prescribed constant level (Song et al. 2012). Therefore, in these systems biodiesel effects are largely eliminated. In spite of that it is known that the operation of a *common rail system* generates a pressure wave in the injector's feed pipe, which can modify the mass flow rate through the injector. Therefore, fuel properties that affect these pressure waves can influence the injection process and, consequently, the total fuel mass injected, especially in multiple injection scenario (Boudy and Seers 2009). It may also be worth noting that the bulk modulus effect appears to be present in *unit injector systems*, but not in *common rail systems*, where a rapid transfer of the pressure wave practically does not occur (Kousoulidou et al. 2012).

Obviously, the situation depends significantly on the selected injection system type. However, minor design modifications, such as an addition of a snubber valve to the fuel transport way, may also influence the result. For example, numerical and experimental investigations on a bus diesel engine M injection system (Table 5.1) show that a change from mineral diesel to rapeseed biodiesel results in a higher injection pressure and earlier needle opening, if the design and initial conditions are kept constant (Kegl and Hribernik 2006; Kegl 2006a). If diesel and biodiesel blends are used, it can be observed that the mean injection pressure increases almost linearly (up to 7 %) with the increasing part of biodiesel in the blend.

However, the addition of a snubber valve to the high pressure path of the fuel changes the situation. Numerical simulation of injection characteristics for rapeseed biodiesel at fuel temperature of 40 °C and at rated condition revealed this influence. A comparison of the pressure after the HP pump p_{p}, the injection pressure p_{inj}, and

Table 5.1 Main specification of the tested M injection system (Kegl 2006a)

Injection model	Direct injection system with wall distribution (M system)
Fuel injection pump type	Bosch PES 6A 95D 410 LS 2542
Pump plunger (diameter × lift)	9.5 × 8 mm
Fuel HP tube (length × diameter)	1,024 × 1.8 mm
Injection nozzle (number × nozzle hole diameter)	1 × 0.68 mm
Needle lift (maximal)	0.3 mm

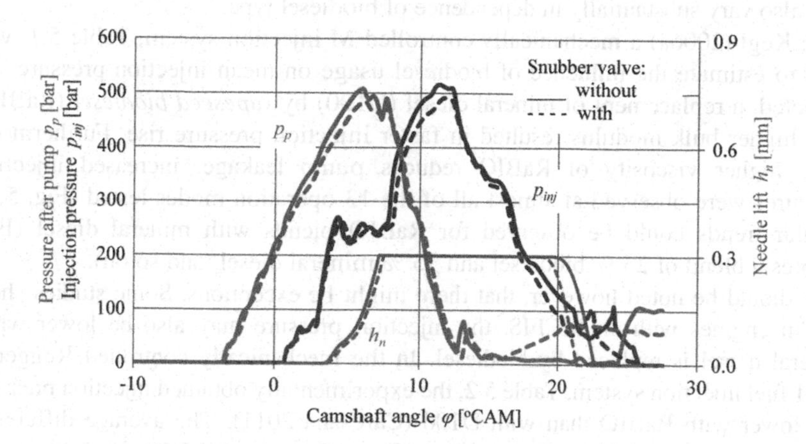

Fig. 5.1 Injection characteristics with and without the snubber valve

the needle lift h_n is shown in Fig. 5.1. When the snubber valve is present, the pressures p_p and p_{inj} are lower by about 20 bar, the injection duration is shorter by about 0.2 ms, and the fuelling is lower by about 10 mm^3/stroke.

5.1.1 Influence of Biodiesel Properties

The physical properties of biodiesel fuels typically differ significantly from those of mineral diesel. These differences lead to various effects, which are most notably present in MCFIS. The most significant effects result from the variations in the following properties:

- *Compressibility* or *bulk modulus*: by using biodiesel, the pressure rise produced by the pump in a MCFIS is achieved faster due to the lower compressibility (higher bulk modulus) of biodiesel.
- *Sound velocity*: when using biodiesel, the fuel transport pressure wave propagates faster toward the injector as a consequence of biodiesel's higher sound velocity; in a MCFIS the delay between the pressure rise after the HP pump and the injection pressure rise is notably shorter.

- *Viscosity*: higher viscosity of biodiesel reduces pump leakage in a MCFIS; this results in increased injection pressure and also contributes to faster and earlier needle opening.

All of these facts result in faster and earlier needle opening when using biodiesel. This fact has been widely accepted to explain the higher temperature peaks and higher rates of NO_x formation when diesel was replaced by biodiesel. Unfortunately, the physical and chemical properties of various biodiesel types may vary significantly, even if they conform to the corresponding standard. Therefore, the effects of biodiesel may also vary substantially in dependence of biodiesel type.

In Kegl (2006a) a mechanically controlled M injection system, Table 5.1, was used to estimate the influence of biodiesel usage on mean injection pressure. As expected, a replacement of mineral diesel (D100) by *rapeseed biodiesel* (RaBIO) with higher bulk modulus resulted in faster injection pressure rise. Furthermore, since higher viscosity of RaBIO reduces pump leakage, increased injection pressures were observed at almost all of the 13 operation modes tested (Fig. 5.2). Similar trends could be observed for RaBIO blends with mineral diesel (B25 denotes a blend of 25 % biodiesel and 75 % mineral diesel, and so on).

It should be noted however, that there might be exceptions. Some studies show that in engines with an MCFIS, the injection pressure may also be lower when mineral diesel is replaced by biodiesel. In the mechanically controlled Ruggerini RF91 fuel injection system, Table 5.2, the experimentally obtained injection pressure was lower with RaBIO than with D100 (Caresana 2011). The average difference observed was around 13 bar for neat RaBIO and 2 bar for the B20 (20 % biodiesel, 80 % mineral diesel) blend. An analysis of the experimental injection pressure traces revealed that just after the needle closing the pressure in the injection tube may fall below the vapor value, which leads to cavitation. In turn, cavitation may be the cause of lower injection pressure in the subsequent injection cycle. The residual pressure obtained with biodiesel was lower than that obtained with mineral diesel (Fig. 5.3).

In Tziourtzioumis and Stamatelos (2012) the influence of fuel properties on variation of the injection pressure was investigated on a HSDI turbocharged engine, equipped with an ECFIS, Table 5.3. The biodiesel used was produced from 40 % rapeseed oil, 30 % soybean oil, and 30 % waste cooking oil.

The results show that the rail pressure increases when biodiesel is added to mineral diesel (Fig. 5.4). This is explained by the ECU algorithm for the computation of the rail pressure. Namely, the rail pressure is computed as a function of the engine speed and fuel delivery per stroke. Since biodiesel has a lower volume heating value than pure diesel, more fuel needs to be injected into the engine cylinder, which causes a higher fuel delivery and thus increased rail pressures for the B70 fuel blend. In Fig. 5.4, the ranges of pressure wave oscillations in the rail tube are given at several engine speeds.

The influence of fuel properties on injection pressure was also investigated numerically for single and multiple injection strategies in an ECFIS (Boudy and Seers 2009; Song et al. 2012). From the results it follows that an increase of fuel viscosity, bulk modulus, or density results in a relatively small reduction of the

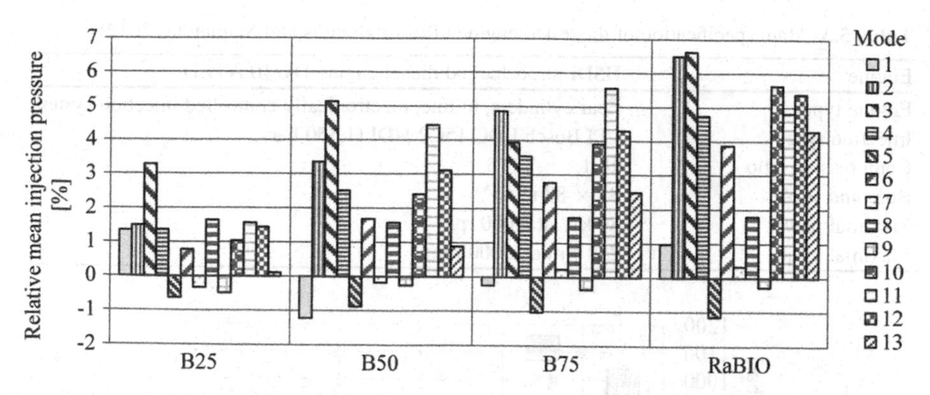

Fig. 5.2 Mean injection pressure at 13 ESC test mode for blends of RaBIO and D100 with respect to D100

Table 5.2 Main specifications of the tested engine (Caresana 2011)

Engine	Ruggerini RF5
Engine type	One cylinder, four stroke, NA, DI, mechanically controlled injection system
Displacement	477 cm^3
Compression ratio	18.5:1
Bore and stroke	90 × 75 mm
Maximal power	8.1 kW at 3,600 rpm
Maximal torque	25 Nm at 2,500 rpm

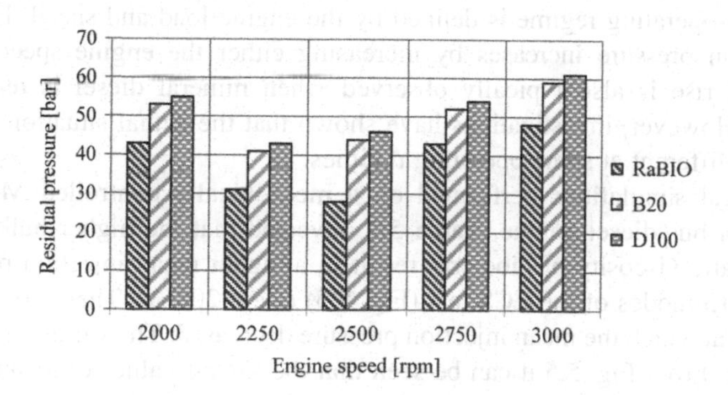

Fig. 5.3 Residual pressure for various tested fuels

pressure wave amplitude. In case of a single injection strategy, the variation in fuel properties are less influential, because the time interval between successive injections is large enough to allow the pressure wave to be damped out before the next injection event. In a multiple injection scenario, the fuel with higher density exhibits a smaller pressure wave amplitude. These differences can be observed even for various biodiesels. For example, the density of RaBIO is higher than that of *palm biodiesel* (PaBIO); consequently the *pressure wave amplitudes* in multiple injection strategy of RaBIO are smaller than those obtained with PaBIO.

Table 5.3 Main specification of the tested engine (Tziourtzioumis and Stamatelos 2012)

Engine	HSDI turbocharged diesel engine DW10 ATED
Engine type	Four cylinders, in line, electronically controlled injection system
Injection type	CPI Bosch EDC 15C2 HDI (1,350 bar)
Compression ratio	18:1
Bore and stroke	85 × 88 mm
Maximal power	80 kW at 4,000 rpm
Maximal torque	250 Nm at 2,000 rpm

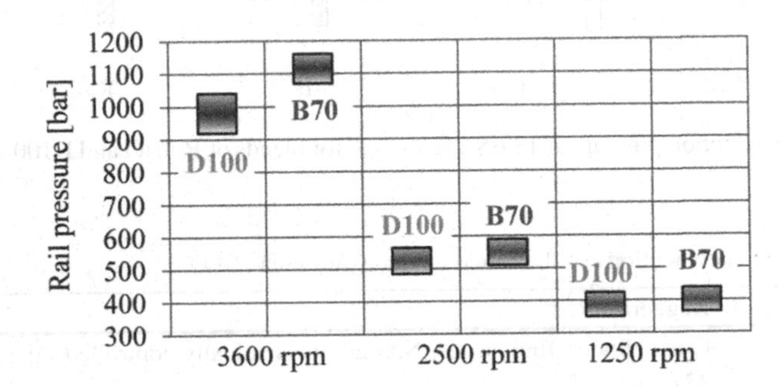

Fig. 5.4 Rail pressure and its oscillation range at various engine speeds

5.1.2 *Influence of Engine-Operating Regime*

The engine-operating regime is defined by the engine load and speed. In general, the injection pressure increases by increasing either the engine speed or load. A pressure rise is also typically observed when mineral diesel is replaced by biodiesel. However, investigations have shown that the actual situation may look somewhat different at some operating regimes.

Numerical simulations performed on a mechanically controlled M injection system of a bus diesel engine, Table 5.1, revealed that the higher bulk modulus and kinematic viscosity of biodiesel result in a higher mean injection pressure at almost all 13 modes of the ECS test (Fig. 5.5) (Kegl 2006a). There are, however, two modes at which the mean injection pressure decreased. The variation was small but notable. From Fig. 5.5 it can be seen that the lowest values of mean injection pressure with RaBIO are obtained at idle (mode 1) and at lower engine speeds and loads (modes 5, 7, 9).

For the RaBIO blends with D100 the situation is quite similar (Fig. 5.2). The mean injection pressure increases almost linearly with the increasing part of biodiesel (up to 7 %) (Kegl 2006a).

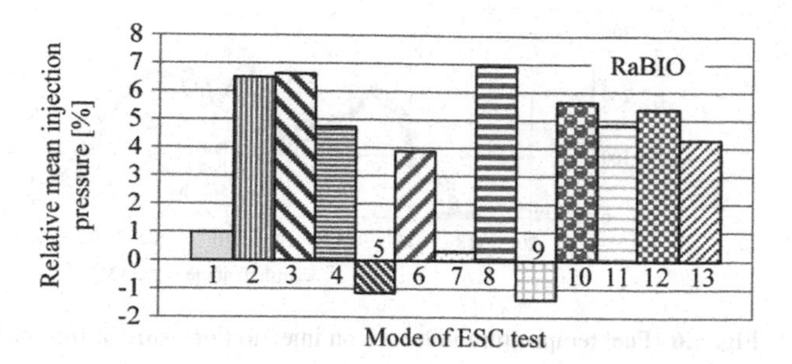

Fig. 5.5 Relative mean injection pressure at the 13 ESC test modes for RaBIO with respect to D100

5.1.3 Influence of Temperature

Fuel temperature influences the bulk modulus, density, kinematic viscosity, and sound velocity of biodiesels. In general, lower fuel temperature means higher bulk modulus, density, and sound velocity. Consequently, a variation in fuel temperature can contribute significantly to variations in all injection characteristics.

At lower fuel temperatures, the injection pressure of biodiesels rises earlier than that of mineral diesel. This is due to the fact that at lower temperatures the difference of bulk modulus between biodiesel and mineral diesel is larger than at higher temperatures.

In a mechanically controlled M fuel injection system, Table 5.1, the influence of fuel temperature on injection pressure was investigated experimentally by using RaBIO (Kegl and Hribernik 2006). The tests were performed at constant pump injection timing of 23°CA BTC at several operating regimes. The fuel temperature was measured at the end of the HP tube, just before the injector. At full load and 500 rpm the maximum injection pressure was reached at higher temperatures (Fig. 5.6). This changed when the speed was increased to 800 rpm (Fig. 5.7) and 1,100 rpm (Fig. 5.8). In the latter cases the pressure maximum was observed at lower temperatures. At lower load regimes, the influence of fuel temperature on the measured injection pressure was observed to be quite similar to that at full load (Kegl and Hribernik 2006).

By lowering the fuel temperature, the bulk modulus and kinematic viscosity of biodiesels typically increase. In general, this means that lower fuel temperature results in higher mean injection pressures. This can be observed at most ESC modes (Fig. 5.9). As one can see, the relative mean injection pressure increases progressively with lower fuel temperature (up to 15 % at −20 °C) (Kegl 2008).

It may be worth noting that some authors report that at higher fuel temperatures (over 80 °C), the variation of bulk modulus and kinematic viscosity does not influence significantly the injection pressure, injection timing, and needle lift history (Yamane et al. 2001).

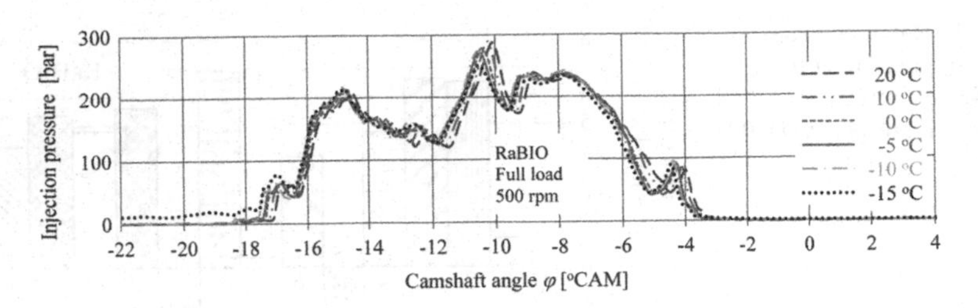

Fig. 5.6 Fuel temperature influence on injection pressure at full load and 500 rpm

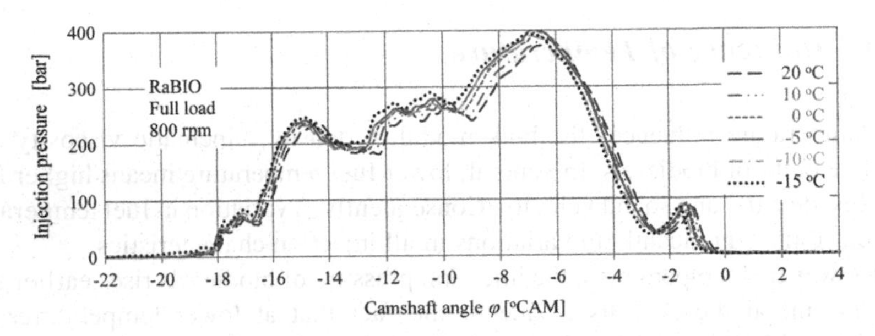

Fig. 5.7 Fuel temperature influence on injection pressure at peak torque

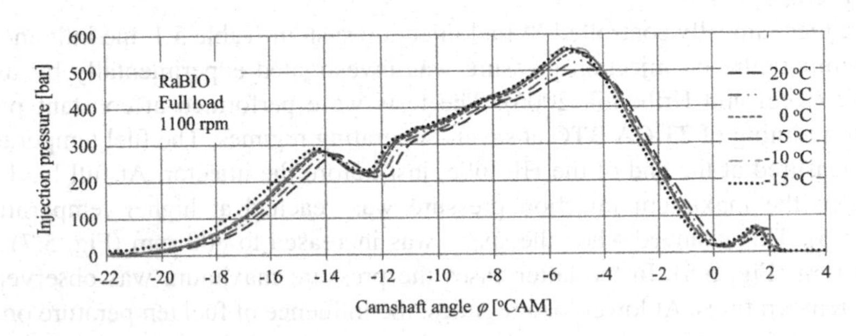

Fig. 5.8 Fuel temperature influence on injection pressure at rated condition

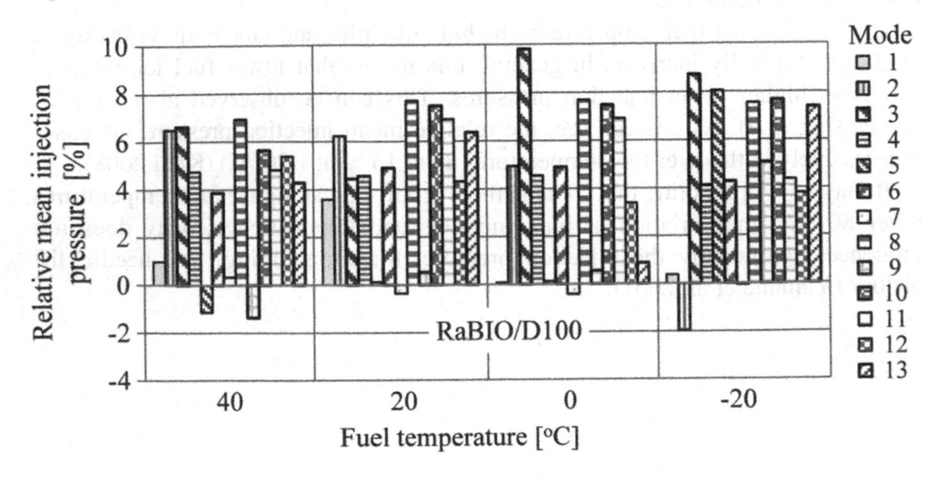

Fig. 5.9 Relative mean injection pressure variation of RaBIO with respect to D100

5.2 Injection Timing

It is well known that injection timing or start of injection is a very important parameter that significantly influences all engine characteristics. This is mainly due to the fact that injection timing influences the mixing quality of the air–fuel mixture and consequently also the combustion process, including harmful emissions. It is generally known that retarded injection decreases the maximal pressure in the cylinder and leads to lower peak rate of heat transfer and consequently to lower combustion noise. Since the delayed injection leads to lower temperatures, the NO_x emissions are also reduced. On the other hand, retarded injection leads to an increase in fuel consumption. Smoke emission may also increase, though trends vary significantly between different engine types. For example, for direct injection diesel engine at high load, the HC emissions are low and vary only modestly with respect to injection timing. At partial loads, the HC emissions are higher and increase as the injection start is shifted significantly from the optimum. This trend is especially evident at idle.

By changing the injection timing and introducing biodiesel fuel, it is necessary to carefully monitor some of the engine characteristics in order to prevent harmful mechanical and thermal stress of the engine. Above all, these characteristics are the *injection pressure*, *in-cylinder pressure*, and *exhaust gas temperature*. All these parameters, as well as ambient pressure and temperature, fuel temperature, cooling water temperature in the inlet and outlet of the engine, intake air pressure and temperature, oil pressure and temperature, intake air mass flow, and oxygen level in the exhaust gases, have to be measured to find the optimal injection timing.

In ECFIS, the start of injection can easily be controlled directly by the ECU, which can be programmed according to various fuel types. When biodiesel replaces diesel, the consequences are not exposed as much as in a MCFIS. In spite of that, they are measurable and have to be taken into account (Song et al. 2012). For example, at the same operating conditions and at the same injection fuelling, *soybean biodiesel* (SoBIO) usage results in a slightly shorter injection delay and consequently in slightly advanced injection timing, compared to mineral diesel (Yoon et al. 2009). The reason for advanced injection timing is mainly due to the higher bulk modulus of biodiesel.

In MCFIS, the start of injection can hardly be controlled directly, because it depends on sophisticated transport phenomena in the pump, high-pressure tube, and the injector. However, the start of injection is closely related to the start of injection pump delivery, which can be set easily to any desired value. For this reason, the start of injection pump delivery or injection pump timing is frequently used as the primary variable to set the injection timing (Kegl 2006b). Compared to mineral diesel, biodiesels exhibit a higher bulk modulus, which results in higher sound velocity. This leads to a more rapid transfer of the pressure wave from the pump to the injector needle and to an earlier needle lift (Boehman et al. 2004; Kegl and Hribernik 2006; Kegl 2006a). Therefore, the injection timing should in general be advanced practically at all operating regimes (Ozsezen et al. 2008).

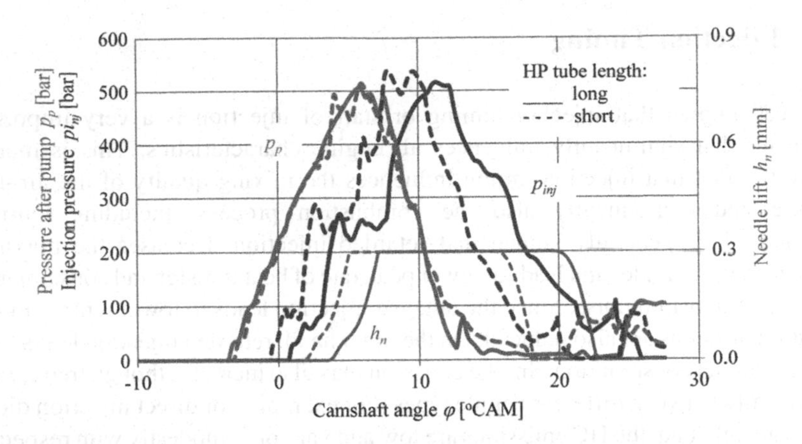

Fig. 5.10 Injection characteristics using RaBIO and various HP tube lengths

It is worth noting that for a MCFIS rather minor design variations, such as the addition of the snubber valve, HP tube length variation, etc., may also lead to significant injection timing variation that may depend on the fuel type. For example, numerical and experimental RaBIO usage investigation on a mechanically controlled M injection system, Table 5.1, revealed that if the HP tube is shortened, e.g., by about 50 %, the pressures p_p and p_{inj} may increase up to 20 bar (Kegl and Hribernik 2006; Kegl 2006a). Furthermore, the injection duration shortens by about 0.4 ms and the fuelling is lower by about 15 mm^3/stroke (Fig. 5.10). Furthermore, the HP tube shortening advanced the injection timing by about 0.3 ms.

5.2.1 Influence of Biodiesel Properties

Fuel injection timing is most notably influenced by the physical properties of the fuel, especially by density, bulk modulus, and kinematic viscosity. Clearly, in mechanically controlled injection systems these effects are much more exposed than in electronically controlled systems.

Biodiesels typically exhibit higher *density, bulk modulus, viscosity*, and *sound velocity*. This leads to faster pressure wave propagation. In MCFIS this leads to earlier needle lift and advanced injection timing (reduced injection delay). Earlier injection typically leads to an increase of the maximum in-cylinder pressure, increased in-cylinder temperature, and higher exhausts emissions (Pandey and Sarviya 2012). Higher viscosity of biodiesels results in reduced fuel losses during the injection process which also contributes to faster evolution of the injection pressure and advanced injection timing. Another point worth notion is the *vapor volume* that depends on the residual pressure in a MCFIS. Biodiesels typically exhibit lower vapor values. This also decreases the injection delay and thus advances the injection timing. This effect is observable for neat biodiesels and for biodiesel blends with mineral diesel.

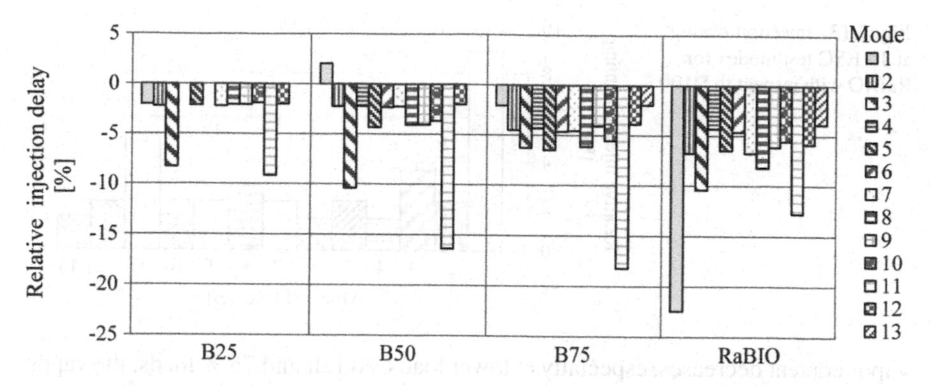

Fig. 5.11 Injection delay at 13 ESC test mode for blends of RaBIO and D100 with respect to D100

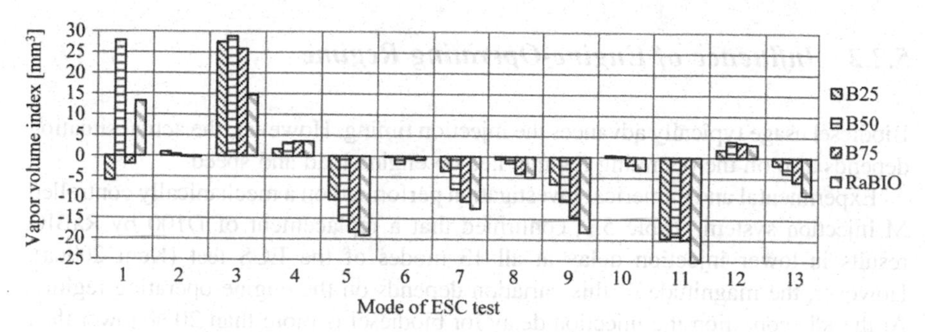

Fig. 5.12 Vapor volume indexes at 13 ESC test modes for blends of RaBIO and D100 with respect to D100

In ECFIS the influence of fuel properties on injection timing is rather minor. Especially for the single injection strategy it can practically be neglected (Boudy and Seers 2009). In the multiple injection strategy, however, variation in the friction coefficient and pressure wave velocity may alter the phasing of the pressure wave arrival and injector opening. For this reason the post injection of biodiesel typically starts earlier.

Experimental and numerical investigation performed on a mechanically controlled M injection system, Table 5.1, confirmed that a replacement of D100 by RaBIO results in advanced injection timing (Kegl 2006a). For the blends, the injection delay decreased practically linearly with the increasing part of RaBIO in the blend (Fig. 5.11). This observation was confirmed for practically all operating modes.

Further numerical simulation also revealed that the injection delay is also influenced by the vapor volume (Kegl 2006a). The vapor contents for various blends of RaBIO and mineral diesel at fuel temperature of 40 °C are shown in Fig. 5.12. Here, the vapor volume index is defined as the vapor volume of the actually tested fuel, multiplied by the fuelling ratio (fuelling of neat mineral diesel divided by the fuelling of actually tested fuel). By increasing the part of RaBIO, the

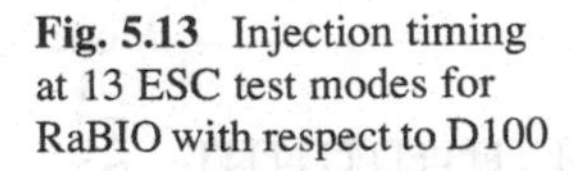
Fig. 5.13 Injection timing at 13 ESC test modes for RaBIO with respect to D100

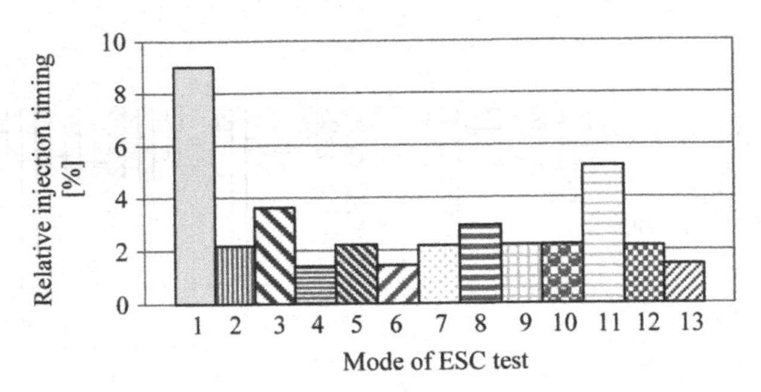

vapor content decreases, especially at lower loads. At full and 75 % loads, the vapor volume indexes are practically the same for all tested fuels.

5.2.2 Influence of Engine-Operating Regime

Biodiesel usage typically advances the injection timing. However, the actual situation depends also on the operating regime, i.e., on engine load and speed.

Experimental and numerical investigation performed on a mechanically controlled M injection system, Table 5.1, confirmed that a replacement of D100 by RaBIO results in lower injection delay at all 13 modes of the ECS test (Kegl 2006a). However, the magnitude of this variation depends on the engine-operating regime. At the idle condition the injection delay for biodiesel is more than 20 % lower than the one obtained with mineral diesel, which results in advanced injection timing by about 9 % (Fig. 5.13). On the other side, the smallest injection timing advancements of less than 2 % were obtained at partial loads and higher engine speeds (modes 4, 6, 13).

5.2.3 Influence of Temperature

The influences of fuel temperature on injection timing can be illustrated by experimental and numerical investigation performed on a mechanically controlled M injection system, Table 5.1 (Kegl and Hribernik 2006). Injection timing was obtained from the needle lift h_n, which was determined at various operating regimes when using RaBIO. The results show that at full load lower fuel temperatures lead to advanced injection timing at 500 rpm (Fig. 5.14) at 850 rpm (Fig. 5.15) and 1,100 rpm (Fig. 5.16). At low load regimes, the influence of fuel temperature on the measured injection timing is quite similar to that at full load.

For the same system, the injection delay and injection timing were investigated at various temperatures and related to the injection delay at fuel temperature of 40 °C. Both, neat D100 and neat RaBIO were investigated. For both fuels, the relative injection delay decreased with decreasing fuel temperatures. This is mostly due to the higher sound velocity and can be observed at practically all operating

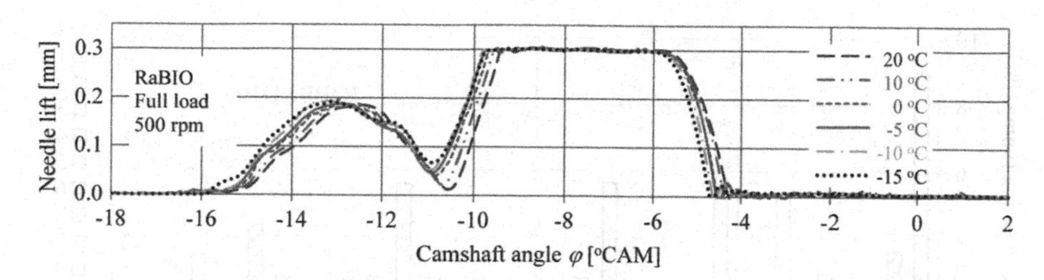

Fig. 5.14 Fuel temperature influence on injection timing at full load and 500 rpm

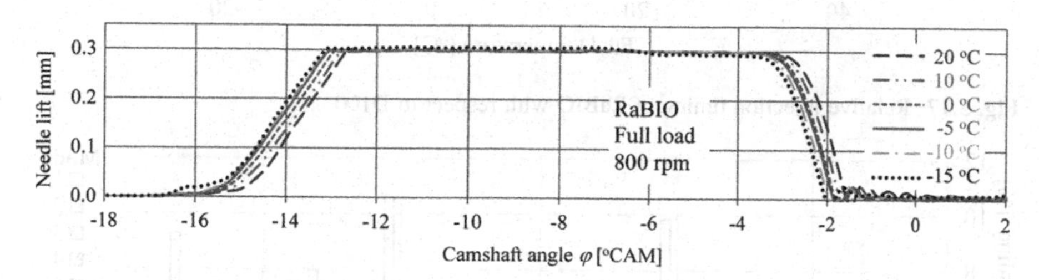

Fig. 5.15 Fuel temperature influence on injection timing at full load and 800 rpm

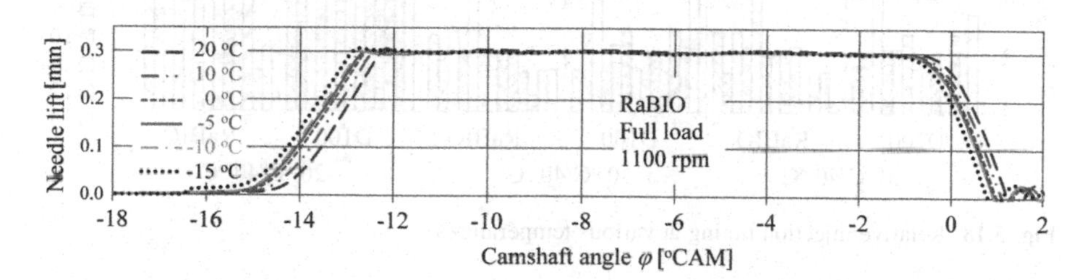

Fig. 5.16 Fuel temperature influence on injection timing at full load and 1,100 rpm

regimes (Kegl 2008). The relative injection timing of RaBIO with respect to D100 at several fuel temperatures is given in Fig. 5.17.

Reduced injection delay means that the injection timing at lower fuel temperatures is advanced. It is known that advanced timing causes the pressure and temperature in the cylinder to rise, leading to increased NO_x emissions. As shown in Fig. 5.18, advanced injection happens with both fuels at all ESC modes.

5.3 Injection Rate

Injection rate history, fuelling at various phases of injection, and the total fuelling depend directly on fuel properties, especially on the viscosity and density. Since viscosity and density of biodiesels may vary substantially; this may lead to

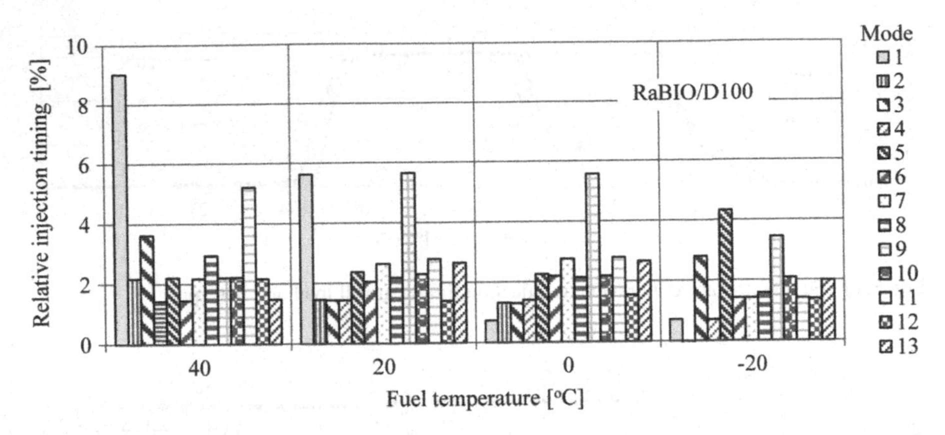

Fig. 5.17 Relative injection timing of RaBIO with respect to D100

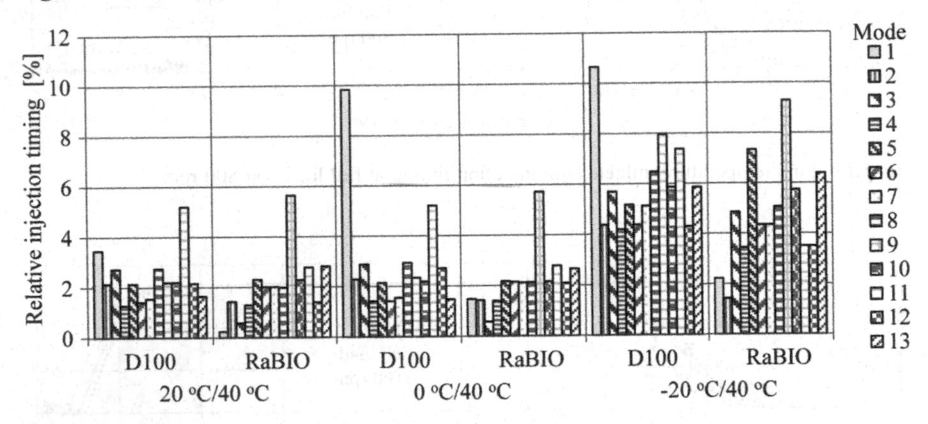

Fig. 5.18 Relative injection timing at various temperatures

significant effects on the injection rate. Furthermore, the density and viscosity of the fuel depend significantly on its temperature. Therefore, fuel temperature also represents a very important parameter.

As with other injection characteristics, biodiesel influence on injection rate is not equally exposed at all fuel injection systems. In MCFIS this influence is rather significant due to transport-related effects and lower injection pressures (Boudy and Seers 2009; Kegl 2008, 2006a). In ECFIS, however, this influence is rather minor. At a first glance, this comes as a surprise since biodiesels typically have higher viscosity than mineral diesel, which tends to retard the fuel flow into the combustion chamber and lower the volume of injected fuel. However, biodiesel fuels have higher density than mineral diesel, which implies that the mass of the injected fuel remains rather constant. These two counteracting effects can explain why there are no significant differences between both fuel injection rates in a high pressure common rail system (Luján et al. 2009a, b; Yoon et al. 2009).

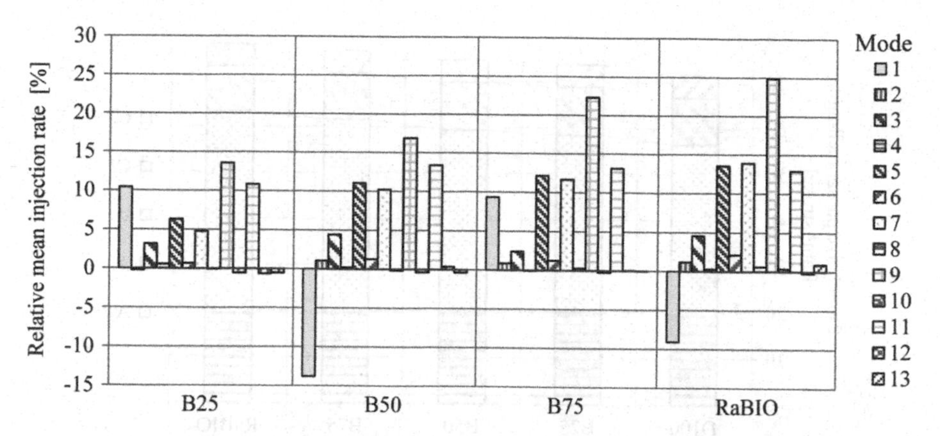

Fig. 5.19 Mean injection rates at 13 ESC test modes for blends of RaBIO and D100 with respect to D100

5.3.1 Influence of Biodiesel Properties

In MCFIS, a general observation is that higher values of fuel density, viscosity, and bulk modulus result in an increased fueling (Boudy and Seers 2009; Kegl 2006a). In ECFIS a higher fuel density may also lead to somewhat higher fueling, but only under special circumstances. In the case of single injection, the fuelling of various biodiesels is practically unchanged with respect to mineral diesel. Fuel density has also a negligible impact during pilot and main injections in the case of triple injection. However, during post-injection the higher density of biodiesel fuels may result in higher injected fuel mass. This effect may be even more emphasized, if the value of the bulk modulus decreases (Boudy and Seers 2009), i.e., if one type of biodiesel is replaced by another one with higher density and lower bulk modulus.

Investigations on a mechanically controlled M fuel injection system, Table 5.1, show that, in general, the mean injection rate increases when D100 is replaced by RaBIO or its blends. This can be observed at all ESC modes, except for neat RaBIO and B50 at idle (Fig. 5.19) (Kegl 2006a).

By taking into account the weighting factors of the 13 modes of the ESC test, the partial fuellings of various injection phases are presented in Fig. 5.20 (Kegl 2006a). The total fuelling consist of the partial fuelling during needle opening phase (A), which is subdivided into A1 (first 10 % of injection) and A2 (from end of A1 till the end of A), open needle phase (B) and needle closing phase (C), which is subdivided into C1 (from start of C till the beginning of C2) and C2 (last 10 % of injection).

The results presented in Fig. 5.20 show that the fuelling during the first 10 % of injection duration (A1) is the highest when using D100; meanwhile the fuelling during the last 10 % of injection duration (C2) is the smallest with B25 and B75. The fuelling during the needle lifting as well as needle closing decreases with increasing content of RaBIO in the blend with D100.

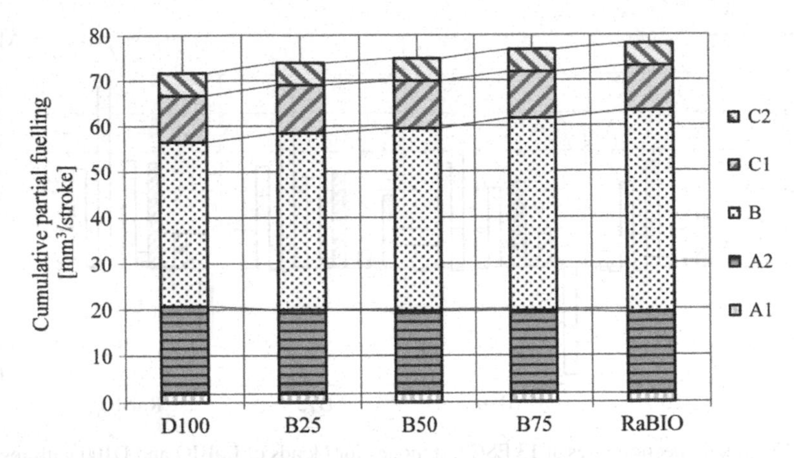

Fig. 5.20 Fuelling in various phases of injection, summed over 13 ESC modes, for various fuels

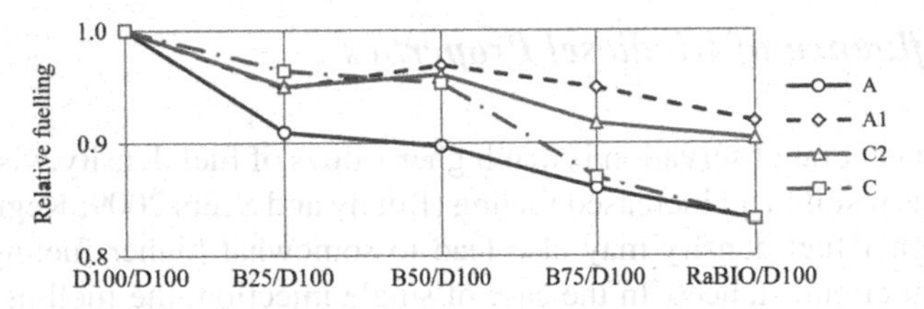

Fig. 5.21 Relative partial fuellings summed over all 13 ESC modes for various blends of RaBIO and D100

In order to enable a comparison between various fuels, the relative partial fuellings, summed over all 13 ESC modes, can be calculated (Kegl 2006a). The most interesting phases (A, A1, C, and C2) are compared in Fig. 5.21. One can see that by increasing the RaBIO content, all of the compared quantities decrease. This offers a good opportunity to reduce NO_x as well as smoke and PM emissions.

5.3.2 *Influence of Engine-Operating Regime*

In general, the mean injection rate is higher at higher engine loads and speeds. However, this may not always be the case. Investigations have shown that this situation may change in a particular operating regime.

In Kegl (2006a) RaBIO blends with D100 were tested in a mechanically controlled M fuel injection system, Table 5.1. It turned out that, in general, the fuelling increases by increasing the part of RaBIO in the blend (Fig. 5.22). Only at idle this was not the case.

The relative mean injection rates of RaBIO with respect to D100 are presented in Fig. 5.23 (Kegl 2006a). One can see that the injection rates increased the most at

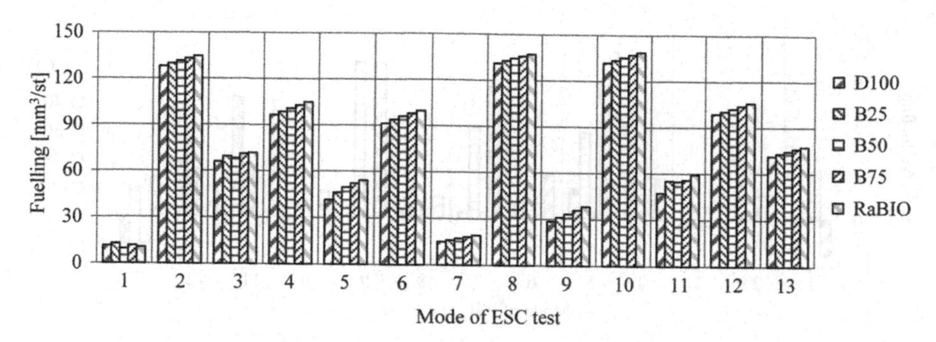

Fig. 5.22 Fuelling of RaBIO blends with D100

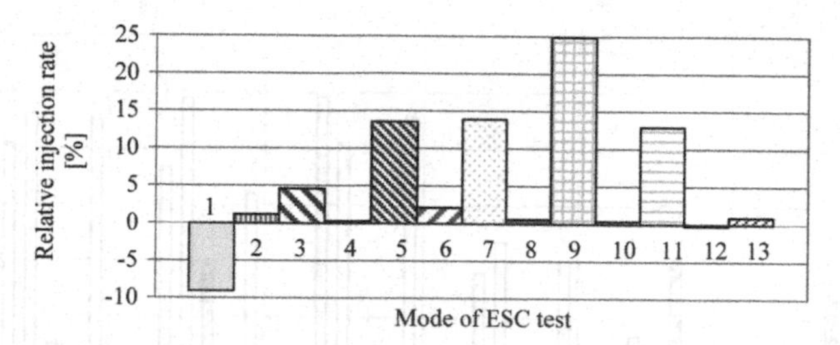

Fig. 5.23 Relative mean injection rate for RaBIO

lower loads (modes 7, 9, 11—25 % load and mode 5—50 % load). On the other side, at idle the mean injection rate is the lowest.

5.3.3 *Influence of Temperature*

In Kegl (2008) the influence of fuel temperature on fuelling per stroke was investigated on the mechanically controlled M fuel injection system, Table 5.1, by using D100 and RaBIO. In general, the fuelling of RaBIO is somewhat higher than that of D100, but the results depend significantly on fuel temperature (Fig. 5.24). The maximum difference (over 20 %) is obtained at lower loads (modes 5, 7, 9, and 11), especially at higher fuel temperatures.

In the mechanically controlled M injection system the fuellings during the initial and final injection phases are closely related to NO_x and smoke emissions. Therefore, it is worth to compare the partial fuellings for D100 and RaBIO at various operating regimes and various fuel temperatures (Kegl 2008). In order to estimate the NO_x emission, the partial fuelling A at various temperatures is shown in Fig. 5.25. One can see that the variations of relative partial fuelling A increase as the temperature becomes lower. For both fuels, D100 and RaBIO, this is especially evident below 0 °C. Both fuels behave similarly, except at three operating regimes (modes 1, 7 and 9) at the temperature of 0 °C. This means that at low load regimes, RaBIO is somewhat more sensitive to temperature. Thus, the relative increase of

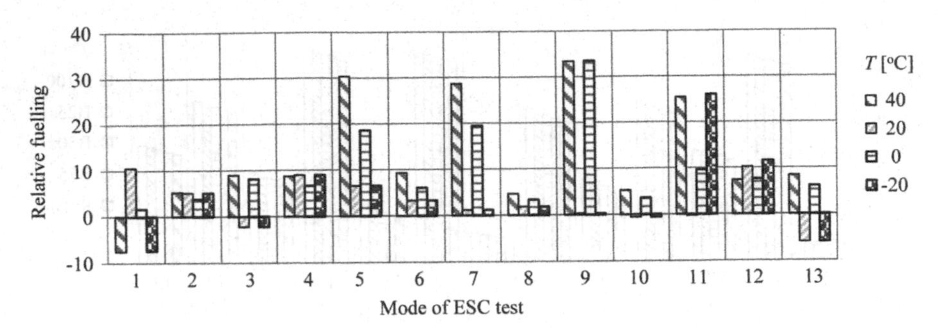

Fig. 5.24 Relative fuelling of RaBIO compared to that of D100 at various fuel temperatures

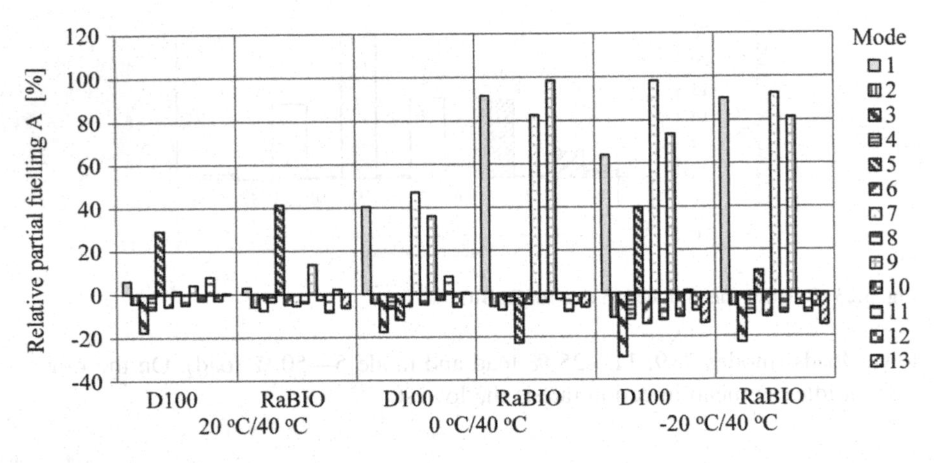

Fig. 5.25 Relative partial fuelling A

NO_x emission of RaBIO may be somewhat higher at these modes. At almost all other modes of the ESC test, the partial fuelling A decreases with lower temperatures.

By analyzing the partial fuelling A1 (Fig. 5.26) one can conclude that the differences between both fuels are relatively small. One can see, however, that the partial fuelling A1 practically always increases as the temperature is reduced. This means that at lower temperatures, higher NO_x emission can be expected for both fuels (Kegl 2008).

To estimate the influence of temperature on smoke emission, the partial fuellings C2 and C are shown in Figs. 5.27 and 5.28 (Kegl 2008). One can see that the relative partial fuelling C2 increases practically at all modes. The increase is relatively large; at some modes it almost reaches 100 %. The differences between RaBIO and D100 are rather small. A similar observation can be made for the relative partial fuelling C. According to the observations in the final phases of injection, one can expect that the influence of fuel temperature on smoke is similar for D100 and RaBIO.

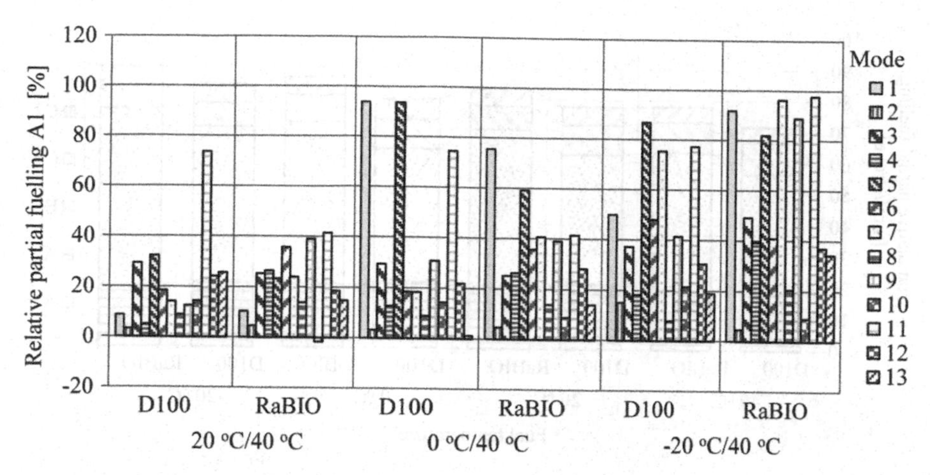

Fig. 5.26 Relative partial fuelling A1

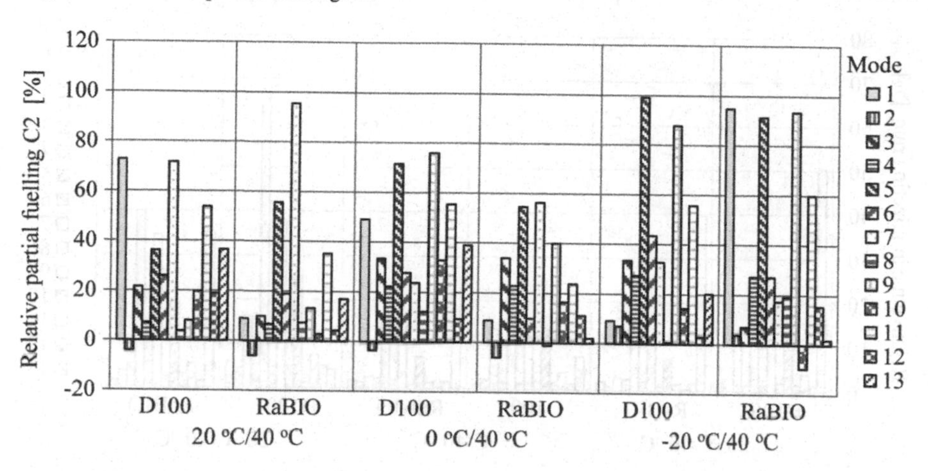

Fig. 5.27 Relative partial fuelling C2

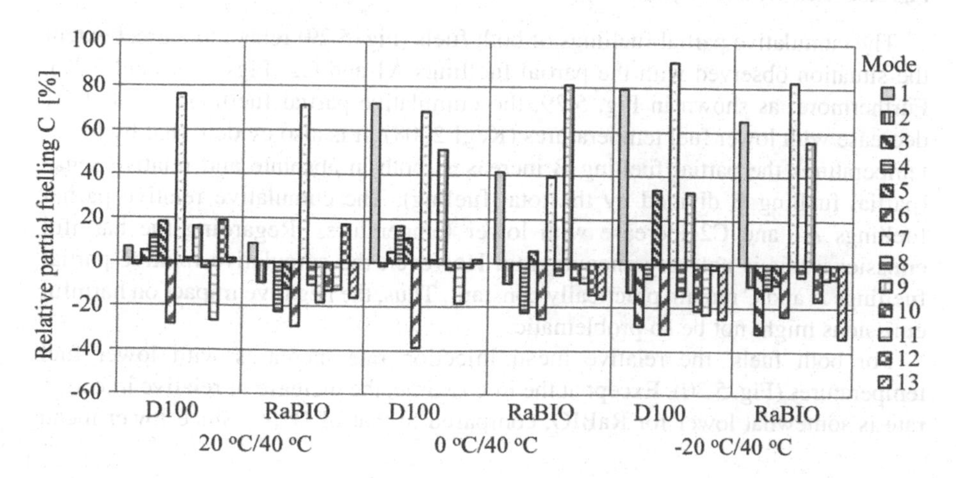

Fig. 5.28 Relative partial fuelling C

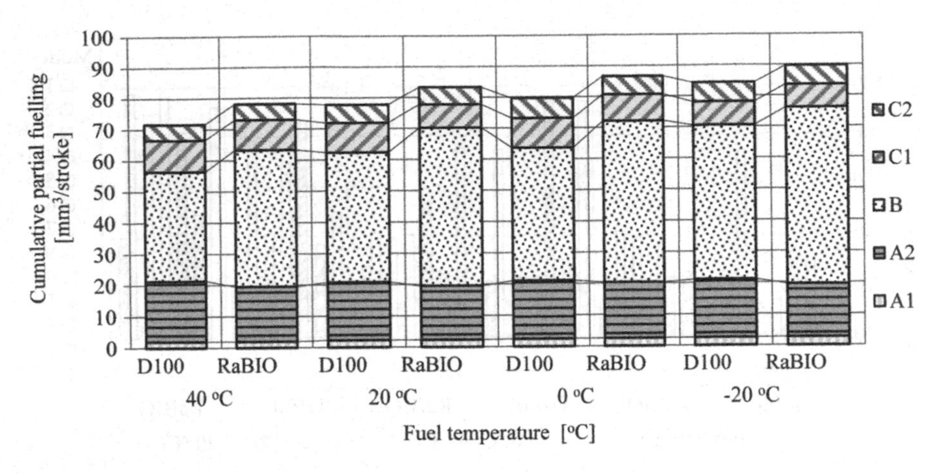

Fig. 5.29 Cumulative partial fuellings

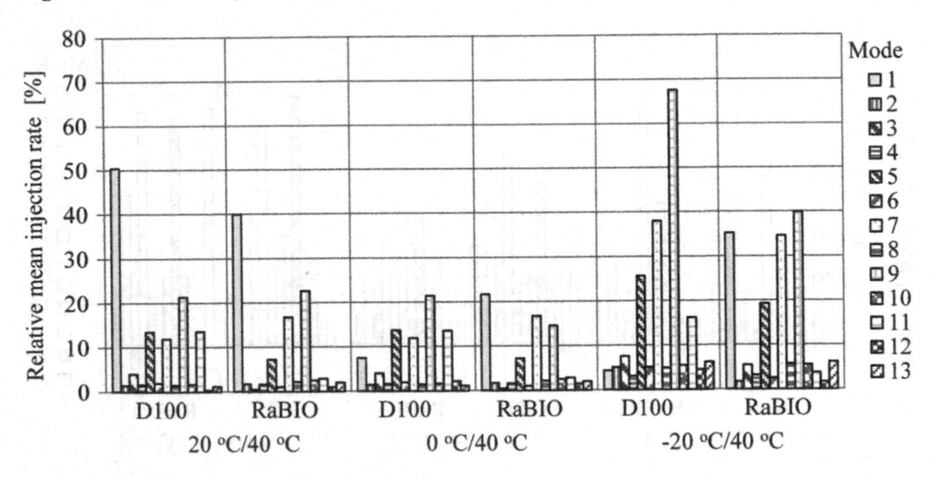

Fig. 5.30 Relative mean injection rate

The cumulative partial fuellings of both fuels (Fig. 5.29) reflect to a great extent the situation observed with the partial fuellings A1 and C2 (Figs. 5.26 and 5.27). Furthermore, as shown in Fig. 5.29, the cumulative partial fuellings A2 and C1 decrease with lower fuel temperatures (Kegl 2008). It is also evident that by lower temperatures the partial fuelling B increases, both in absolute and relative sense (partial fuelling B divided by the total fueling). The cumulative relative partial fuellings A1 and C2 increase with lower temperatures. Regarding the harmful emissions, this is not very encouraging. However, the cumulative relative partial fuellings A and C remain practically constant. Thus, the negative impact on harmful emissions might not be so problematic.

For both fuels, the relative mean injection rate increases with lower fuel temperatures (Fig. 5.30). Except at the idle regime, the increase of relative injection rate is somewhat lower for RaBIO, compared to that of D100. Since lower mean

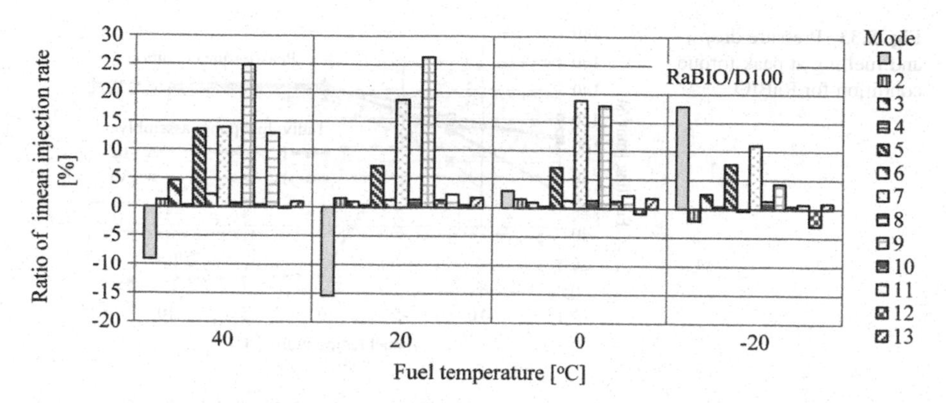

Fig. 5.31 Ratio of mean injection rate

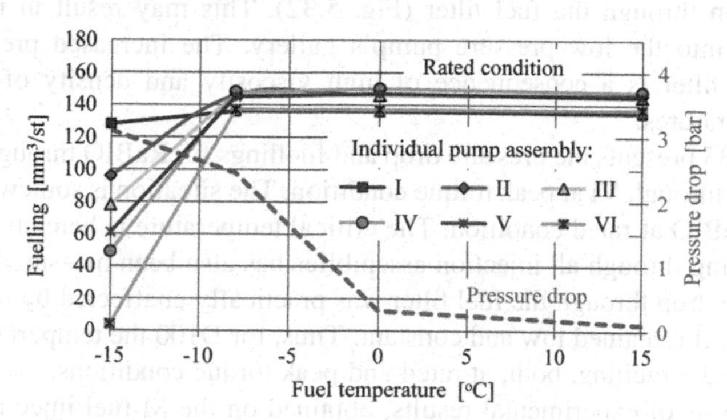

Fig. 5.32 Pressure drop and fuelling at rated condition for RaBIO

injection rate leads to higher smoke emission, it looks like RaBIO would give higher smoke emission, especially at the lowest temperature (Fig. 5.30). To clarify the situation, the ratio of mean injection rate of RaBIO and D100 is shown in Fig. 5.31. As shown, RaBIO offers higher mean injection rates at almost all ESC modes and at all temperatures (Kegl 2008).

Experimental investigations also show that low temperature can have a significant influence on unequal fuelling with respect to individual assemblies in the injection pump of a multi cylinder engine. This influence was investigated experimentally for the whole M fuel injection system of a 6-cylinder engine with injection assemblies I through VI (Kegl 2008). Experiments were done at two engine operating regimes (rated and peak torque conditions). The tested fuels were neat RaBIO and D100.

At rated condition, the fuellings of RaBIO for individual assemblies (I through VI) are presented in Fig. 5.32. One can see that above −7 °C the differences of fuellings through individual assemblies are within acceptable limits. However, at temperatures lower than −7 °C, these differences raise unacceptably. Only for the injection assembly I, being located the closest to the fuel inlet into the low pressure pump's gallery, the fuelling remains acceptable. For other assemblies, the situation becomes worse, as the distance between the gallery inlet and assembly inlet

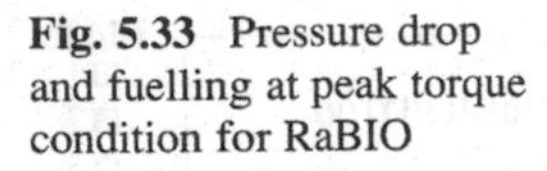

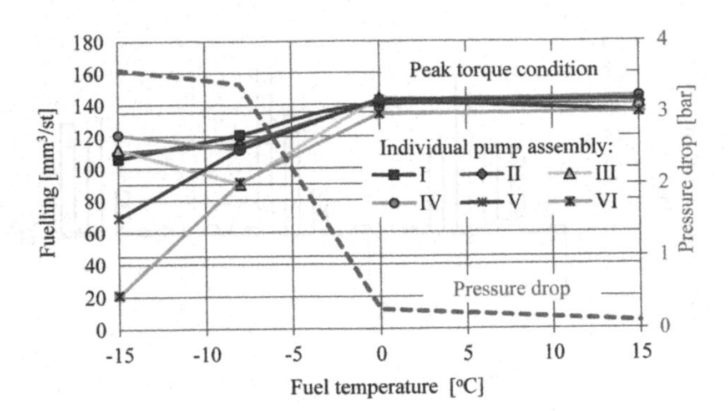

Fig. 5.33 Pressure drop and fuelling at peak torque condition for RaBIO

becomes larger. This observation can probably be explained by the increased pressure drop through the fuel filter (Fig. 5.32). This may result in insufficient fuel supply into the low pressure pump's gallery. The increased pressure drop through the filter is a consequence of high viscosity and density of RaBIO at lower temperatures.

Figure 5.33 presents the pressure drop and fuellings for RaBIO through injection assemblies I through VI at peak torque condition. The situation is somewhat similar to that of RaBIO at rated condition. The critical temperature is here about −3 °C.

The fuelling through all injection assemblies has also been measured for D100. The pressure drop through the fuel filter was practically unaffected by temperature changes, i.e., it remained low and constant. Thus, for D100 the temperature had no influence on the fuelling, both, at rated and peak torque conditions.

The analysis of experimental results, obtained on the M fuel injection system, shows that the distribution of RaBIO fuellings becomes very unequal among individual injection assemblies as the fuel temperature falls below the critical one (−7 °C at rated and −3 °C at peak torque regime). Obviously, the viscosity and the density of RaBIO increase to such extent that the fuel supply to the individual injection assemblies becomes critical. This is caused by the pressure drop through fuel filter and by increased flow resistance through the low pressure pump's gallery. The heating of RaBIO would be necessary to avoid these negative effects.

5.4 Discussion

The analysis of injection characteristics reveals that, in general, the fuelling, injection duration, injection timing, mean injection rate, and injection pressure increase under most considered operating regimes when mineral diesel is replaced by biodiesel. The higher sound velocity and bulk modulus of biodiesels lead to reduced injection delay and advanced injection timing. Regarding the partial fuellings at various phases of the injection process, the numerical results depend significantly on the operating regime. However, summing the results over all 13 modes of the ESC test (and taking into account the corresponding weighting factors) the fuelling at the beginning of injections (fuelling during needle lifting and

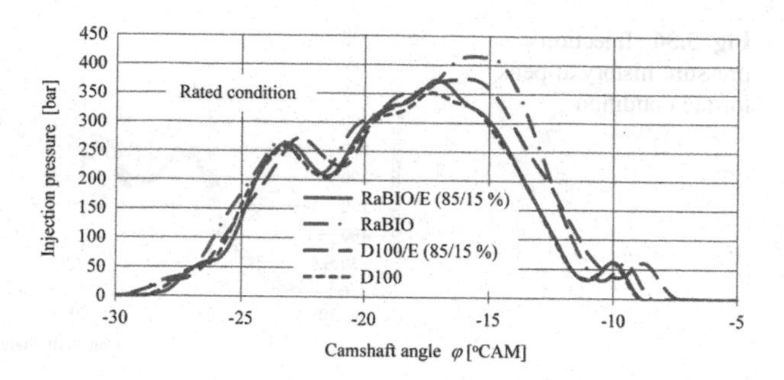

Fig. 5.34 Injection pressure history at rated condition

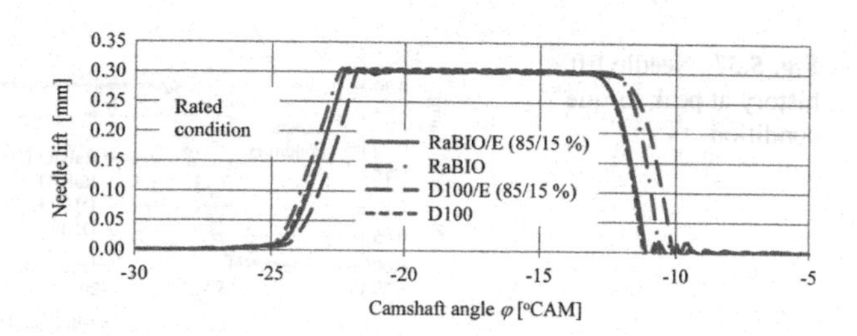

Fig. 5.35 Needle lift history at rated condition

during first 10 % of injection period) as well as the fuelling at the end of injections (fuelling during needle closing and during last 10 % of injection period) are lower when biodiesel (RaBIO) replaces mineral diesel.

In a mechanically controlled injection system the high mean injection pressure and mean injection rate of biodiesels offer a potential to reduce harmful smoke and NO_x emissions. This can be achieved by retarding the injection pump timing accordingly. It has to be pointed out that by proper modification of the injection pump timing most harmful engine emissions can be reduced, while the specific fuel consumption and other engine performances remain within acceptable limits.

Biodiesels obviously offer an interesting and potentially quite beneficial fuel alternative. Therefore, it may also be worth to investigate if it is possible to improve the injection characteristics by blending biodiesels with appropriate additives, like bioethanol.

First of all, it has to be pointed out that the injection characteristics obtained with the blends of RaBIO and bioethanol can be matched close to those obtained with mineral diesel (Torres et al. 2011). The experimental results, obtained at rated conditions in a bus MAN diesel engine with mechanically controlled M injection system, are shown in Fig. 5.34.

It can be seen that the addition of bioethanol decreases the maximal injection pressure. It may be worth noting that the maximal pressure decreases more, when bioethanol is added to biodiesel than to mineral diesel. Furthermore, one can see that the injection pressure history of the blend of 15 % bioethanol and 85 % RaBIO (RaBIO/E 85/15 %) is very close to that of neat mineral diesel.

From Fig. 5.35, where the needle lift history is shown, it can be observed that biodiesel injection timing is advanced with respect to that of mineral diesel and that

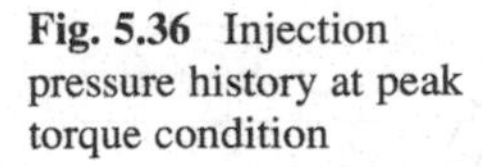

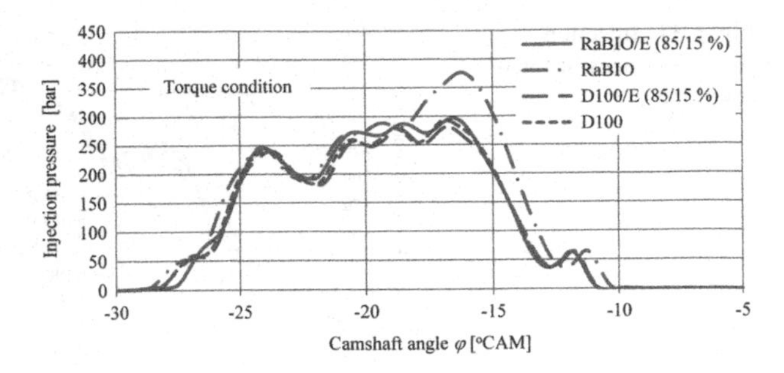

Fig. 5.36 Injection pressure history at peak torque condition

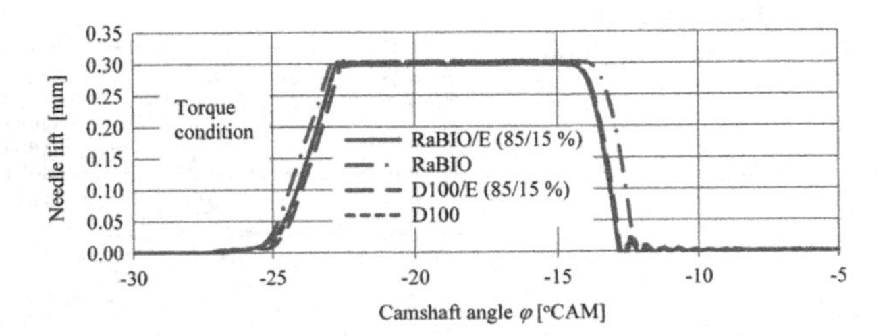

Fig. 5.37 Needle lift history at peak torque condition

biodiesel injection duration is longer. The addition of bioethanol retards the injection timing of diesel and biodiesel fuel. It has to be pointed out that this influence is more significant in biodiesel. Furthermore, the needle lift history of RaBIO/E (85/15 %) is practically the same as that of mineral diesel. This means that the injection timing, injection duration, and injection delay of RaBIO/E (85/15 %) are very close to those of D100.

The influence of bioethanol in biodiesel and diesel fuels at peak torque condition is similar to that obtained at rated condition. From Figs. 5.36 and 5.37, the injection characteristics comparison confirms that the RaBIO/E (85/15 %) blend delivers the injection pressure and needle lift histories very close to those of D100.

On the basis of experimental investigations, it can be summarized that for all injection characteristics, the influence of bioethanol in biodiesel is much more significant that in mineral diesel. Furthermore, bioethanol addition to biodiesel has a beneficial effect on the injection characteristics because it can bring them close to those of mineral diesel (Torres et al. 2011). It seems that the RaBIO/E (85/15 %) blend gives the best results. Namely, at most operating regimes the injection characteristics of this blend are the closest to those of mineral diesel or even seem to be better.

It is known that, compared to mineral diesel, RaBIO exhibits higher fuelling at most engine operating conditions. Bioethanol addition to biodiesel decreases the fuelling and can bring it close to that of mineral diesel. On the other hand, bioethanol addition doesn’t modify substantially mineral diesel fuelling or increases it just slightly.

Injection timing and consequently injection delay are significantly influenced by the type of injection system (e.g., in-line or common rail injection system). When using a low pressure injection system, injection timing of biodiesel is retarded by bioethanol addition and injection delay increases due to lower density of bioethanol. The variation in injection timing caused by bioethanol addition is thus expected to decrease biodiesel's NO_x emissions. Injection timing is advanced slightly when increasing bioethanol content in mineral diesel and injection delay decreases due to bioethanol's lower viscosity.

An increment of bioethanol concentration in biodiesel leads to a decrement in injection duration. Meanwhile, the duration doesn't change or increases slightly when bioethanol is added to mineral diesel.

In general, mean injection rate varies almost negligible when bioethanol is added to diesel fuel because higher fuelling is compensated by longer injection duration. For biodiesel blends, the lower fuelling has more influence than the shorter injection duration, caused by bioethanol addition, which leads to a decrease in mean injection rate.

In most cases studied, bioethanol addition to biodiesel decreases the maximum injection pressure more than bioethanol addition to mineral diesel. In case of bioethanol–biodiesel blends it is expected that bioethanol offers a possibility to reduce NO_x emissions with respect to neat biodiesel.

References

Boehman, A. L., Morris, D., Szybist, J., & Esen, E. (2004). The impact of bulk modulus of diesel fuels on fuel injection timing. *Energy & Fuels, 18*, 1877–1882.

Boudy, F., & Seers, P. (2009). Impact of physical properties of biodiesel on the injection process in a common-rail direct injection system. *Energy Conversion and Management, 50*, 2905–2912.

Caresana, F. (2011). Impact of biodiesel bulk modulus on injection pressure and injection timing. The effect of residual pressure. *Fuel, 90*, 477–485.

Kegl, B. (2006a). Numerical analysis of injection characteristics using biodiesel fuel. *Fuel, 85*, 2377–2387.

Kegl, B. (2006b). Experimental investigation of optimal timing of the diesel engine injection pump using biodiesel fuel. *Energy & Fuels, 20*, 1460–1470.

Kegl, B. (2008). Biodiesel usage at low temperature. *Fuel, 87*, 1306–1317.

Kegl, B., & Hribernik, A. (2006). Experimental analysis of injection characteristics using biodiesel fuel. *Energy & Fuels, 20*, 2239–2248.

Kousoulidou, M., Ntziachristos, L., Fontaras, G., Martini, G., Dilara, P., & Samaras, Z. (2012). Impact of biodiesel application at various blending ratios on passenger cars of different fueling technologies. *Fuel, 98*, 88–94.

Luján, J. M., Bermúdez, V., Tormos, B., & Pla, B. (2009a). Comparative analysis of a DI diesel engine fuelled with biodiesel blends during the European MVEG-A cycle: Performance and emissions (II). *Biomass and Bioenergy, 33*, 948–956.

Luján, J. M., Tormos, B., Salvador, F. J., & Gargar, K. (2009b). Comparative analysis of a DI diesel engine fuelled with biodiesel blends during the European MVEG-A cycle: Preliminary study (I). *Biomass and Bioenergy, 33*, 941–947.

Ozsezen, A. N., & Canakci, M. (2010). The emission analysis of an IDI diesel engine fueled with methyl ester of waste frying palm oil and its blends. *Biomass and Bioenergy, 34*, 1870–1878.

Ozsezen, A. N., Canakci, M., & Sayin, C. (2008). Effect of biodiesel from used frying palm oil on the performance, injection, and combustion characteristics of an indirect injection diesel engine. *Energy & Fuels, 22*, 1297–1305.

Pandey, R. K., & Sarviya, R. M. (2012). Impact of alternative fuel properties on fuel spray behavior and atomization. *Renewable and Sustainable Energy Reviews, 16*, 1762–1778.

Payri, R., Salvador, F. J., Martí-Aldaraví, P., & Martínez-López, J. (2012). Using one-dimensional modeling to analyse the influence of the use of biodiesels on the dynamic behavior of solenoid-operated injectors in common rail systems: Detailed injection system model. *Energy Conversion and Management, 54*, 90–99.

Postrioti, L., Grimaldi, C.N., Ceccobello, M., Gioia, R.D. (2004). Diesel common rail injection system behavior with different fuels. *SAE Technical Paper Series* 2004-01-0029

Song, H., Tompkins, B. T., Bittle, J. A., & Jacobs, T. J. (2012). Comparisons of NO emissions and soot concentrations from biodiesel-fuelled diesel engine. *Fuel, 96*, 446–453.

Torres-Jimenez, E., Dorado, M. P., & Kegl, B. (2011). Experimental investigation on injection characteristics of bioethanol–diesel fuel and bioethanol–biodiesel blends. *Fuel, 90*, 1968–1979.

Tziourtzioumis, D., & Stamatelos, A. (2012). Effects of a 70 % biodiesel blend on the fuel injection system operation during steady-state and transient performance of a common rail diesel engine. *Energy Conversion and Management, 60*, 56–67.

Usta, N. (2005). Use of tobacco seed oil methyl ester in a turbocharged indirect injection diesel engine. *Biomass and Bioenergy, 28*, 77–86.

Yamane, K., Ueta, A., & Shimamoto, Y. (2001). Influence of physical and chemical properties of biodiesel fuels on injection, combustion and exhaust emission characteristics in a direct injection compression ignition engine. *International Journal of Engine Research, 2*, 249–261.

Ye, P., & Boehman, L. (2010). Investigation of the impact of engine injection strategy on the biodiesel NO_x effect with a common-rail turbocharged direct injection diesel engine. *Energy & Fuels, 24*, 4215–4225.

Ye, P., & Boehman, A. L. (2012). An investigation of the impact of injection strategy and biodiesel on engine NOx and particulate matter emissions with a common-rail turbocharged DI diesel engine. *Fuel, 97*, 476–488.

Yoon, S. H., Suh, H. K., & Lee, C. S. (2009). Effect of spray and EGR rate on the combustion and emission characteristics of biodiesel fuel in a compression ignition engine. *Energy & Fuels, 23*, 1486–1493.

Zhang, Y., & Boehman, A. L. (2007). Impact of biodiesel on NO_x emissions in a common rail direct injection diesel engine. *Energy & Fuels, 21*, 2003–2012.

Chapter 6
Effects of Biodiesel Usage on Fuel Spray Characteristics

The spatial and temporal distribution of a fuel spray in a diesel engine has a determining effect on noise and exhaust emissions, fuel consumption, and engine performance (Battistoni and Grimaldi 2012; Kuti et al. 2013; Park et al. 2008). The most important fuel spray characteristics are *spray penetration*, *spray angle*, and *Sauter mean diameter*. These characteristics depend to a great extent on the injector type, fuel composition, and engine operating regime (Kousoulidou et al. 2012; Lin and Lin 2011; Mancaruso et al. 2011).

Spray development depends significantly on the processes occurring within the injector orifice, turbulence and cavitation being perhaps the most important. Once the fuel exits the orifice, spray development is determined by the fuel atomization process. Unfortunately, all the mechanisms acting in this phase are still not fully understood. Because various mechanisms are in action and because the magnitude of their influence may depend significantly on several circumstances, it is very difficult to make concise and unique conclusions. In other words, what is observed under some circumstances may change substantially under somewhat modified situation.

The *injector type influence* is largely determined by its control and geometric parameters. Among the control parameters, the injection pressure is probably the most important; its influence will be addressed along with the effects of engine operating regime. On the other side, the most important geometric parameters are those related to the injector nozzle geometry. Roughly speaking, there are two main orifice geometries: the more common cylindrical and the conical orifice. A major difference between these two types is the presence of cavitation phenomena, which is much more exposed in the cylindrical orifice. The most important parameters of the nozzle are related to the (one or more) *orifice diameters*, its *length*, and the *shape* of the *orifice inlet* (sharp or rounded edges).

In general, smaller orifice diameters (combined with higher injection pressures) lead to better atomization and surface evaporation of the spray. For example, several investigations show that the usage of high injection pressure and a micro-hole nozzle represents an effective method to improve spray atomization and mixture preparation processes of biodiesel fuels (Kuti et al. 2013). On the other

B. Kegl et al., *Green Diesel Engines*, Lecture Notes in Energy 12,
DOI 10.1007/978-1-4471-5325-2_6, © Springer-Verlag London 2013

side, the shape and dimension of the nozzle hole have a notable influence on the cavitation phenomena, fuel exit velocity, and the discharge coefficient. In general, an increase of the orifice inlet radius leads to increased mean exit velocity and discharge coefficient near the orifice wall, while the region of the cavitation contracts.

Since fuel composition has also an important influence on spray characteristics, one can expect that a replacement of mineral diesel by biodiesel will also have significant effects. In fact, biodiesel usage usually results in longer spray tip penetration and narrower cone angles; this is mostly attributed to poorer atomization characteristics of biodiesels. Furthermore, compared to mineral diesel, the liquid lengths of biodiesels are higher due to higher boiling temperature and vaporization heat of biodiesel fuels.

Fuel spray development, especially in its early phase, depends heavily on the injection rate and spray momentum, which is the energy of the spray, delivered at the nozzle exit (Desantes et al. 2009). These quantities vary in dependence on the engine operating regime. This means that spray development has to be investigated at various engine loads and speeds.

6.1 Spray Penetration

Spray tip penetration is most significantly affected by the fuel *density*, *viscosity*, and *surface tension* (Kegl 2008; Pogorevc et al. 2008; Yoon et al. 2009). This influence varies in dependence on the engine operating regime.

6.1.1 Influence of Biodiesel Properties

In general, biodiesels have higher density, viscosity, surface tension, and bulk modulus than mineral diesel. Higher density increases the momentum of the injected fuel. Furthermore, in mechanically controlled fuel injection systems, higher density and higher bulk modulus also mean higher injection pressures. This means that, in general, higher density and bulk modulus will result in higher penetration lengths. Higher viscosity and surface tension result in poorer fuel vaporization. This is another reason for higher penetration lengths. On the other hand, higher viscosity means higher friction between the fuel and the injector nozzle surface. Under some circumstances this may result in lower penetration lengths, for example, in the case of some biodiesel blends with mineral diesel.

Obviously, one can expect that biodiesel usage will prolong the penetration lengths. This may lead to serious problems, if the liquid part of the spray reaches the combustion chamber wall and the piston surface. Heavy carbon deposits on the walls, piston ring sticking and breaking, and dilution of the lubricating oil may be the consequences (Pandey et al. 2012).

Table 6.1 Investigation data (Kuti et al. 2013)

Injection system/injector	Common rail for DI diesel engine
Nozzle type	Single hole
Length of nozzle hole	1.2 mm
Nozzle hole diameter	0.08 mm and 0.16 mm
Injection pressure	2,000 bar
Injection duration	1.5 ms
Ambient temperature	885 K
Ambient pressure	40 bar

In the following, the effects of variable fuel properties on fuel spray tip penetration are discussed by using several types of biodiesel in various injectors of mechanically and electronically controlled fuel injection systems.

In Kuti et al. (2013) spray tip penetration of *palm biodiesel* (PaBIO) was investigated numerically and experimentally in a diesel engine with a *common rail injection system* (Table 6.1). The experimental conditions were similar to real engine conditions. The constant volume vessel was filled with nitrogen gas.

The results revealed that a change from mineral diesel (D100) to PaBIO results in longer spray tip penetration at both tested nozzle hole diameters (Fig. 6.1). Furthermore, it was evident that the spray tip penetration is higher for the nozzle hole with larger diameter. This is true for both fuels, D100 and PaBIO. It was observed that after an initial spray development period, the tip of the liquid phase fuel region stopped penetrating and fluctuated about a mean axial location as a result of turbulence. The PaBIO produced longer liquid phase lengths than D100. This implies that D100 evaporated more quickly than PaBIO. One reason for that is obviously the higher boiling point of PaBIO, which is characterized by the distillation temperature. Because of the high boiling point property, PaBIO is less volatile than D100. As a result of less volatility, PaBIO produces longer liquid phase lengths and the energy required to heat and vaporize PaBIO is higher. Since the entrainment rate of energy into the spray is limiting the vaporization process, the requirement for more energy to heat and vaporize the less volatile fuel translates to a longer spray tip penetration.

In Battistoni and Grimaldi (2012) an extensive experimental and numerical research of the influence of various *nozzle hole types* on D100 and *soybean biodiesel* (SoBIO) spray penetration has been done. The numerical part was based on 3D simulation of the unsteady multiphase flow in the injector (Table 6.2).

The results show that higher flow rates were achieved by the conical nozzle. This is because cavitation phenomena in a conical orifice are far less exposed than in a cylindrical orifice. Consequently, the penetration lengths obtained with the conical orifice are notably higher than those obtained with the cylindrical orifice. This can be observed for both fuels, D100 and SoBIO. What is quite interesting here is that the penetration length differences are much larger for D100 than for SoBIO. Figure 6.2 shows the relative variations of the penetration lengths when the cylindrical orifice is replaced by a conical one. Although for SoBIO the conical

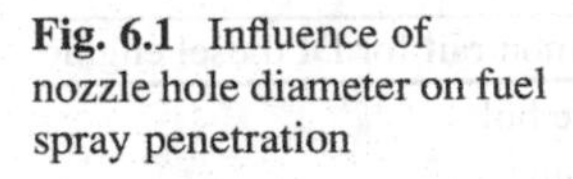

Fig. 6.1 Influence of nozzle hole diameter on fuel spray penetration

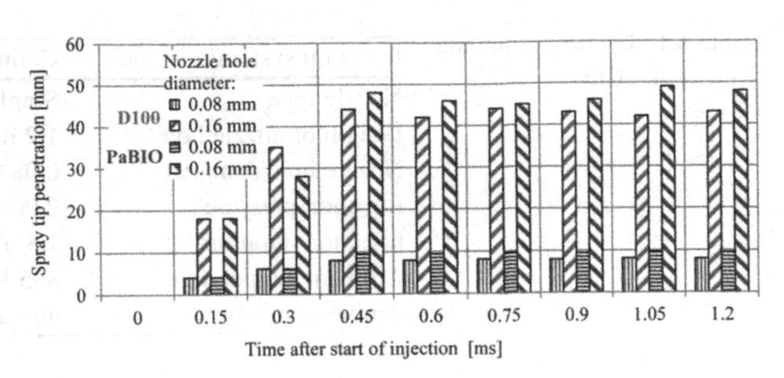

Table 6.2 Investigation data (Battistoni and Grimaldi 2012)

Injection system/injector	Common rail Bosch CRI1
Nozzle type	Mini-sac, five holes: cylindrical and conical orifice shape
Length of nozzle hole	0.7 mm
Nozzle hole diameter	0.13 mm
Injection pressure	1,350 bar
Energizing duration	0.5 ms
Ambient pressure	10 bar

Fig. 6.2 Influence of nozzle hole type on fuel spray penetration

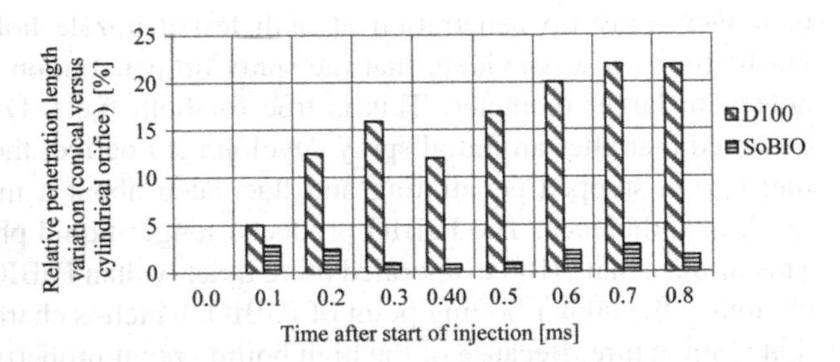

shape results in a slightly higher penetration length, the relative length variation with respect to the length, obtained with the cylindrical orifice, is rather minor.

In Kegl (2008) and Pogorevc et al. (2008) fuel spray penetration of *rapeseed biodiesel* (RaBIO) in a bus diesel engine with a mechanically controlled M injection system (Table 6.3) was investigated numerically and experimentally. The fuel was injected through a one-hole nozzle into a glass chamber at room temperature and atmospheric pressure and filmed with a high-speed digital camera.

The fuel spray development for D100 and RaBIO at partial load of 50 % and 500 rpm is shown in Fig. 6.3. It is evident that RaBIO spray is longer than that of mineral diesel by about 5 %. Partially, this is because RaBIO atomizes poorly compared to mineral diesel due to its higher surface tension, density, and viscosity. On the other hand, RaBIO usage results in a stepper rise and higher injection pressure, which also influences the development of the spray.

In Wang et al. (2010, 2011) the liquid lengths of PaBIO and *waste cooking biodiesel* (WcBIO) were investigated under simulated diesel engine conditions in a constant volume chamber (Table 6.4). A common rail injection system was used to

Table 6.3 Investigation data (Kegl 2008)

Injection system/injector	Direct injection system with wall distribution (M system)
Nozzle type	Single hole
Length of nozzle hole	1.95 mm
Nozzle hole diameter	0.68 mm
Needle opening pressure	175 bar
Needle lift (maximum)	0.3 mm

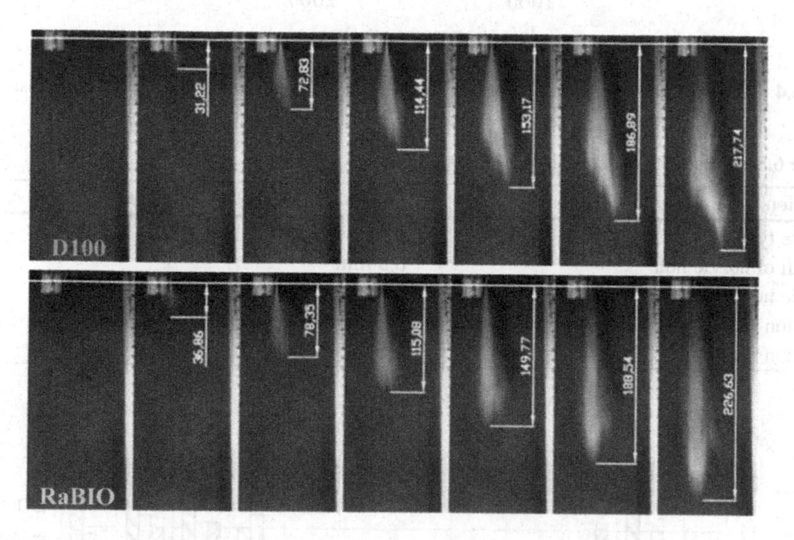

Fig. 6.3 Influence of fuel properties on spray penetration at full load

Table 6.4 Investigation data (Wang et al. 2011)

Injection system/injector	Common rail injector
Nozzle type	Single hole
Nozzle hole diameter	0.16 mm
Injection pressure	1,000, 2,000, 3,000 bar
Injection duration	1.5 ms
Ambient pressure	Up to 40 bar
Ambient temperature	885 K

ensure injection pressures of 1,000, 2,000, and 3,000 bar. The spray chamber with adjustable pressure was filled with nitrogen (ambient) gas. The ambient temperature was fixed at 295 K. The tested ambient densities, controlled by the pressure, were 15 and 30 kg/m^3.

The liquid lengths of D100, PaBIO, and WcBIO are shown in Fig. 6.4. It is evident that biodiesel fuels exhibit longer liquid lengths than mineral diesel. The highest penetration is obtained with PaBIO. Furthermore, the differences of liquid lengths increase as the injection pressure rises.

In Gao et al. (2009) fuel spray tip penetration of *jatropha biodiesel* (JaBIO), WcBIO, and PaBIO was investigated by using a high-speed camera. A

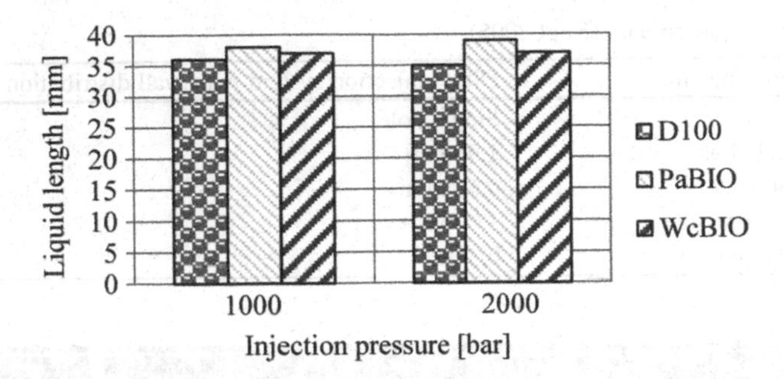

Fig. 6.4 Influence of various fuels on liquid lengths of the spray at various injection pressures

Table 6.5 Investigation data (Gao et al. 2009)

Injection system/injector	In-line mechanically controlled injection system
Nozzle type	Single hole
Length of nozzle hole	0.8 mm
Nozzle hole diameter	0.18 mm
Injection pressure—maximal	430 bar
Injection duration	1.6 ms

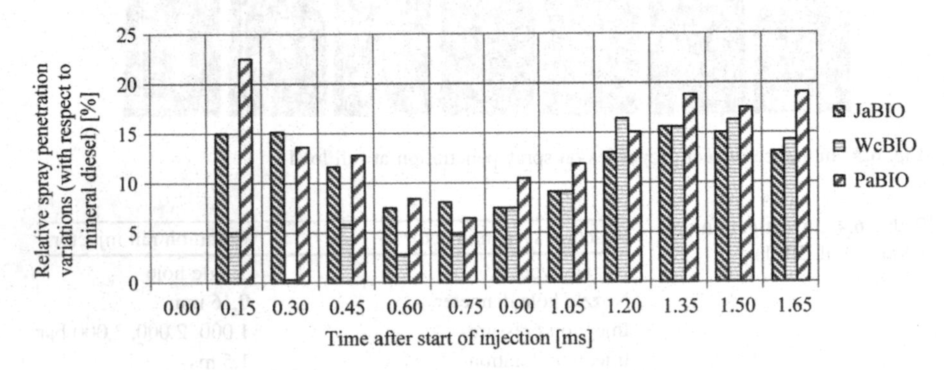

Fig. 6.5 Influence of various fuels on spray penetration history

mechanically controlled fuel injection system with a single-hole injector was used to inject the fuel into a constant volume chamber (Table 6.5).

Among all tested fuels, PaBIO has the highest viscosity. It is known that the increase in fuel viscosity retards the breakup of the spray jet, which results in larger spray droplets. This reduces the total resistance or drag. For this reason, the spray penetration of PaBIO is the highest practically during the whole injection period. Figure 6.5 shows the relative variations of penetration lengths, obtained when D100 was replaced by the tested biodiesels. Interestingly, these variations are the smallest in the period of 0.6–0.8 ms after the start of injection. Before and after this period, the variations are larger.

According to the investigations done so far, one can say that at the same injection pressure, fuelling, injection velocity, and engine operation regime, the spray evolutions of mineral diesel and biodiesels show similar trends. However, because of poorer atomization of biodiesel droplets, biodiesels typically exhibit a slightly higher spray tip penetration than mineral diesel. Furthermore, compared to mineral diesel, biodiesel droplets have a higher ratio of momentum to drag force. This also contributes to longer spray tip penetration of biodiesels (Yoon et al. 2009).

6.1.2 Influence of Engine Operating Regime

A change of the engine operating regime is typically accompanied by a change of the injection pressure. This also affects spray penetration lengths and this influence may depend on biodiesel type (Agarwal and Chaudhury 2012; Kegl 2008; Som et al. 2010).

In general, higher engine loads and speeds are accompanied by higher injection pressures which result in higher penetration lengths. But this influence may vary in dependence on the actual operating regime. For example, even at low load (but high engine speed) the biodiesel spray tip penetration may be prolonged up to 15 % with respect to that of mineral diesel.

Ambient pressure and temperature represent two parameters that are not directly related to the engine operating regime. Nevertheless, they influence significantly fuel spray development and it may be of benefit to consider them in the context of engine operating regimes. In general, spray tip penetration and spray width increase as the ambient temperature rises at constant pressure. This is because higher temperature decreases the density of the ambient gas. On the other hand, by keeping the temperature constant and lowering the ambient pressure, the evaporation of the fuel is improved and a decrease of spray tip penetration can be observed.

In Agarwal and Chaudhury (2012) the influence of *engine loads* on the spray penetration of *karanja biodiesel* (KaBIO) and D100 was investigated. A simple mechanically controlled fuel pump with a fuel delivery pressure set at 200 bar and a mechanical injector were used (Table 6.6). The fuel was injected into a customized constant volume spray visualization chamber at various chamber pressures, which simulated various engine loads.

The results show that the spray tip penetration decreases as the chamber pressure increases for both tested fuels. Higher pressure means higher air density in the chamber. This increases the drag force on the droplets at the spray tip. KaBIO exhibited higher penetration lengths at all tested engine loads (Fig. 6.6). This is because the density and viscosity of KaBIO are higher than those of D100 and D100 atomizes more rapidly than KaBIO.

In Som et al. (2010) the influence of injection pressure on spray penetration history was investigated numerically for SoBIO and D100. A common rail injection system was utilized (Table 6.7).

Table 6.6 Investigation data (Agarwal and Chaudhury 2012)

Injection system/injector	In-line mechanically controlled injection system
Nozzle type	Three-nozzle hole
Nozzle hole diameter	0.29 mm
Injection pressure	200 bar

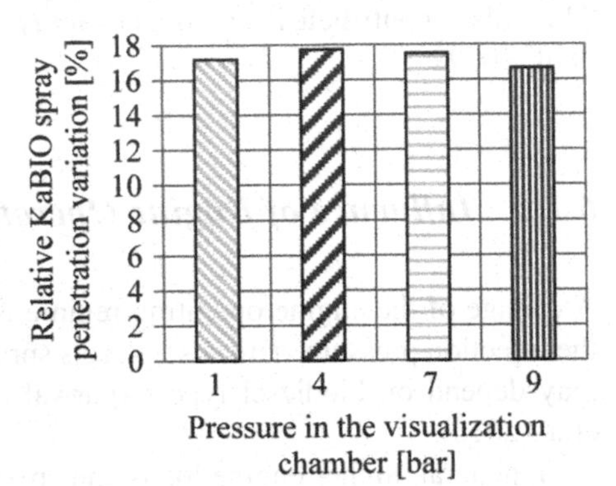

Fig. 6.6 Influence of the ambient pressure on spray penetration of KaBIO with respect to D100

Table 6.7 Investigation data (Som et al. 2010)

Injection system/injector	Common rail injector Detroit diesel
Nozzle type	Mini-sac nozzle with six cylindrical holes
Nozzle hole diameter	0.169 mm
Injection pressure	1,100 bar and 1,300 bar
Injection duration	3 ms
Ambient density	34 kg/m^3
Ambient temperature	300 K

From the results it follows that higher injection pressure yields higher penetration lengths. The relative penetration variation of SoBIO is higher than that of D100 (Fig. 6.7).

In Pogorevc et al. (2008) spray development of RaBIO was investigated at various engine operating regimes (Table 6.3). The effect of pump speed variation at full load is shown in Fig. 6.8. It is evident that lower engine speeds result in a notably shorter spray tip penetrations.

Similar trends were observed at other engine loads. Figure 6.9 illustrates the effect of variable pump speed at partial load (PL) of 50 %.

The obtained experimental results are summarized in Fig. 6.10. It can be seen that engine load influences the spray tip penetration less than engine speed. This can be observed for both fuels, RaBIO and D100.

In Chen et al. (2013) the influence of *injection pressure* on spray tip penetration of WcBIO and D100 was investigated experimentally. The fuels were stored at room temperature of 25 °C and injected into air at ambient temperature and pressure by using a common rail injector with a single hole nozzle (Table 6.8).

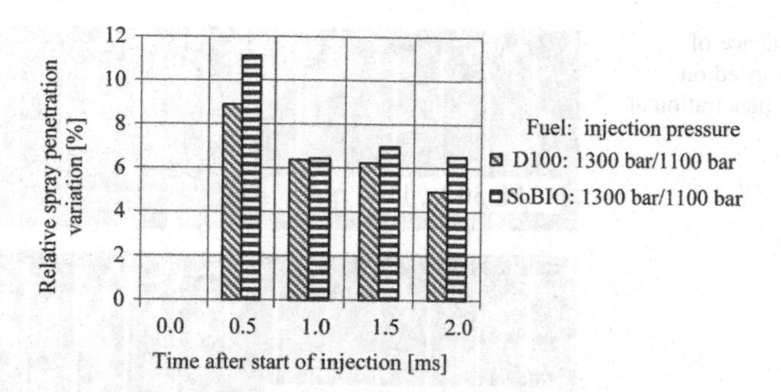

Fig. 6.7 Influence of injection pressure on spray penetration of SoBIO and D100

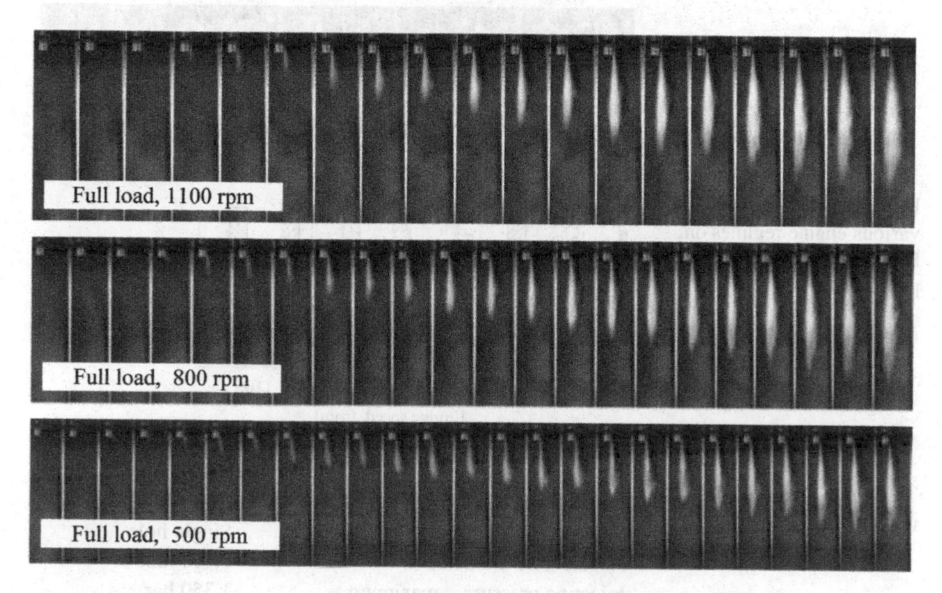

Fig. 6.8 Influence of engine pump speed on RaBIO spray penetration at full load

Figure 6.11 illustrates the obtained spray tip penetration histories. It is evident that tip penetrations of WcBIO and D100 fuels increase with higher injection pressure. This trend can be easily observed by looking at D100 penetration under 300 bar. At 1.5 ms after injection start, the spray tip penetration is 100 mm compared with 140 mm under 1,000 bar. It is interesting that WcBIO shows significantly longer penetrations than D100 at low injection pressures (e.g., 300 bar), but this difference diminishes at higher injection pressures. By regarding the history of the spray development, one can also see that WcBIO penetrates faster than D100 in the early stage of injection. In the later stage D100 catches up and the difference between WcBIO and D100 reduces gradually.

In Park et al. (2009a) the influence of *fuel temperatures* and *ambient gas conditions* on SoBIO spray penetration was investigated by using a common rail injection system (Table 6.9).

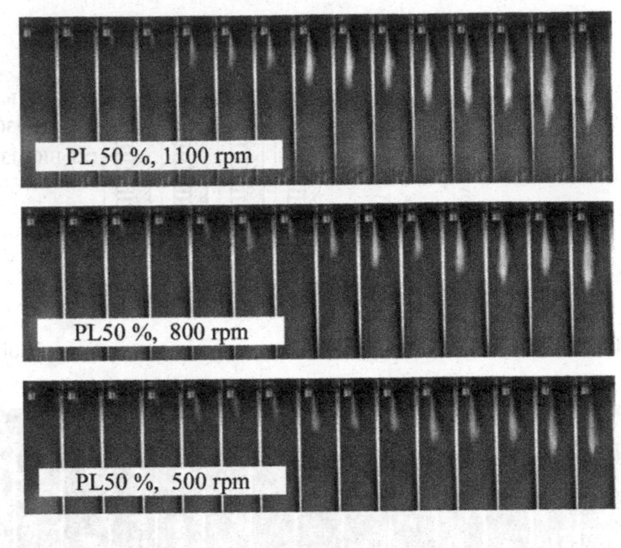

Fig. 6.9 Influence of engine pump speed on RaBIO spray penetration at PL 50 %

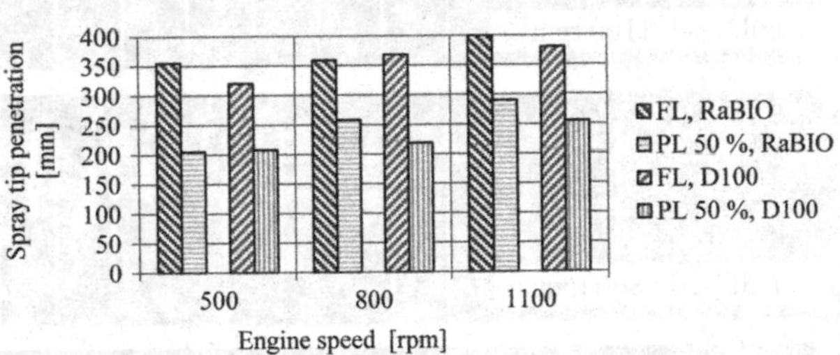

Fig. 6.10 Influence of various engine regimes on RaBIO and D100 spray penetration

Table 6.8 Investigation data (Chen et al. 2013)

Injection system/injector	Common rail injector
Nozzle type	Single hole
Nozzle hole diameter	0.140 mm
Injection pressure—maximum	1,350 bar

The experimentally and numerically obtained results show that the spray tip penetration increases with the increase of the *injection pressure*, which leads to an increase of the initial spray momentum. Furthermore, the spray tip penetration is only slightly affected by the variation of the *fuel temperature*, which causes a variation of fuel properties. On the other hand, increased *ambient air temperature* results in higher spray liquid tip penetration. This is because the ambient air density decreases with rising ambient air temperature. The influence of fuel temperature T_{fuel} and ambient temperature T_{amb} on SoBIO spray penetration under 1,200 bar injection pressure and 20 bar of ambient gas pressure is shown in Fig. 6.12.

In Lee et al. (2005) the SoBIO blend with D100 was tested under various engine regimes, simulated by various injection pressures. The 20 % blend B20 was injected into atmospheric ambient conditions by a common rail injection system (Table 6.10).

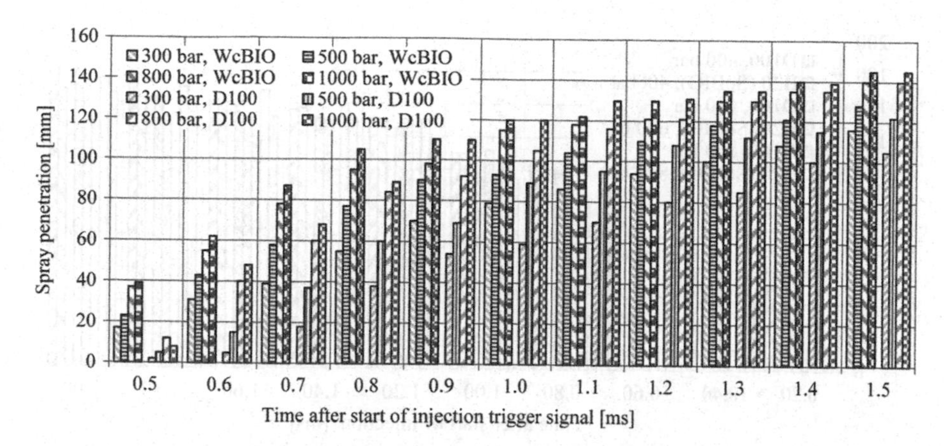

Fig. 6.11 Effect of injection pressure on fuel spray penetration history

Table 6.9 Investigation data (Park et al. 2009a)

Injection system/injector	Common rail injector
Nozzle type	Six cylindrical holes
Length of nozzle hole	0.8 mm
Nozzle hole diameter	0.126 mm
Injection pressure	600 bar and 1,200 bar
Energizing duration	1 ms
Ambient pressure	20 bar

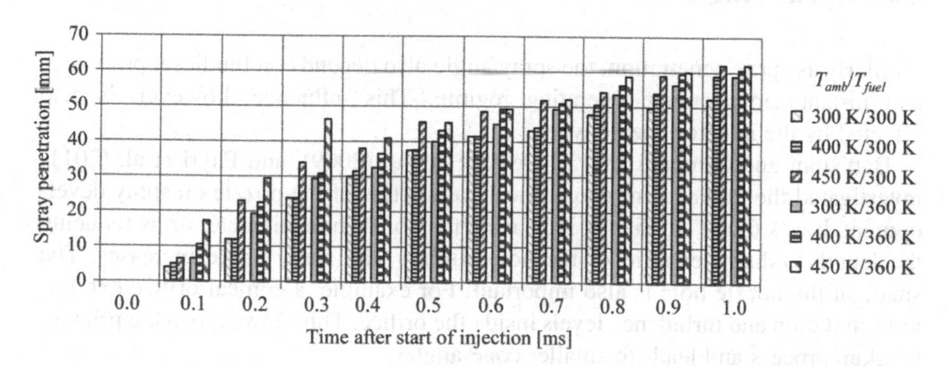

Fig. 6.12 Influence of fuel and ambient air temperature on SoBIO spray penetration history

Table 6.10 Investigation data (Lee et al. 2005)

Injection system/injector	Common rail injector
Nozzle type	Single hole
Nozzle hole diameter	0.3 mm
Injection pressure	400, 600, and 800 bar
Energizing duration	1.0 ms
Ambient pressure	1 bar
Ambient temperature	295 K

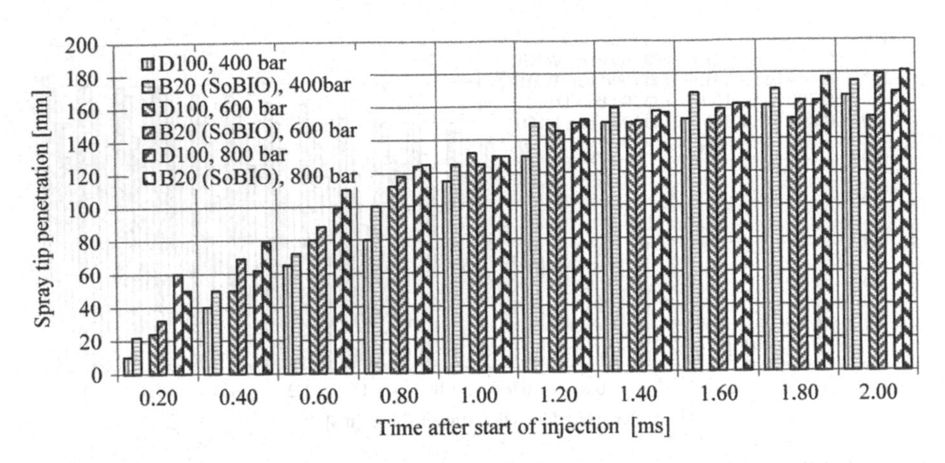

Fig. 6.13 Influence of injection pressure on spray tip penetration history

In Fig. 6.13 the spray tip penetration history of the B20 blend is compared to that of D100 at various injection pressures. Obviously, the spray tip penetration increases at higher injection pressures because of injection velocity rise. It should be noted, however, that the differences between B20 and D100 are rather small.

6.2 Spray Angle

Similarly as spray penetration, the spray angle also depends on biodiesel properties and instantaneous engine operating regime. This influence, however, is also affected by the injector nozzle type.

Battistoni and Grimaldi (2012), Desantes et al. (2009), and Payri et al. (2011) investigated the influence of geometrical parameters of the nozzle on spray development. It was found out that by increasing the nozzle hole diameter or by reducing the length-to-diameter ratio of the hole slightly, the spray angle increases. The shape of the nozzle hole is also important. For example, a conical orifice exhibits low cavitation and turbulence levels inside the orifice. This slows down the primary breakup process and leads to smaller cone angles.

In Battistoni and Grimaldi (2012) the influence of various *nozzle hole types* on D100 and SoBIO spray angle was analyzed. For this purpose, numerical simulation of unsteady multiphase flow inside the injector was employed by using the Eulerian–Eulerian approach. The spray was computed by using the discrete droplet model in the Lagrangian framework. The results show that D100 and SoBIO produce similar cone angles when using a cavitating *cylindrical nozzle hole with sharp edges*. On the other hand, D100 spray angle was significantly smaller than that of SoBIO when a *conical nozzle hole with rounded edges* was used. In this case, SoBIO produces a large cloud after the secondary breakup that increases significantly its global spray cone angle. This result contrasts with the usually observed trend of narrower angles obtained with biodiesels. Maybe this can be partially attributed to possible inaccuracy of the numerical models, but it surely demonstrates the strong influence of the orifice shape on the spray. Another interesting result was that D100 appeared to be much more sensitive to the orifice shape since it showed very different breakup characteristics: with the rounded conical hole, D100 produced a faster and denser spray which remained compact for a long time. On the other hand, the cylindrical and highly cavitating hole enhanced the D100 spray breakup significantly.

6.2.1 Influence of Biodiesel Properties

The fuel spray angle is most significantly influenced by the following properties: *density, viscosity*, and *surface tension* (Kegl 2008; Pandey et al. 2012; Pogorevc et al. 2008; Park et al. 2010; Som et al. 2010; Yoon et al. 2009). In general, biodiesels exhibit higher values of these properties than mineral diesel. Consequently, biodiesels usually form a narrower spray under most operating regimes. Usually, this is also true for the biodiesel blends with mineral diesel (Desantes et al. 2009).

In Wang et al. (2011, 2010) the influence of fuels properties on the spray angle was investigated experimentally under simulated diesel engine conditions (Table 6.4). The results show that by using either PaBIO or WcBIO the spray angles were always smaller than in the case of D100 (Fig. 6.14). One can also see that the differences between both biodiesels decrease as the injection pressure increases.

In Pogorevc et al. (2008) the fuel spray angle was investigated numerically and experimentally (Table 6.3). Numerical simulation was performed by using the AVL Fire program and the Euler–Lagrangian approach. To calculate the *primary breakup* of the spray, the *diesel core injection model* was chosen, while the *secondary breakup* was simulated by using the *wave model*.

Figure 6.15 shows a comparison of the filmed and simulated spray development at pump rotational speed of 500 rpm and full load for D100 and RaBIO. It can be seen that the RaBIO spray is narrower than the D100 spray due to higher spray velocities as a result of higher injection pressure, which is caused by higher viscosity and density of biodiesel.

In Gao et al. (2009) the fuel spray angles of JaBIO, WcBIO, and PaBIO were investigated and compared to that of mineral diesel (Table 6.5). For mechanically controlled injection systems, the spray cone angle history usually shows a gradual increase of the angle during the early phase of injection. This is mainly because the accompanying increase of the injection pressure causes the spray to spread around more intensively, leading to a gradual increase in the cone angle. In the later phase of injection, the cone angle begins to decrease and then stabilizes at a certain value, which depends on the fuel. Namely, in the later phase the outer spray droplets become smaller and diffuse easily, which leads to a reduction of the spray angle.

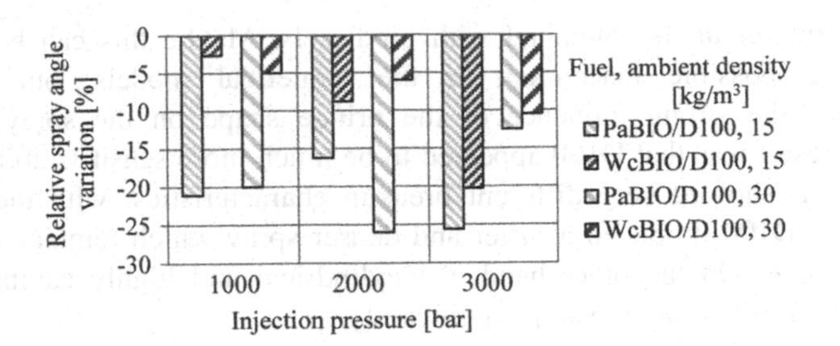

Fig. 6.14 Influence of injection pressure and ambient density on spray angle of PaBIO and WcBIO with respect to D100

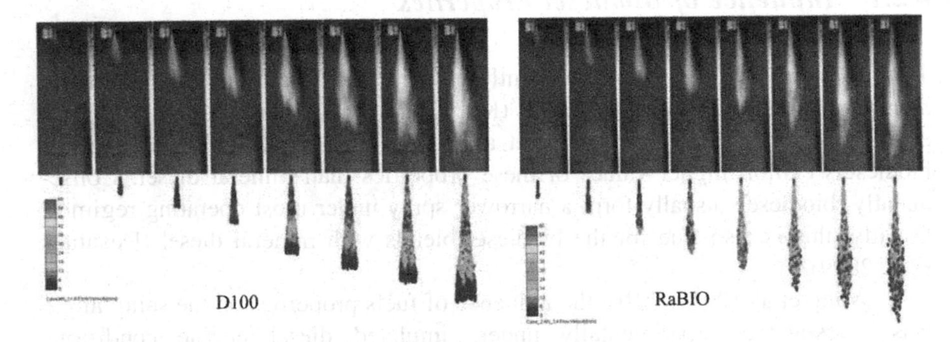

Fig. 6.15 Filmed and simulated spray development at full load

From Fig. 6.16, which shows the relative spray angle variations with respect to D100, one can see that among all tested fuels this behavior is the most exposed for JaBIO. Furthermore, one can see that the stabilized spray angles of all tested biodiesels are lower than that of mineral diesel. The lowest angle is obtained with PaBIO, which has the highest viscosity.

In Valentino et al. (2011) the cone angles of RaBIO and SoBIO blends with D100 were investigated experimentally. The tests were performed on a turbo-charged, water-cooled, 4-valve DI diesel engine, equipped with a common rail injection system (Table 6.11). The results show clearly the dependency of the spray angle on fuel viscosity and ambient gas density.

The spray cone angles are measured at 500 ms after the start of injection. The tested fuels are D100 and its 50 % blends with SoBIO (B50, SoBIO) and with RaBIO (B50, RaBIO). The measurements were carried out on the main injection and averaged over seven sprays and ten shots. The spray cone angle for each tested fuel increased by raising the ambient density. The highest spray cone angles were obtained with D100 which exhibits a viscosity of 3.20 mm^2/s. Somewhat lower angles were observed with SoBIO B50 with 3.46 mm^2/s. RaBIO B50 with the largest viscosity of 3.67 mm^2/s yielded the smallest spray cone angles. Figure 6.17 illustrates nicely the dependence of the spray cone angle on fuel viscosity. One can be seen that lower viscosities result in higher cone angles at any operating condition.

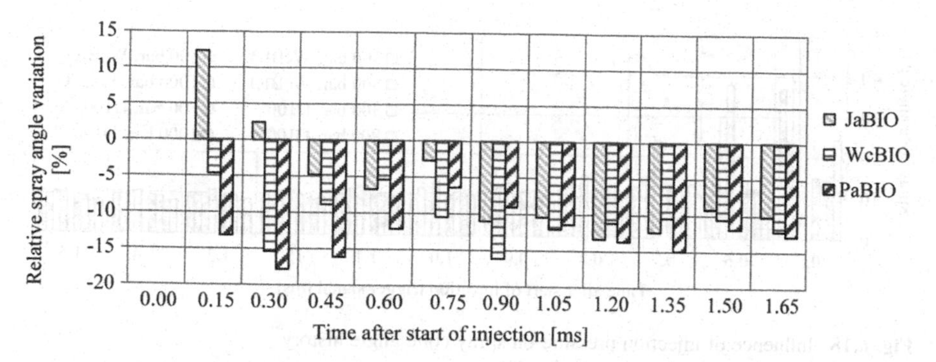

Fig. 6.16 Influence of various biodiesel fuels on spray angle

Table 6.11 Investigation data (Valentino et al. 2011)

Injection model	Bosch CRI 2.2 MW
Nozzle type	Mini-sac seven holes
Nozzle hole diameter	0.136 mm
Injection pressure—maximum	1,600 bar

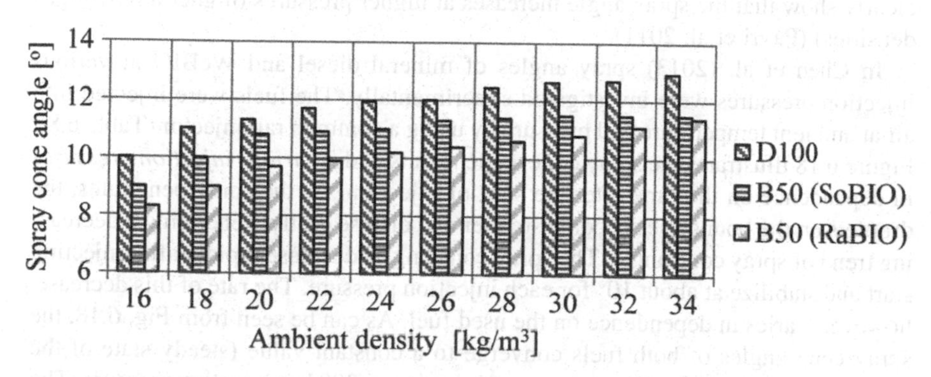

Fig. 6.17 Influence of ambient density on spray cone angle for various fuels

6.2.2 *Influence of Engine Operating Regime*

A change of the engine operating regime is typically accompanied by a change of the injection pressure. This may also affect the spray angle, depending also on the fuel type (Payri et al. 2011). Lower injection pressures may produce a considerably smaller spray angles than higher pressures. This may be mostly related to the flow regime inside the injection orifice, since at lower injection pressure the flow regime may be closer to a laminar one with low turbulence. On the other hand, high injection pressures increase the turbulence inside the orifice and may cause extensive cavitation. When the fuel exits the nozzle, the radial turbulence velocities eject some fuel away from the main stream and thus promote atomization.

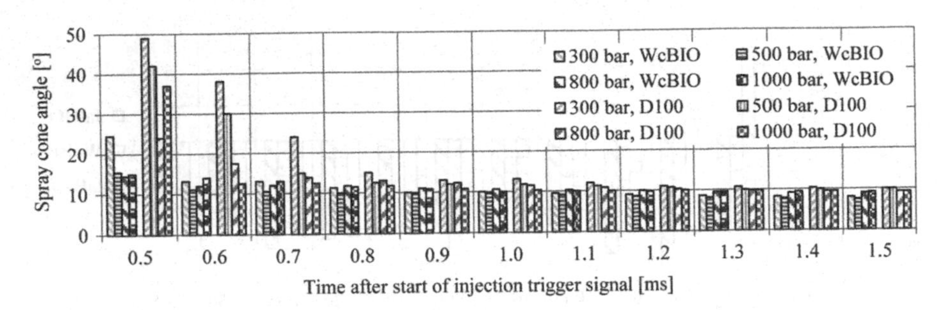

Fig. 6.18 Influence of injection pressure on spray cone angle history

Besides the injection pressure, the ambient pressure and temperature may play an important role. Unfortunately, the spray dependency on these quantities is quite sophisticated. In the literature one can find experimental results (high injection pressure, cylindrical orifice) indicating that high ambient temperature (lower ambient gas density) increases the spray angle. On the other hand, for example, the experimental results on non-cavitating (conical) nozzle holes with rounded edges clearly show that the spray angle increases at higher pressures (higher ambient gas densities) (Payri et al. 2011).

In Chen et al. (2013) spray angles of mineral diesel and WcBIO at various injection pressures were investigated experimentally. The fuels were injected into air at ambient temperature and pressure by using a common rail injector (Table 6.8). Figure 6.18 illustrates the spray cone angle results under *various injection pressures* in dependence on the time after the start of injection. As the spray penetrates, the droplets on the boundaries become smaller and diffuse easily, generating a decreasing trend of spray cone angle. The spray cone angles decrease rapidly after injection start and stabilize at about 10° for each injection pressure. The rate of this decrease, however, varies in dependence on the used fuel. As can be seen from Fig. 6.18, the spray cone angles of both fuels converge to a constant value (steady state of the spray) at around 0.8 ms after the start of injection at 300 bar injection pressure. The steady state is achieved sooner, if the injection pressure increases, for example, at 0.7 ms for 500 and 800 bar. These results show that the differences between various fuels diminish after achieving the steady state. The smaller cone angle of biodiesel after the injection start can probably be mostly attributed to higher *viscosity* of WcBIO, compared to D100.

In Agarwal and Chaudhury (2012) the influence of *engine loads* on the spray cone angle of D100 and KaBIO was investigated by using a simple mechanical control fuel pump and an injector (Table 6.6). The spray cone angle increased as the pressure in the chamber increased. This was observed for all test fuels. At low loads (e.g., at 1 bar of ambient pressure) the ambient air density is lower which results in lower spray cone angles. As the engine load increases (the pressure in chamber increases) the shear resistance exerted by ambient air rises which leads to an increase of the spray angle. The spray cone angle is 10.7° at 1 bar and 16° at 9 bar. Under identical experimental conditions, the spray cone angle of KaBIO is

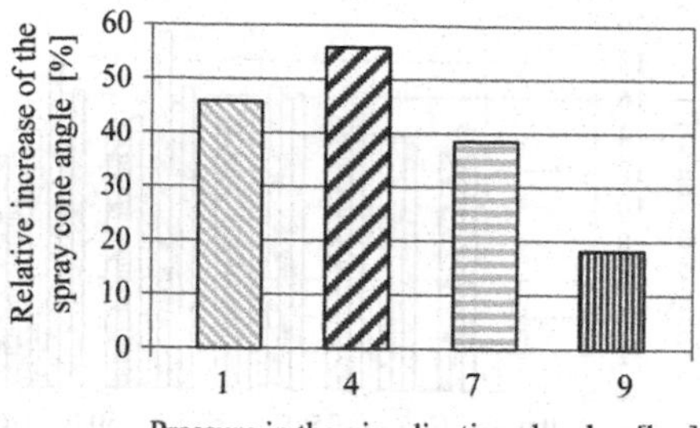

Fig. 6.19 Relative increase of the spray cone angle of KaBIO with respect to D100 under various ambient pressures

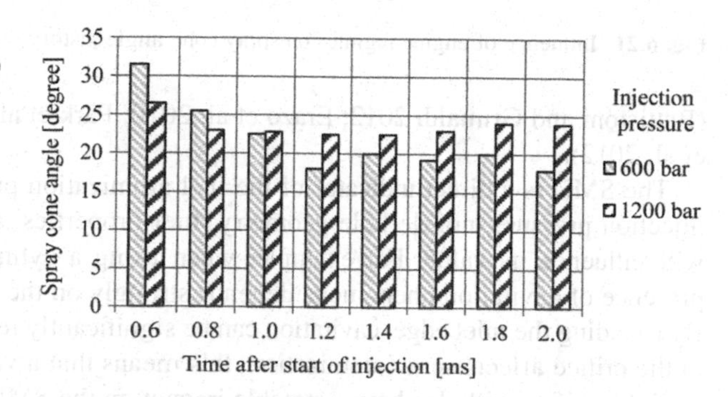

Fig. 6.20 Influence of injection pressure on SoBIO spray cone angle history

higher than that of D100 at all pressure conditions because of different fuel densities (Fig. 6.19).

In Park et al. (2011) the influence of injection pressure on spray angle of SoBIO in a naturally aspirated DI diesel engine with a common rail injection system was investigated. The fuel was injected through a nozzle with six holes with a diameter of 0.128 mm at ambient pressure of 30 bar. At 600 bar of injection pressure, the spray cone angle decreased sharply at the early stage of injection and became stable about 1.2 ms after the start of injection (Fig. 6.20). On the other hand, at 1,200 bar of injection pressure, the variation of the spray cone angle was rather minor.

For a bus diesel MAN engine with a mechanically controlled M fuel injection system (Table 6.3), the experimentally obtained comparison of spray cone angle at various engine speeds at full load is given in Fig. 6.21. It can be seen that the spray cone angles of D100 are higher than those of RaBIO at all engine speeds. For both tested fuels the smallest spray cone angles were observed at 800 rpm (near the peak torque condition).

6.3 Sauter Mean Diameter

The Sauter mean diameter (SMD) of the spray is closely related to the combustion process and consequently to all engine characteristics. Its value influences significantly noise and exhaust emissions, fuel consumption, and engine performance

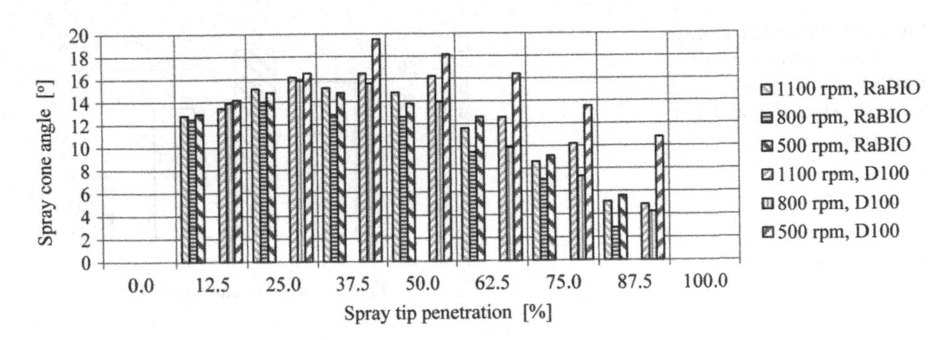

Fig. 6.21 Influence of engine regimes on spray cone angle history

(Battistoni and Grimaldi 2012; Erazo et al. 2010; Park et al. 2008; Shervani-Tabar et al. 2012).

The SMD is a direct indicator of the fuel atomization process. This means that injection pressure, nozzle hole geometry, fuel properties, and ambient conditions will influence its value. For example, when using a cylindrical nozzle hole, the presence of cavitation phenomena depends strongly on the shape of the inlet edge. By rounding the inlet edge, cavitation can be significantly reduced. Since cavitation in the orifice affects spray atomization, this means that a variation of the radius of the inlet orifice will also have a notable impact on the SMD.

6.3.1 Influence of Biodiesel Properties

Fuel atomization and combustion characteristics depend on the following fuel properties: viscosity, surface tension, density, latent heat of vaporization, thermal conductivity, specific heat capacity, boiling point, and heat of combustion. However, *viscosity* and *surface tension* are those that influence fuel atomization and SMD the most (Choi and Oh 2012; Desantes et al. 2009; Pandey et al. 2012; Park et al. 2009b; Wang et al. 2010, 2011).

In general, as the viscosity of the fuel increases, the diameters of fuel droplets increase also. Biodiesels have roughly about 50 % higher viscosity than mineral diesel, which means that larger droplets will form in the spray. The operation of the fuel injector may become less accurate and more deposit formation might be observed on the injector or in the combustion chamber. Higher surface tension also makes the spray harder to break up into smaller droplets. Since surface tension of biodiesels in general is higher than that of mineral diesel, this is another factor promoting larger Sauter mean diameters of biodiesels.

In a fuel spray, the SMD varies along the central axis of the spray and with the radial distance from the axis (Erazo et al. 2010). Experiments with *canola biodiesel* (CaBIO) and D100 revealed that the droplet size may increase with increasing radial distance at all axial locations. This was due to the swirl imparted by the

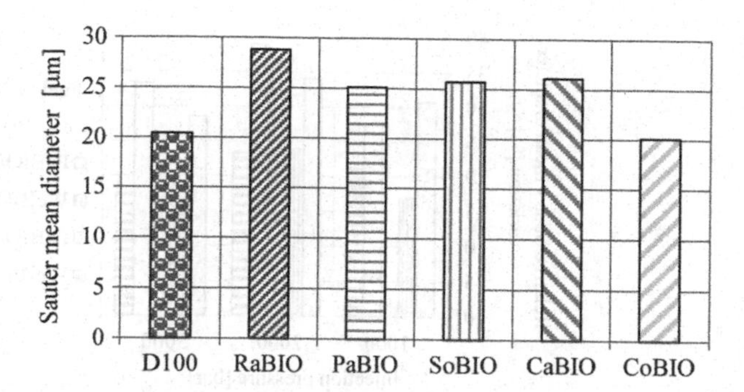

Fig. 6.22 Influence of biodiesel type on SMD

injector, by which the large drops were thrown to the spray edge and took longer to evaporate and burn. The result was an increase of the SMD at the outer edge of the spray. The droplet SMD of CaBIO was smaller than that of D100 at the nozzle exit, implying faster vaporization rates for CaBIO. Furthermore, the droplet sizes in D100 and CaBIO sprays were comparable in the near-injector region and further downstream of the injector. At an axial distance of 30 mm from the injector, however, the SMD of CaBIO was larger than that of the D100 spray. This might be due to the smaller drops of CaBIO evaporating faster than the D100 drops, leaving only the larger drops to remain at this axial location.

In Battistoni and Grimaldi (2012) the influence of various types of nozzle holes on SMD, by using D100 and SoBIO, was investigated by numerical simulation (Table 6.2). The numerical results for the SMD show that SoBIO produces quite larger droplet diameters and its breakup process is slower than that of D100. The main reasons stated are the lower velocities and the lower Weber numbers of SoBIO. The orifice shape had a negligible effect on the SMD, meaning that under given circumstances only the fuel properties determine the average stable non-evaporating droplet size.

A comparison of the SMDs, obtained on a direct fuel injection system for D100, RaBIO, PaBIO, SoBIO, CaBIO, and *coconut biodiesel* (CoBIO) at 80 °C, is given in Fig. 6.22 (Ejim et al. 2007). RaBIO has the largest SMD, which differs more than 40 % from that of D100. CoBIO has the smallest SMD, which is very close to that of D100.

The influence of fuel properties on the SMD was also investigated by (Wang et al. 2010, 2011). The experiments were performed under simulated diesel engine conditions (Table 6.4). The tested ambient densities were 15 and 30 kg/m^3. At both ambient densities the SMD of PaBIO yielded significantly larger Sauter diameters than WcBIO (Fig. 6.23). The difference between both tested biodiesels was larger at higher injection pressure. Both biodiesels always produced larger diameters than D100.

In Chen et al. (2013) spray properties of various fuels, WcBIO being one of them, were investigated experimentally (Table 6.8). The results showed that the SMD of WcBIO increases with the distance from the injector nozzle tip due to momentum loss. This was observed for axial distances of 6 cm and upwards.

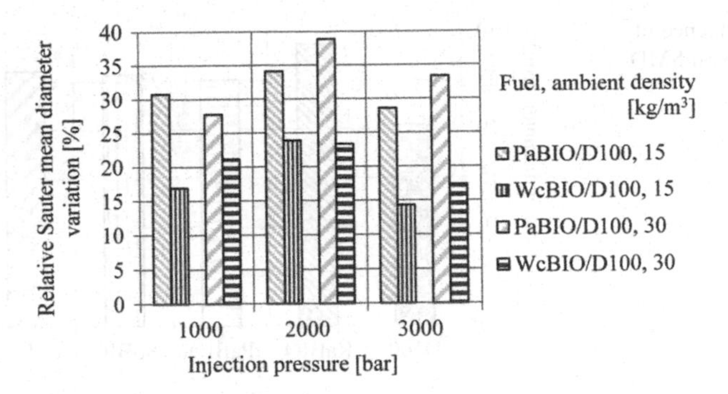

Fig. 6.23 Effect of injection pressure and ambient density on SMD of PaBIO and WcBIO, with respect to D100

Table 6.12 Investigation data (Park et al. 2009b)

Injection system/injector	Common rail injector
Nozzle type	Single hole
Length of nozzle hole	0.80 mm
Nozzle hole diameter	0.30 mm
Injection pressure—maximal	600 bar
Energizing duration	1.2 ms
Ambient temperature	293 K
Ambient pressure	1 bar

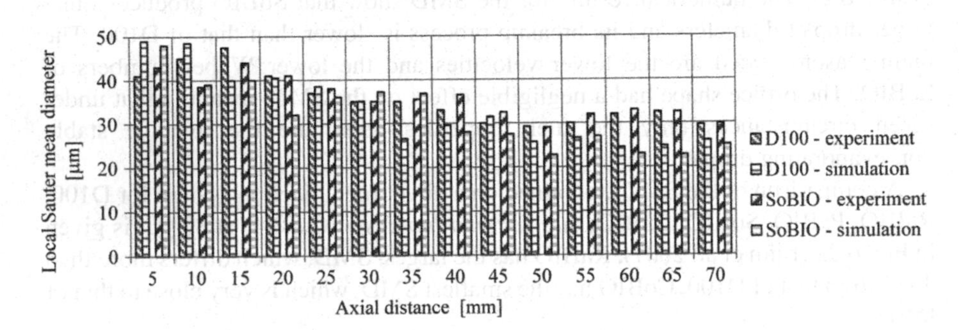

Fig. 6.24 Local SMD history

In Park et al. (2009b) spray atomization characteristics of SoBIO under various ambient pressures were investigated experimentally and numerically. A common rail injector (Table 6.12) was used to inject SoBIO and D100 into a high pressure chamber, filled with nitrogen gas.

The results obtained at 600 bar injection pressure and ambient pressure of 1 bar show that local SMD decreases along the spray axis at the distances from 1 cm up to 5 cm. From 5 cm upwards, it begins to increase slightly. In comparison to D100, the SoBIO exhibited a slightly larger droplet size. However, the difference in SMD between both fuels was relatively small (Fig. 6.24).

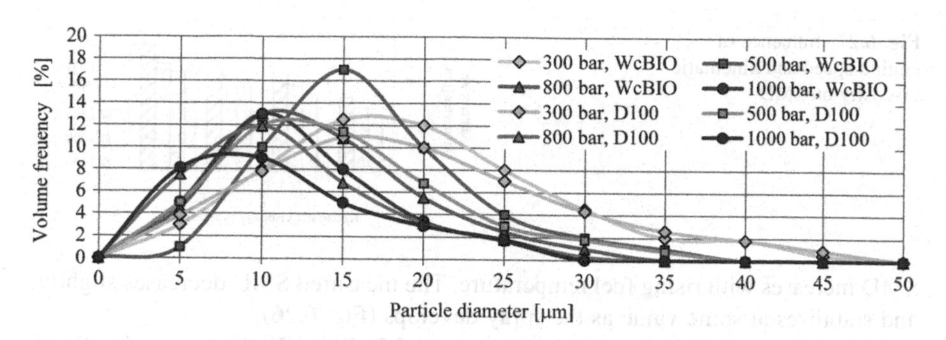

Fig. 6.25 Influence of injection pressure on SMD for various fuels

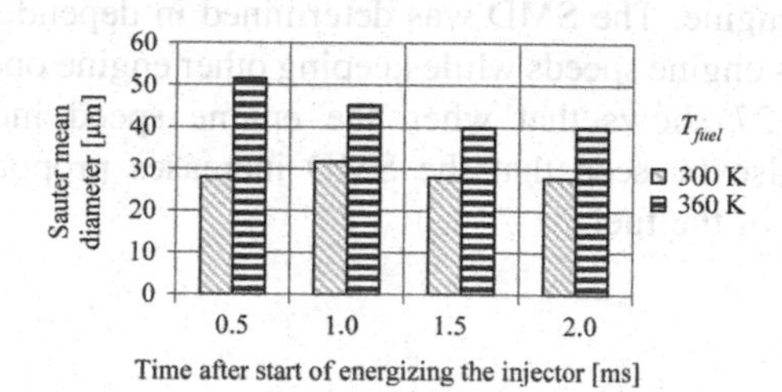

Fig. 6.26 Influence of fuel temperature on SMD history of SoBIO

6.3.2 *Influence of Engine Operating Regime*

A change of the engine operating regime is typically accompanied by a change of the injection pressure, which also affects the SMD. In general, higher injection pressure produces smaller droplets. In this context it may be worth to pay attention also on the fuel temperature, since its variations also influence the SMD.

In Chen et al. (2013) the influence of *variable injection pressure* on the SMD of WcBIO and D100 was investigated experimentally. The fuels at room temperature were injected into air by using a common rail injector (Table 6.8). The injection pressures were 300, 500, 800, and 1,000 bar.

Figure 6.25 illustrates the droplet size distribution (volume frequency) results under *various injection pressures*. All distributions of WcBIO and D100 fuels show a trend towards smaller droplet volumes as the injection pressure increases. D100 consistently exhibits smaller droplet volumes than WcBIO, although at 300 bar the difference is rather small.

In Park et al. (2009a) the effects of *fuel temperature* and *ambient gas temperature* on SoBIO spray atomization behavior were investigated (Table 6.9). It was observed that an increase of fuel temperature results in an increase of overall and local SMD (measured along the spray axis). This is because at higher temperatures small droplets seem to evaporate and vanish very quickly while this process is considerably slower for large droplets. As a result, the SMD increases. The effects of fuel temperature on the overall SMD at ambient temperature of 300 K and ambient pressure of 1 bar are shown in Fig. 6.26. It is evident that the overall

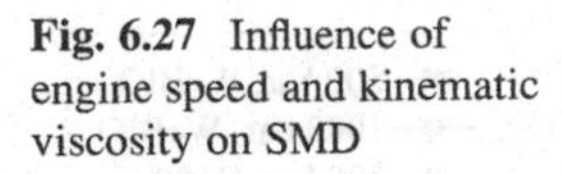
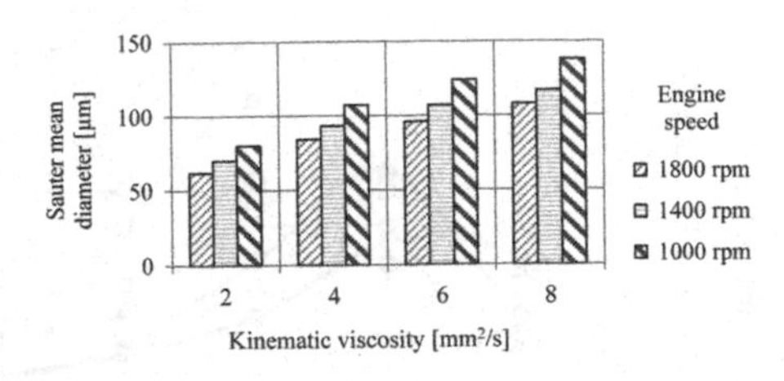

Fig. 6.27 Influence of engine speed and kinematic viscosity on SMD

SMD increases with rising fuel temperature. The measured SMD decreases slightly and stabilizes at some value as the spray develops (Fig. 6.26).

In Tesfa et al. (2010) spray development of RaBIO, WcBIO, *corn biodiesel* (CrBIO), and their blends with D100 was investigated numerically for a turbocharged DI diesel engine. The SMD was determined in dependence on kinematic viscosity for various engine speeds while keeping other engine operating conditions constant. Figure 6.27 shows that when the engine speed increases the SMD decreases. It can also be seen that the SMD increases proportionally with the kinematic viscosity of the fuel.

6.4 Discussion

A replacement of mineral diesel by biodiesel influences the fuel spray development in a rather sophisticated way. This is because this influence depends on many factors, starting from those related to the injection system to those related to biodiesel properties and engine operating conditions. One of the most important parameters of the injection system is the geometry of the injector's nozzle holes. For example, the spray angle may be affected significantly by varying only the rounding radius of the inlet edge of a cylindrical nozzle hole. A reduction of this radius typically increases the turbulences and cavitation inside the orifice, which leads to a wider spray angle. Of course, all such modifications require caution and need to be implemented carefully. For example, too much cavitation may lead to damages of the nozzle.

The task to predict the biodiesel spray development becomes even harder if one takes into account that biodiesels come from a variety of sources and that this variety is also reflected in their properties. A lot of experimental work is therefore needed. In order to reduce the related cost, numerical simulation becomes more and more important. Adequate software, such as AVL Fire or KIVA-3 V, can deliver remarkable results and can be used to reduce experimental work. However, it should be noted that such simulation software typically requires many coefficients of the underlying mathematical models that have to be carefully determined in order to get usable results. This is not an easy task and typically has to be accompanied by experiments.

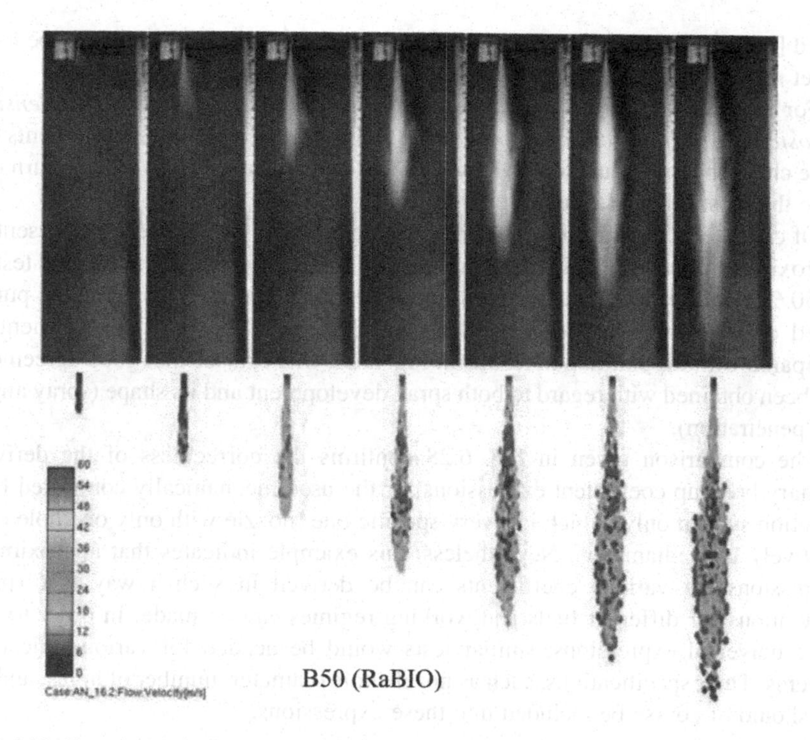

Fig. 6.28 Experimental and simulated spray development at peak torque condition for B50 RaBIO obtained with the approximated primary breakup coefficients

For example, for the primary diesel breakup model (Sect. 2.2.2) the coefficients C_i, $i = 1$, 2, and 3 are needed. One possibility is to build approximation formulas, for example, in the form:

$$C_i = \chi_1^{a_1} \times \chi_2^{a_2} \times \chi_3^{a_3} \ldots \tag{6.1}$$

where a_i are unknown coefficients to be determined and χ_i represents either a fuel physical property, an injection characteristics, or a working regime parameter. The influence of all χ_i parameters on each primary breakup coefficient may be analyzed experimentally and the results may be used to get the coefficients a_i. Under the conditions, given in Pogorevc et al. (2008), the resulting approximations for C_1, C_2, and C_3 are

$$\begin{aligned} C_1 &= \rho_f^{7.808} \times \mu_f^{-2.276} \times \sigma_f^{14.676} \times t_{\text{inj}}^{-31.348} \times s_q^{-9.864} \times n^{-8.391} \\ C_2 &= \rho_f^{-11.507} \times \mu_f^{-0.912} \times \sigma_f^{-23.985} \times q_{\text{inj}}^{38.725} \times p_{\text{ave}}^{15.9889} \times n^{-12.420} \\ C_3 &= \rho_f^{13.447} \times \mu_f^{2.369} \times \sigma_f^{25.764} \times t_{\text{inj}}^{-69.983} \times p_{\text{ave}}^{-23.873} \times n^{-4.015} \end{aligned} \tag{6.2}$$

where ρ_f is fuel density [kg/m^3], μ_f is fuel viscosity [mPas], σ_f is fuel surface tension [N/mm], t_{inj} stands for injection time [ms], p_{ave} is average injection pressure [MPa], $s_q = p_{\text{ave}}/p_{\max}$ is squareness, q_{inj} represents fuelling [mm^3/cycle], and n is pump speed [$\min^{-1}$]. In some limited context, such approximate models can also be used to get more insight into the influence of some parameters.

For example, in the presented approximate model one can see that fuel *density*, *viscosity*, and *surface tension* appear in the expressions for all three coefficients C_i. If we check the individual terms, however, fuel density and surface tension turn out to be the most influential fuel properties.

Of course, the validity of such approximations has to be verified. The presented approximation model was inserted into the AVL Fire solver steering file and tested on 50 % blend of diesel and rapeseed biodiesel fuel (B50 RaBIO) at engine pump speed of 800 rpm (Pogorevc et al. 2008). The recorded spray development is compared with the simulation results in Fig. 6.28. One can see that good agreement has been obtained with regard to both spray development and its shape (spray angle and penetration).

The comparison given in Fig. 6.28 confirms the correctness of the derived primary breakup coefficient expressions for the used mechanically controlled fuel injection system only, which is a very specific one (nozzle with only one hole of a relatively large diameter). Nevertheless, this example indicates that approximate expressions for various coefficients can be derived in such a way that spray simulations for different fuels and working regimes can be made. In order to get more universal expressions, similar tests would be needed for various injection systems. Their specifications, such as nozzle hole diameter, number of holes, and so on, should of course be included into these expressions.

References

Agarwal, A. K., & Chaudhury, V. H. (2012). Spray characteristics of biodiesel/blends in a high pressure constant volume spray chamber. *Experimental Thermal and Fluid Science, 42*, 212–218.

Battistoni, M., & Grimaldi, N. (2012). Numerical analysis of injector flow and spray characteristics from diesel injectors using fossil and biodiesel fuels. *Applied Energy, 97*, 656–666.

Chen, P. C., Wang, W. C., Roberts, W. L., & Fang, T. (2013). Spray and atomization of diesel fuel and its alternatives from a single-hole injector using a common rail fuel injection system. *Fuel, 103*, 850–861.

Choi, S., & Oh, Y. (2012). The spray characteristics of unrefined biodiesel. *Renewable Energy, 42*, 136–139.

Desantes, J. M., Payri, R., García, A., & Manin, J. (2009). Experimental study of biodiesel blends effects on diesel injection process. *Energy & Fuels, 23*, 3227–3235.

Ejim, C. E., Fleck, B. A., & Amirfazli, A. (2007). Analytical study for atomization of biodiesels and their blends in a typical injector: Surface tension and viscosity effects. *Fuel, 86*, 1534–1544.

Erazo, J. A., Parthasarathy, R., & Gollahalli, S. (2010). Atomization and combustion of canola methyl ester biofuel spray. *Fuel, 89*, 3735–3741.

Gao, Y., Deng, J., Li, C., Dang, F., Liao, Z., Wu, Z., & Li, L. (2009). Experimental study of the spray characteristics of biodiesel based on inedible oil. *Biotechnology Advances, 27*, 616–624.

Kegl, B. (2008). Effects of biodiesel on emissions of a bus diesel engine. *Bioresource Technology, 99*, 863–873.

Kousoulidou, M., Ntziachristos, L., Fontaras, G., Martini, G., Dilara, P., & Samaras, Z. (2012). Impact of biodiesel application at various blending ratios on passenger cars of different fueling technologies. *Fuel, 98*, 88–94.

Kuti, O. A., Zhu, J., Nishida, K., Wang, X., & Huang, Z. (2013). Characterization of spray and combustion processes of biodiesel fuel injected by diesel engine common rail system. *Fuel, 104*, 838–846.

Lee, C. S., Park, S. W., & Kwon, S. I. (2005). An experimental study on the atomization and combustion characteristics of biodiesel blended fuels. *Energy & Fuels, 19*, 2201–2208.

Lin, Y. S., & Lin, H. P. (2011). Spray characteristics of emulsified castor biodiesel on engine emissions and deposit formation. *Renewable Energy, 36*, 3507–3516.

Mancaruso, E., Sequino, L., & Vaglieco, B. (2011). First and second generation biodiesels spray characterization in a diesel engine. *Fuel, 90*, 2870–2883.

Pandey, R. K., Rehman, A., & Sarviya, R. M. (2012). Impact of alternative fuel properties on fuel spray behavior and atomization. *Renewable and Sustainable Energy Reviews, 16*, 1762–1778.

Park, S. H., Kim, H. J., & Lee, C. S. (2010). Fuel spray and exhaust emission characteristics of an undiluted soybean oil methyl ester in a diesel engine. *Energy & Fuels, 24*, 6172–6178.

Park, S. H., Kim, H. J., Suh, H. K., & Lee, C. S. (2009a). Experimental and numerical analysis of spray-atomization characteristics of biodiesel fuel in various fuel and ambient temperatures conditions. *International Journal of Heat and Fluid Flow, 30*, 960–970.

Park, S. H., Kim, H. J., Suh, H. K., & Lee, C. S. (2009b). A study on the fuel injection and atomization characteristics of soybean oil methyl ester (SME). *International Journal of Heat and Fluid Flow, 30*, 108–116.

Park, S. H., Suh, H. K., & Lee, C. S. (2008). Effect of cavitating flow on the flow and fuel atomization characteristics of biodiesel and diesel fuels. *Energy & Fuels, 22*, 605–613.

Park, S. H., Yoon, S. H., & Lee, C. S. (2011). Effects of multiple-injection strategies on overall spray behavior, combustion, and emissions reduction characteristics of biodiesel fuel. *Applied Energy, 88*, 88–98.

Payri, R., Salvador, F. J., Gimeno, J., & Novella, R. (2011). Flow regime effects on non-cavitating injection nozzles over spray behavior. *International Journal of Heat and Fluid Flow, 32*, 273–284.

Pogorevc, P., Kegl, B., & Škerget, L. (2008). Diesel and biodiesel fuel spray simulations. *Energy & Fuels, 22*, 1266–1274.

Shervani-Tabar, M. T., Parsa, S., & Ghorbani, M. (2012). Numerical study on the effect of the cavitation phenomenon on the characteristics of fuel spray. *Mathematical and Computer Modelling, 56*, 105–117.

Som, S., Longman, D. E., Ramírez, A. I., & Aggarwal, S. K. (2010). A comparison of injector flow and spray characteristics of biodiesel with petrodiesel. *Fuel, 89*, 4014–4024.

Tesfa, B., Mishra, R., Gu, F., & Powles, N. (2010). Prediction models for density and viscosity of biodiesel and their effects on fuel supply system in CI engines. *Renewable energy, 35*, 2752–2760.

Valentino, G., Allocca, L., Iannuzzi, S., & Montanaro, A. (2011). Biodiesel/mineral diesel fuel mixtures: Spray evolution and engine performance and emissions characterization. *Energy, 36*, 3924–3932.

Wang, X., Huang, Z., Kuti, O. A., Zhang, W., & Nishida, K. (2010). Experimental and analytical study on biodiesel and diesel spray characteristics under ultra-high injection pressure. *International Journal of Heat and Fluid Flow, 31*, 659–666.

Wang, X., Huang, Z., Kuti, O. A., Zhang, W., & Nishida, K. (2011). An experimental investigation on spray, ignition and combustion characteristics of biodiesels. *Proceedings of the Combustion Institute, 33*, 2071–2077.

Yoon, S. H., Suh, H. K., & Lee, C. S. (2009). Effect of spray and EGR rate on the combustion and emission characteristics of biodiesel fuel in a compression ignition engine. *Energy & Fuels, 23*, 1486–1493.

Chapter 7
Effects of Biodiesel Usage on Engine Performance, Economy, Tribology, and Ecology

Biodiesel usage influences directly the injection and combustion processes and consequently also the *engine performance*, *ecology*, and *economy characteristics*. This influence is determined by the chemical and physical properties of biodiesel and depends on many parameters related to the engine, operating conditions, and so on.

According to most investigations, engine power and torque, particulate matter (PM), carbon monoxide (CO), and unburned hydrocarbons (HC) in general decrease when mineral diesel is replaced by biodiesel. On the other hand, nitrogen oxides (NO_x) typically increase. Of special interest is the variation of PM and NO_x emissions which is attributed both to the difference in chemical characters of mineral diesel and biodiesel (which affects combustion kinetics) and to different physical properties, which affect fuel spray characteristics (Kousoulidou et al. 2012; Giakoumis 2012; Kegl 2008).

Biodiesel fuels have lower energy content than mineral diesel. Thus, if the efficiency is kept constant, the engine fuel consumption will be higher when mineral diesel is replaced by biodiesel. Furthermore, biodiesel usage can contribute to the formation of deposits, the degradation of materials, or the plugging of filters. This depends mainly on their degradability, their glycerol content, their cold flow properties, and other fuel quality specifications (Lapuerta et al. 2008a).

In spite of all potential problems related to biodiesel usage, biodiesel fuels exhibit an interesting potential to improve engine characteristics and reduce harmful emissions. In order to get as much insight as possible, these topics need to be addressed with regard to *biodiesel properties*, *engine type*, and *engine operating conditions* (including *ambient conditions*). In this context it may be worth noting that most modern engines have electronically controlled high pressure direct fuel injection systems, which are more sensitive to fuel quality than mechanically controlled injection systems. Since modern engines are optimized for mineral diesel, it is reasonable to compare all engine characteristics, obtained by usage of biodiesel fuels, to those obtained with mineral diesel (Canakci 2007).

B. Kegl et al., *Green Diesel Engines*, Lecture Notes in Energy 12,
DOI 10.1007/978-1-4471-5325-2_7, © Springer-Verlag London 2013

7.1 Engine Performance, Economy, and Tribology Characteristics

In general, investigation results show that the *engine power* and *torque* will decline when biodiesel replaces mineral diesel. This is especially evident for neat biodiesel fuels, because biodiesels have lower energy content than diesel. It is interesting to note, however, that the results reported are not completely uniform. For example, some investigations reveal that because of power recovery the observed power loss is actually lower than expected due to lower heating value of biodiesel (Xue et al. 2011). Anyhow, most of researches agree that lower energy content of biodiesels has evident consequences, which are also reflected in higher *fuel consumption*.

The influence of biodiesel fuels on the engine tribology characteristics depends to a great extent on the raw material used for biodiesel production and consequently its properties. Compared to mineral diesel, biodiesels typically exhibit higher viscosity and lower volatility, which can lead to injector coking and trumpet formation on the injectors. Furthermore, one can observe various influences on carbon deposits, oil ring sticking, and thickening and gelling of the engine lubricant oil, after the engine has been operated on biodiesel for a longer time period. Therefore, many biodiesel investigations are related to wear of engine components like piston, piston ring, cylinder liner, bearing, crankshaft, cam tappet, valves, and injectors (Dorado et al. 2003; Demirbas 2006; Giannelos et al. 2005). On the other hand, biodiesels typically exhibit better lubrication properties than mineral diesel. Therefore, biodiesels and their blends with mineral diesel may reduce long-term engine wear. In some tests, the engine wear was reduced to less than half of what was observed in the engine running on current low sulfur diesel fuel (Demirbas 2009).

7.1.1 *Influence of Biodiesel Properties*

Biodiesel chemical and physical properties influence significantly the injection process, fuel spray development, and combustion. Especially in a mechanically controlled injection system a replacement of mineral diesel by biodiesel results in higher injection pressure, the injection timing and start of combustion are advanced, the ignition delay is shorter, and the in-cylinder pressure and temperature rise earlier. Furthermore, the maximum firing temperature, the exhaust gas temperature, and heat release rate are smaller than to those obtained with mineral diesel. Shorter ignition delay of biodiesel results in an advanced combustion, longer expansion period, and lower exhaust gas temperature. Clearly, according to all this, biodiesel properties indirectly influence all engine characteristics. But apart from this, they also have a direct influence on engine performance, economy, and tribology characteristics (Kegl 2008, 2011; Ozsezen et al. 2008; Sayin and Gumus 2011;

Xue et al. 2011). In this context, the following properties are perhaps the most exposed:

- *Energy content*: biodiesels typically exhibit lower energy content than mineral diesel. Thus, biodiesel usage typically results in lower effective engine power and higher fuel consumption. Of course, this effect is the most exposed for neat biodiesel and declines by blending biodiesel with mineral diesel. Furthermore, it should be noted that the energy content varies in dependence on biodiesel source, its production processes, and quality.
- *Density*: higher density of biodiesel means that if the injected fuel volume is kept constant, the mass of injected fuel and consequently the fuel consumption are also higher.
- *Lubricity and solvent action*: in general, biodiesels exhibit good lubricity properties and good solvent action. Therefore, compared to mineral diesel, biodiesel usage may result in lower carbon deposits and wear of the key engine parts. The durability of the engine may further be improved due to the lower soot formation when using biodiesel. Good lubricity properties also positively affect the engine effective power.

In Kegl (2006) a *DI diesel bus* MAN 2566 MUM engine (Table 7.1) was tested in order to investigate the influence of *rapeseed biodiesel* (RaBIO) usage on engine performance. The results were compared to those obtained with mineral diesel (D100).

The histories of *heat release rate*, *in-cylinder pressure*, and *in-cylinder temperature* are shown in Figs. 7.1 and 7.2, along with the corresponding injection pressure and needle lift histories. It is evident that the maximums of the in-cylinder pressure, in-cylinder temperature, and heat release rate appear earlier when using RaBIO. The lower in-cylinder temperature of RaBIO can be attributed to the fact that, compared to D100, RaBIO exhibits higher latent heat of vaporization and lower heating value. Thus, more heat is needed for RaBIO vaporization, while the energy released by RaBIO is lower than that from the same mass of D100. As a result, the in-cylinder gas temperature can be lower for RaBIO. The start of injection and the start of combustion at peak torque and at rated conditions are also shown in Figs. 7.1 and 7.2. The start of injection is indicated by the point of needle lifting. The start of combustion is marked by a rapid increase of in-cylinder gas pressure and by the point of heat release start.

In Pehan et al. (2009) the same engine (Table 7.1) was used in order to get some insight into the influence of RaBIO on *tribology characteristics* of the engine. Firstly, the most important pump plunger surfaces were analyzed before and after biodiesel usage. Then, biodiesel deposits at the injectors and in the combustion chambers were examined. The deposits in the injector nozzle holes were investigated by considering the measured discharge coefficients. Before using RaBIO, the engine was run with mineral diesel.

The influence of RaBIO usage on *pump plunger surface* is illustrated in Fig. 7.3. The surface area shown in the figure is positioned close to the top of the pump plunger. This area has been selected for examination since it has a very important

Table 7.1 Engine specifications (Kegl 2006)

Engine model	MAN 2566 MUM
Engine type	4 stroke, 6 cylinders in line, water cooled
Displacement	11,413 cm^3
Bore and stroke	125 mm × 155 mm
Compression rate	17.5:1
Maximal power at engine speed	162 kW at 2,200 rpm
Maximal torque at engine speed	158 Nm at 1,600 rpm
Injection model	Mechanically controlled M direct injection system
Injector opening pressure	175 bar
Injection pump timing	23°CA BTC

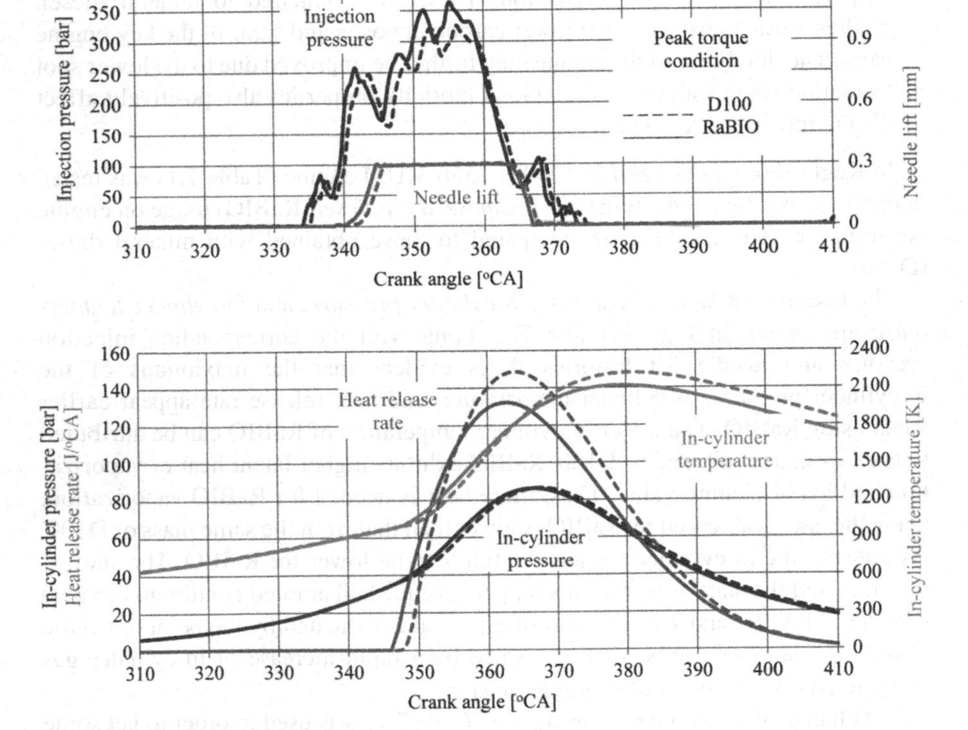

Fig. 7.1 Combustion characteristics for RaBIO (*solid*) and D100 (*dashed*) fuels at peak torque condition

influence on the injection pressure. It turned out that under the microscope the surface looked always pretty the same, regardless of the fuel used.

In order to obtain the *surface roughness* parameters, five measurements were performed on both plunger skirt and head for each parameter. It turned out that the influence of RaBIO usage is rather minor for the *plunger skirt* (Fig. 7.4). On the contrary, the roughness parameters of the *plunger head* exhibited significant changes after RaBIO usage (Fig. 7.5).

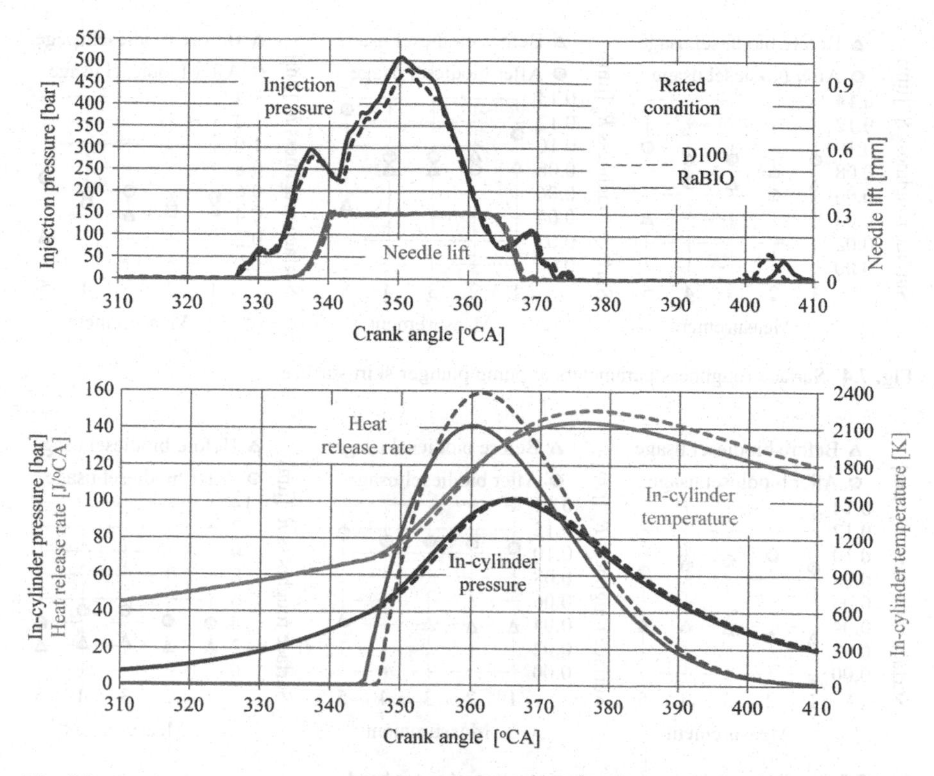

Fig. 7.2 Combustion characteristics for RaBIO (*solid*) and D100 (*dashed*) fuels at rated condition

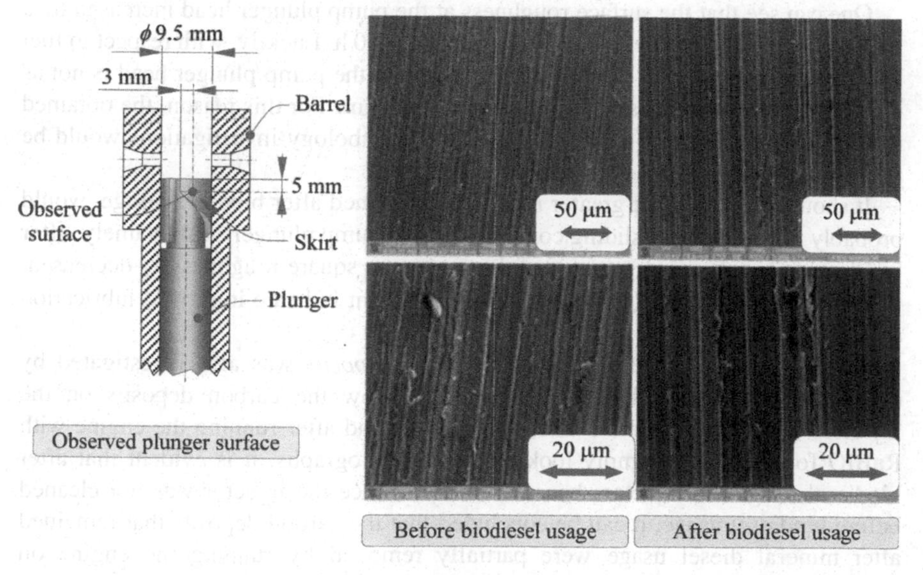

Fig. 7.3 Pump plunger skirt surface before and after RaBIO usage

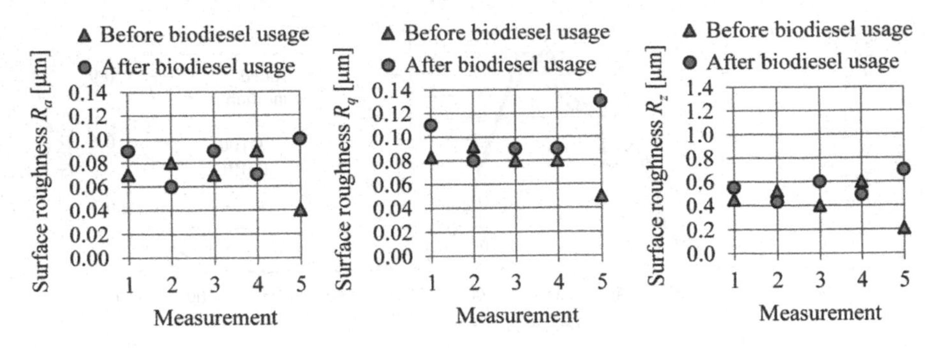

Fig. 7.4 Surface roughness parameters at pump plunger skirt surface

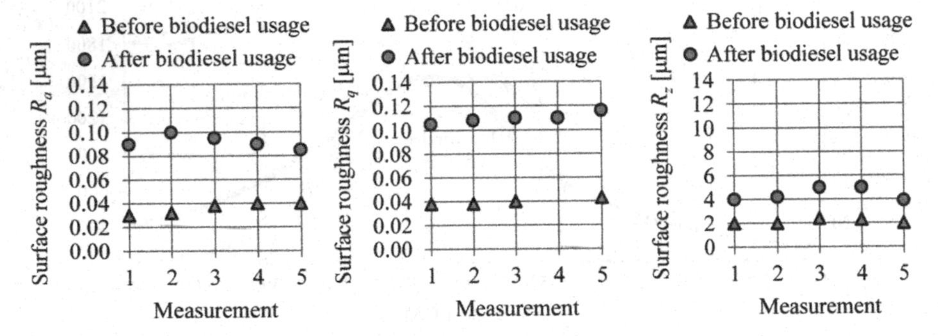

Fig. 7.5 Surface roughness parameters at pump plunger head

One can see that the surface roughness at the pump plunger head increased by a factor of two when biodiesel was used for about 110 h. Luckily, with respect to fuel leakage in the HP pump, the surface roughness at the pump plunger head is not as important as the roughness at the pump plunger skirt. For this reason, the obtained results are not alarming, although some further tribology investigations would be necessary to evaluate the situation more precisely.

It should be noted that greater roughness, obtained after biodiesel usage, would probably not worsen the sliding conditions at the pump plunger skirt. Namely, after biodiesel usage the average value of the root mean square roughness R_q decreased; in fact, it dropped from $0.45a$ to $0.4a$, which might indicate improved lubrication conditions.

The influence of RaBIO usage on *carbon deposits* was also investigated by endoscopic inspection. Figures 7.6 and 7.7 show the carbon deposits on the injectors of cylinders 3 and 5 after D100 usage and after running the engine with RaBIO for 110 h. By simply looking at the photographs, it is evident that after biodiesel usage the injectors became cleaner. Since the injector was not cleaned before biodiesel usage, it can be concluded that the carbon deposits that remained after mineral diesel usage were partially removed by running the engine on biodiesel. This is obviously due to the physical and chemical properties of the

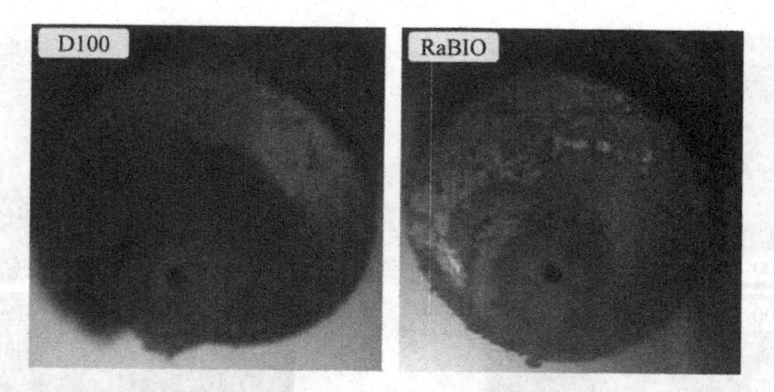

Fig. 7.6 Carbon deposits on the injector of cylinder 3

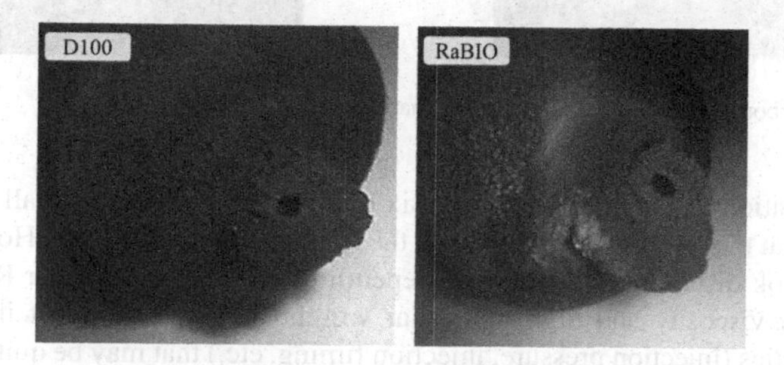

Fig. 7.7 Carbon deposits on the injector of cylinder 5

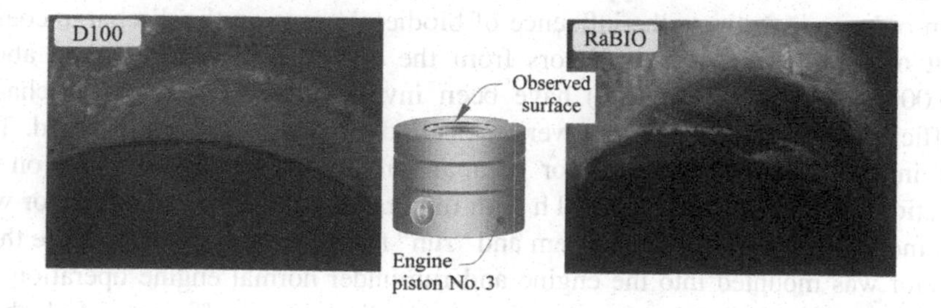

Fig. 7.8 Carbon deposits in the combustion chamber of cylinder 3

tested biodiesel, which is known to have excellent solvent action and can also loosen old deposits.

Furthermore, the combustion chambers were observed by using videoscopy. It was found out that at some places of the combustion chamber the carbon deposits increased, while at some other places they decreased. Figure 7.8 shows the carbon deposits on one side of the combustion chamber of the third cylinder. The situation in chamber 5 (at two positions) is shown in Fig. 7.9. One can say that the carbon deposits in the combustion chambers vary in dependence on the fuel used.

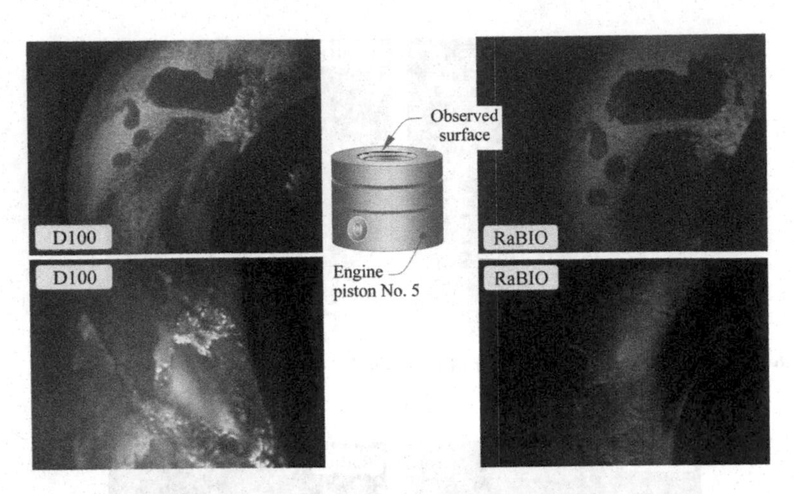

Fig. 7.9 Carbon deposits in the combustion chamber of cylinder 5

Examination and comparison of all six combustion chambers (in all cylinders) revealed that the situation is pretty much the same in all six chambers. However, the deposits look differently distributed—depending on whether D100 or RaBIO was used. High viscosity and high molecular weight of biodiesel result in injection characteristics (injection pressure, injection timing, etc.) that may be quite different to those of D100. This may lead to different distribution of the deposits. However, although biodiesel is known to exhibit poor atomization and low volatility, it seems that the total amount of the deposits did not increase by using RaBIO.

In order to investigate the influence of biodiesel usage on the discharge coefficient of the injector, three injectors from the engine (that has been run about 500,000 km on mineral diesel) have been investigated. At first, the discharge coefficients of all three injectors were measured by using a calibration fluid. The first injector (that served only for comparison purposes) was mounted on an injection system and "run" for 110 h with mineral diesel. The second injector was also mounted on the injection system and "run" for 110 h with biodiesel. The third injector was mounted into the engine and run under normal engine operation for 110 h by using biodiesel. After that the nozzle discharge coefficients of all three injectors were measured again by using the calibration fluid.

The analysis of the obtained results showed that the influence of biodiesel usage on the nozzle discharge coefficient is rather minor (Fig. 7.10). There were minor differences between the measured discharge coefficients, but these differences practically vanished at higher needle lifts. The results show that after biodiesel usage the coefficients in the injection system are lower than before usage of biodiesel for the most of the needle lifts. This was a somewhat surprising result. Namely, since biodiesel has a good solvent action, it was expected that the old deposits in the injector nozzle would be reduced, resulting in an increase of the discharge coefficient. Therefore, the same experiment was repeated with a new (clean) injector and the result was as expected: the discharge coefficient was not decreased after biodiesel usage. As it looks, the only reasonable explanation seems

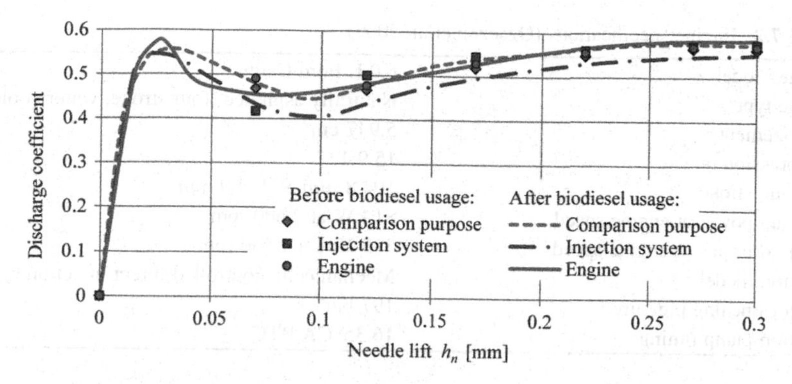

Fig. 7.10 Influence of biodiesel usage on nozzle discharge coefficient

to be that biodiesel caused the old deposits in the nozzle to swell up. It is possible that after a longer period of biodiesel usage the old deposits would begin to decay (at least partially). In any case, the experimental time period (110 h) is obviously not enough to reduce the deposits. Finally, it may be worth noting that even if the coefficient variations are not so dramatic, one must recognize that many investigations are done primarily on the basis of numerical simulation of the fuel injection processes. Therefore, taking into account the influence of small nozzle discharge coefficient variations on the numerically obtained injection characteristics might prove to be of benefit when high accuracy results are needed.

Most investigations confirm that biodiesel usage reduces engine effective power due to lower *energy content* of biodiesel (Buyukkaya 2010; Carraretto et al. 2004; Choi and Oh 2006; Hazar 2009; Kaplan et al. 2006; Karabektas 2009; Xue et al. 2011). However, the results reported show some fluctuations. Some investigators report that the actually observed power loss is lower than expected and that this is a consequence of power recovery (Murillo et al. 2007; Ozsezen et al. 2009; Oğuz et al. 2007). For example, a power loss of about 7 % was observed on a 3-cylinder, naturally aspirated, submarine diesel engine at full load, when mineral diesel was replaced by biodiesel; the energy content of used biodiesel fuels was about 13.5 % lower than that of mineral diesel. Furthermore, by testing RaBIO, *soybean biodiesel* (SoBIO), and *palm biodiesel* (PaBIO) on a 4-stroke, 30 kW TUMOSAN diesel engine (without any engine modification), it was shown that, compared to D100, there were no significant differences in engine power (Oğuz et al. 2007).

In Ozsezen et al. (2009) a naturally aspirated DI diesel engine (Table 7.2) was tested under full load. The tested fuels were D100, PaBIO, and *canola biodiesel* (CaBIO). As one can see from Fig. 7.11, D100 always delivered the highest engine *effective power*, but the differences were rather minor. Furthermore, the maximum brake power values obtained with either PaBIO or CaBIO were quite similar. Larger differences were observed in the *specific fuel consumption* (Fig. 7.12). The differences between mineral diesel and both biodiesels agreed approximately with the differences in fuel heating values.

Table 7.2 Engine specifications (Ozsezen et al. 2009)

Engine model	6.0 L Ford Cargo
Engine type	Naturally aspirated, four stroke, water cooled
Displacement	5,947 cm^3
Compression rate	15.9:1
Bore and stroke	104.8 mm × 114.9 mm
Maximal power at engine speed	81 kW at 2,600 rpm
Maximal torque at engine speed	335 Nm at 1,500 rpm
Injection model	Mechanically controlled direct injection system
Injector opening pressure	197 bar
Injection pump timing	16.35°CA BTC

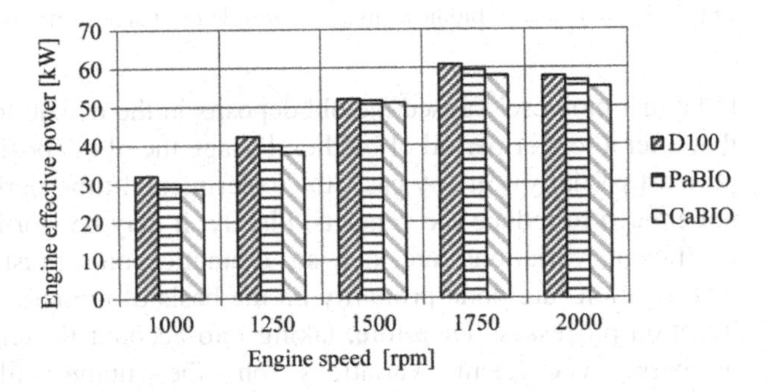

Fig. 7.11 Engine effective power obtained with various fuels

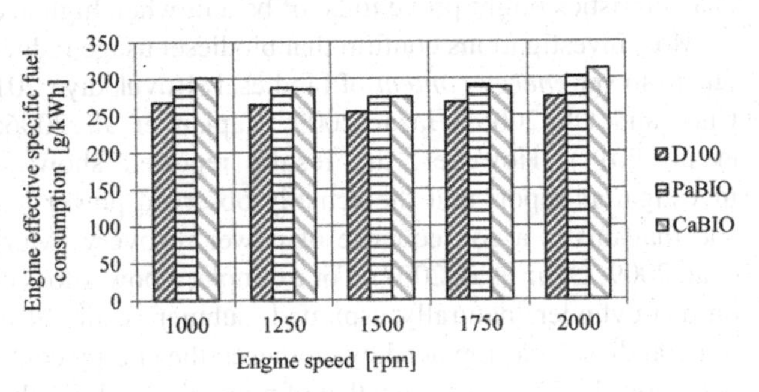

Fig. 7.12 Engine effective specific fuel consumption obtained with various fuels

The *effective specific fuel consumption* may vary significantly in dependence on biodiesel source (Ozsezen et al. 2009; Reyes and Sepúlveda 2006; Lin et al. 2009; Sahoo et al. 2009), which determines the *energy content*, *density*, and *viscosity* of biodiesel. A comparison of biodiesels from various sources, performed at constant conditions, revealed that the highest value of the effective specific fuel consumption was obtained with PaBIO, which has particularly low energy content and shorter *carbon chains*, compared to other biodiesels (Lin et al. 2009). It may be worth noting that the explanations for the increased fuel consumption of biodiesel are not very uniform among the authors. In some investigations, the most exposed reason is the higher density of biodiesel, which causes the higher mass injection for the same volume at the same injection pressure (Buyukkaya 2010; Qi et al. 2009;

Godiganur et al. 2009, 2010; Puhan et al. 2005a). Some other authors interpret the increase in fuel consumption to be a consequence of combination of biodiesel properties. For example, the following reasons are often highlighted: lower energy content and higher density (Utlu and Koçak 2008; Carraretto et al. 2004; Tsolakis et al. 2007), combined effect of higher viscosity and lower energy content of biodiesel (Aydin and Bayindir 2010; Ramadhas et al. 2005; Qi et al. 2010), and interaction of higher density, higher viscosity, and lower energy content of biodiesel (Lin et al. 2009; Song and Zhang 2008).

7.1.2 Influence of Engine Type

The influence of biodiesel usage on engine characteristics may depend significantly on various parameters related to engine design, control, and so on (Hazar 2009; Karabektas 2009; Pandey et al. 2012; Sayin and Gumus 2011). Of those parameters, the following ones might be the most important:

- *Compression ratio*: specific fuel consumption and thermal efficiency are considerably improved by increasing the compression ratio; higher compression ratio raises the density of air charge in the cylinder; higher air density leads to higher spray angles which results in an increase of the amount of air entrainment in the spray; this contributes to more complete combustion.
- *Injection timing*: in order to get minimal specific fuel consumption and maximal thermal efficiency, the injection timing has to be set optimally for each individual biodiesel fuel.
- *Naturally aspirated and turbocharged operation*: effective engine power and torque obtained with mineral diesel are higher than those obtained with biodiesel; this holds true for both naturally aspirated and turbocharged operation; by applying a turbocharger, the benefit of increased effective power and torque is more evident in biodiesel usage; because of higher fuel density and lower energy content, biodiesel shows slightly higher specific fuel consumption for both operations in comparison with mineral diesel; for both fuel types the specific fuel consumption decreases in the turbocharged operation, compared to naturally aspirated operation.
- *Coated and uncoated engine*: for biodiesel and diesel, the effective power of a coated engine is typically higher than that of the uncoated engine; this may be explained by an increase in the temperature of the combustion chamber elements, due to a thermal barrier effect caused by the ceramic coating; higher temperatures improve combustion.

In Sayin and Gumus (2011) the influence of *compression ratio* and *injection pump timing* on the brake specific fuel consumption and brake thermal efficiency was investigated experimentally on a *naturally aspirated DI diesel* engine (Table 7.3). The used fuel was biodiesel from a commercial supplier. Here and in the following, the term *relative variation* will be used to denote the relative difference Δp of a parameter p, calculated as

Table 7.3 Engine specifications (Sayin and Gumus 2011)

Engine model	Lombardini 6 LD 400
Engine type	Naturally aspirated, four stroke, one cylinder
Displacement	395 cm^3
Compression rate	18:1
Bore and stroke	86 × 68 mm
Maximal power at engine speed	7.5 kW at 3,600 rpm
Maximal torque at engine speed	21 Nm at 2,200 rpm
Injection model	Mechanically controlled direct injection system
Injector opening pressure	200 bar
Injection pump timing	20°CA BTC

$$\Delta p[\%] = \frac{p_{\mathrm{BIO}} - p_{\mathrm{D100}}}{p_{\mathrm{D100}}} \times 100, \qquad (7.1)$$

where p_{BIO} is the parameter obtained with biodiesel and p_{D100} is the same parameter obtained with mineral diesel usage.

Among other quantities, the brake specific fuel consumption (BSFC) and the brake thermal efficiency (BTE) were measured and compared to the same quantities obtained with mineral diesel. By increasing the *compression ratio* from 17:1 to 19:1 the relative differences in BSFC and BTE both tend to decrease (Fig. 7.13). It is evident that with higher compression ratios the difference between biodiesel and mineral diesel declines.

By increasing the *pump injection timing* from 15 to 25°CA BTC the relative difference in BSFC also showed some variation (Fig. 7.14). This variation, however, was not monotonic. On the other hand, the relative BTE difference was practically unaffected by various injection pump timings.

In Kegl (2006) a *DI diesel bus* MAN 2566 MUM engine (Table 7.1) was tested in order to investigate the influence of *pump injection timing* on injection and in-cylinder pressures. The fuels investigated were RaBIO and D100.

The injection pressure obtained with RaBIO was higher than that obtained with D100. The opposite was observed for the peak in cylinder pressure (Fig. 7.15). By reducing the pump injection timing, the peak injection pressure location of RaBIO was shifted towards the top dead center while the peak in-cylinder pressures decreased. This indicates that retarded pump injection timing considerably influences the air–fuel mixing, start of combustion, and consequently all engine performances.

The influence of *pump injection timing* on engine performance is shown in Fig. 7.16, which presents the relative differences of several parameters. The compared quantities are the effective power, effective specific fuel consumption, thermal efficiency, and temperature of exhaust emissions. The presented results show that for RaBIO the minimal specific fuel consumption g_e is obtained with pump injection timing of $\alpha_i = 19°$CA TDC at peak torque condition and $\alpha_i = 20°$CA TDC at rated condition (Fig. 7.16). This can be explained with the nature of fuel injection of the employed engine. The M injection system with its single-hole injection

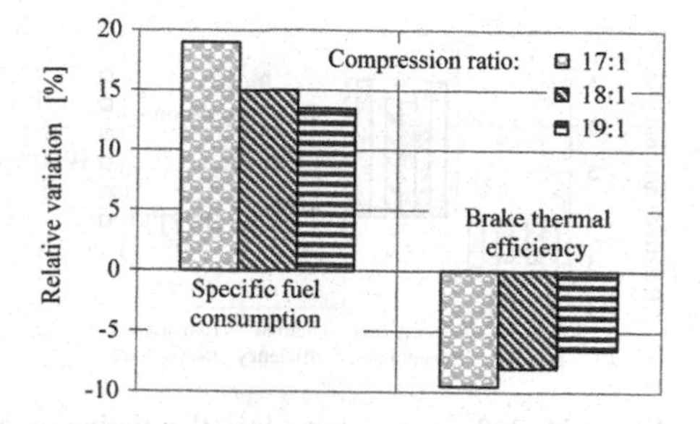

Fig. 7.13 Relative variations of BSFC and BTE for various compression ratios

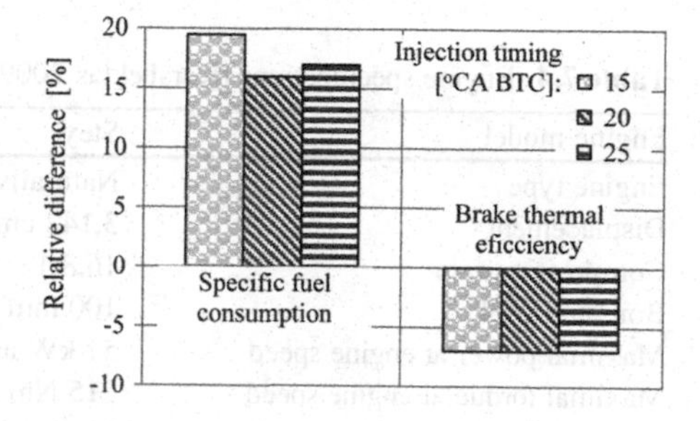

Fig. 7.14 Relative variations of BSFC and BTE for various injection timings

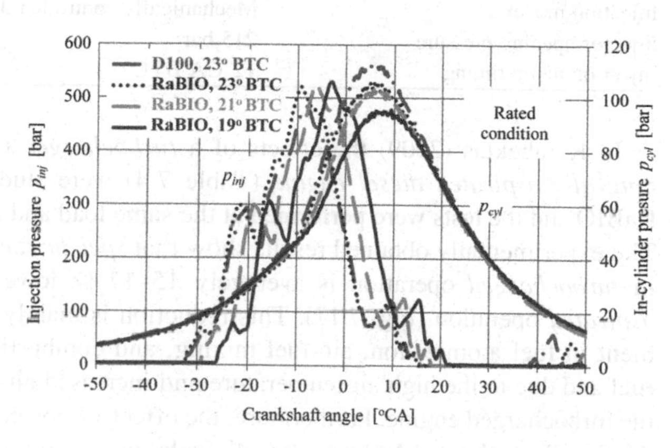

Fig. 7.15 Injection and in-cylinder pressures for various pump injection timings

nozzle is oriented so that most of the fuel is deposited on the piston bowl walls. The use of a bowl-in-piston combustion chamber results in a substantial swirl amplification at the end of the compression process. The air swirl increases as the piston approaches the top center, influencing significantly the fuel–air mixing rate. It is known, however, that optimum (and not maximum) swirl level gives minimum specific fuel consumption. Obviously, the optimum swirl for RaBIO fuel was obtained at $\alpha_i = 19$°CA TDC. At this setting the maximum cylinder pressure is lower by about 15 bar compared to D100 with standard $\alpha_i = 23$°CA TDC (Fig. 7.15). The temperatures of exhaust gases are at the lowest levels at $\alpha_i = 20$°CA TDC and the thermal efficiency is optimal at $\alpha_i = 20$ °CA TDC for rated condition and at $\alpha_i = 19$°CA TDC for peak torque condition.

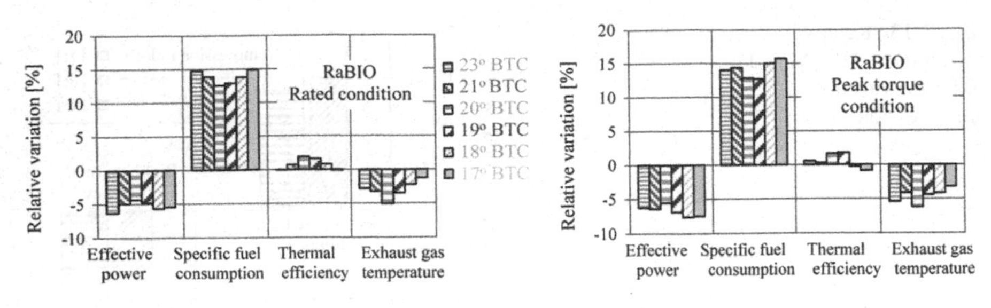

Fig. 7.16 Influence of pump injection timing on engine performance

Table 7.4 Engine specifications (Karabektas 2009)

Engine model	Steyr
Engine type	Naturally aspirated, water cooled, 4 stroke, 4 cylinders
Displacement	3,142 cm^3
Compression rate	16.8:1
Bore and stroke	100 mm × 100 mm
Maximal power at engine speed	51 kW at 2,400 rpm
Maximal torque at engine speed	215 Nm at 1,400 rpm
Injection model	Mechanically controlled direct injection system
Injector opening pressure	215 bar
Injection pump timing	12°CA BTC

In Karabektas (2009) the effects of a *turbocharger system* installation into a *naturally aspirated diesel engine* (Table 7.4) were studied. The fuel used was RaBIO and the tests were performed at the same load and at various *engine speeds*. The experimentally obtained results show that *specific fuel consumption* of RaBIO in *turbocharged* operation is averagely 15–17 % lower than that of *naturally aspirated* operation (Fig. 7.17). This reduction is mainly caused by the improvement in fuel atomization, air–fuel mixing, and combustion characteristics of the fuel and due to the high air temperature and increased air charge in the cylinder of the turbocharged engine. Furthermore, the effective power reached its peak value at the speed of about 2,400 rpm for all fuels and engine operations. The effective power obtained with D100 is higher than that obtained with RaBIO for both naturally aspirated and turbocharged operations. In naturally aspirated operation, the mean reduction in the effective power is about 5 % when RaBIO replaces D100. Due to the fact that the energy content of RaBIO is about 12 % lower than that of D100, both the effective torque and power decline. However, the differences are relatively small in most cases. Figure 7.18 also shows that the difference in the effective power between D100 and RaBIO reduces in turbocharged operation. The effective power obtained with RaBIO is on average 3 % lower than that of D100 in turbocharged operation, due to better combustion, resulting from increased air supply.

In Hazar (2009) the influence of low heat rejection engine on engine characteristics was investigated. The investigated fuel was CaBIO. Two versions

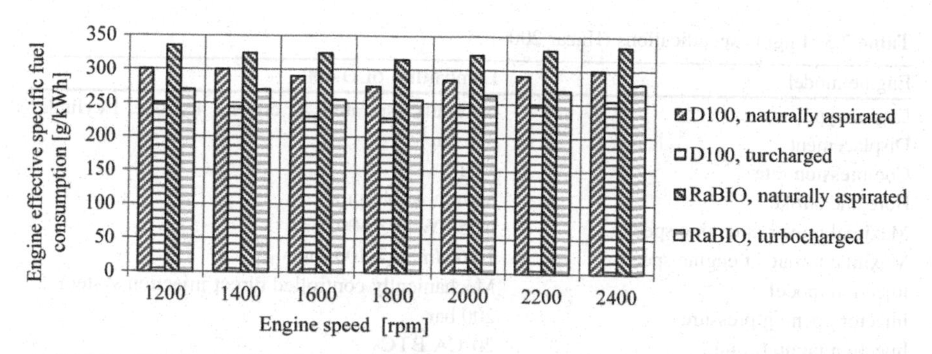

Fig. 7.17 Effective specific fuel consumption of the naturally aspirated and turbocharged engine

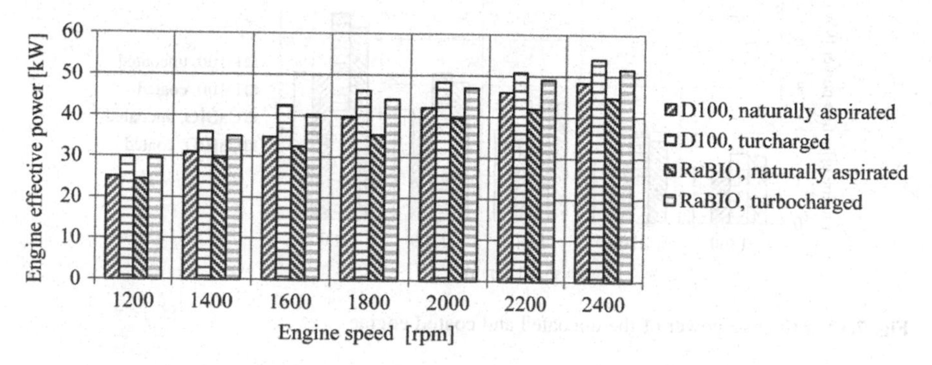

Fig. 7.18 Effective power of the naturally aspirated and turbocharged engine

of the tested engine (Table 7.5) were utilized. The first one was the regular, *uncoated*, version. The second, *coated*, version was modified as follows: the cylinder head, exhaust, and inlet valves were coated with the ceramic material $MgO–ZrO_2$ by the plasma spray method; the piston surface was coated with ZrO_2.

Increased engine power and decreased specific fuel consumption were observed for either CaBIO or D100 fuelled coated engine, compared to the uncoated version. The average increase of power in the coated engine was about 8.4 % for D100 and 3.5 % for CaBIO (Fig. 7.19). This increase in power value may be explained by the increase in the temperature of the combustion chamber elements due to a thermal barrier effect, caused by the ceramic coating. Higher temperatures improve the combustion.

In Haşimoğlu et al. (2008) the effects of *sunflower biodiesel* (SuBIO) usage on engine performance of an *uncoated* and *coated* engine (Table 7.6) were investigated. In the coated version the cylinder head and valves were coated with plasma-sprayed yttria-stabilized zirconia ($Y_2O_3ZrO_2$) with a thickness of 0.35 mm over a 0.15 mm thickness of NiCrAl bond coat.

Table 7.5 Engine specifications (Hazar 2009)

Engine model	Lombardini 6LD 400
Engine type	Naturally aspirated, air cooled, 4 stroke, 1 cylinder
Displacement	395 cm^3
Compression rate	18:1
Bore and stroke	86 mm × 68 mm
Maximal power at engine speed	6.25 kW at 3,600 rpm
Maximal torque at engine speed	16.6 Nm at 2,400 rpm
Injection model	Mechanically controlled direct injection system
Injector opening pressure	200 bar
Injection pump timing	20°CA BTC

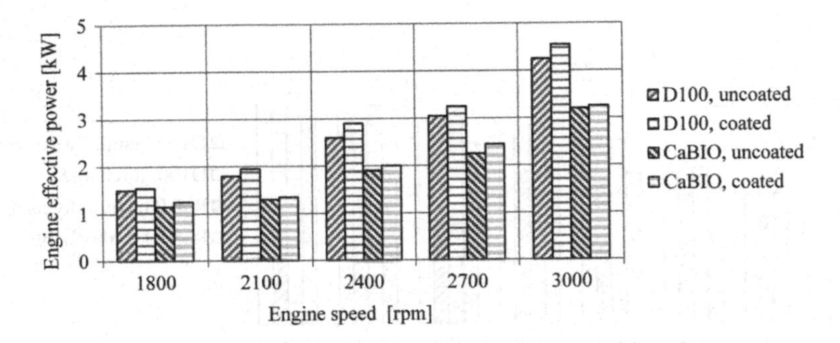

Fig. 7.19 Effective power of the uncoated and coated engine

Table 7.6 Engine specifications (Haşimoğlu et al. 2008)

Engine model	Mercedes-Benz/OM364A
Engine type	Turbocharged diesel engine, four cylinders
Displacement	3,972 cm^3
Compression rate	17.25:1
Bore and stroke	97.5 mm × 133 mm
Maximal power at engine speed	66 kW at 2,800 rpm
Maximal torque at engine speed	266 Nm at 1,400 rpm
Injection model	Mechanically controlled direct injection system
Injector opening pressure	200 bar
Injection pump timing	18°CA BTC

The variations of engine torque in dependence on engine speed are shown in Fig. 7.20. By using D100, the coated engine torque was higher by about 7 % at all engine speeds. By using SuBIO, however, the coated engine torque at low and at high engine speeds was up to 10 % lower than that of the uncoated engine. At medium engine speeds, the coated engine torque was higher up to 6 %. The reason for increased torque of the coated engine, compared to the uncoated version, was the increase of exhaust gas energy. This leads to improved volumetric efficiency, because of the increased turbocharger outlet pressure. Furthermore, it was observed

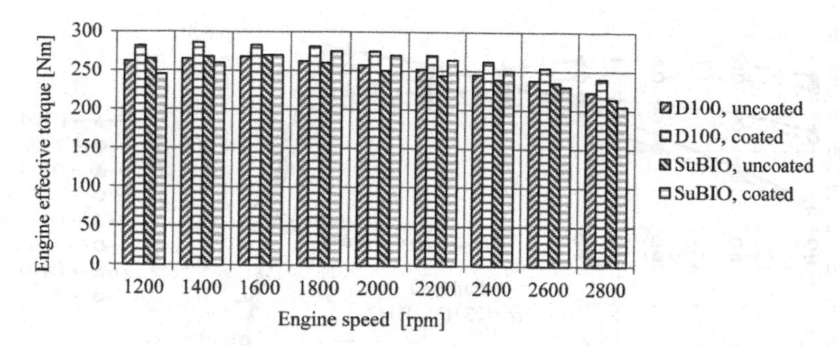

Fig. 7.20 Effective torque of the uncoated and coated engine

that the specific fuel consumption of the coated engine, when using SuBIO, was lower by approximately 4 %, compared to the uncoated version.

7.1.3 Influence of Engine Operating Regime

The influence of biodiesel fuels on engine performance may depend to some extent on the engine operating regime (Kegl 2008, 2011; Meng et al. 2008; Raheman and Phadatare 2004; Ramadhas et al. 2005; Qi et al. 2010; Zhu et al. 2010; Xue et al. 2011):

- *Engine load*: with increased load, the effective specific fuel consumption of biodiesel decreases since the brake power increases faster than the fuel consumption; full engine load delivers the maximal difference in engine power between biodiesel and mineral diesel; by reducing the engine load, the power delivered by biodiesel becomes similar to the one delivered by mineral diesel; engine load has a variable (non-monotonic) effect on exhaust gas temperature.
- *Engine speed*: the basic trends of engine power and specific fuel consumption in dependence on engine speed are similar for biodiesel and diesel, but there is of course some offset between the biodiesel and diesel curves.

In Kegl (2006) the influence of *engine speeds* on engine parameters was investigated on a bus engine (Table 7.1). The engine was run at partial (PL) and full (FL) loads and the injection pump timing used was the one that is prescribed for D100. The engine speeds were varied from 1,000 up to 2,500 rpm. As one can see from Fig. 7.21, the effective torque M_e and power P_e decreased by about 5 % when D100 was replaced by RaBIO. Furthermore, the effective specific fuel consumption g_e (for the actual fuel mass) increased by about 10–15 % in the whole engine speed range. On the other hand, the temperatures of exhaust gases $T_{g,e}$ declined by about 30 °C, which is probably mostly due to the lower calorific value of RaBIO.

The effective engine power varied practically negligible at lower loads at practically all engine speeds (Fig. 7.22).

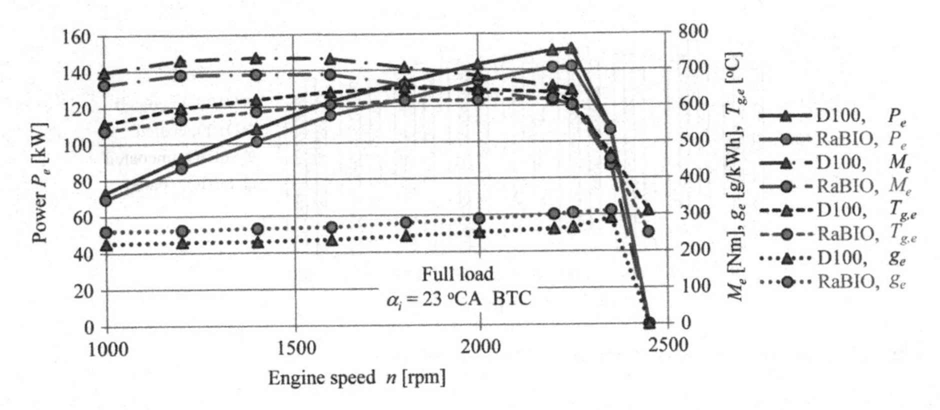

Fig. 7.21 Engine parameters at full load

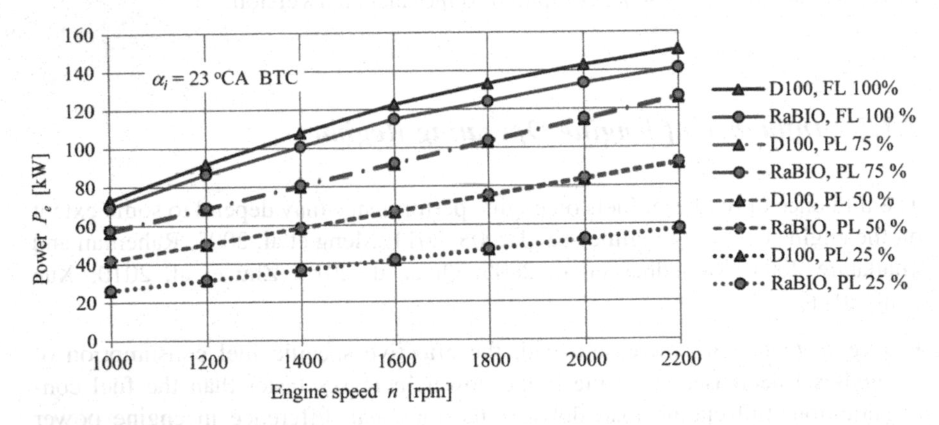

Fig. 7.22 Engine power at various engine regimes

The comparison of temperature $T_{g,e}$ of exhaust gases at FL and PL shows that $T_{g,e}$ at 25 %, 50 %, and 75 % load is higher for RaBIO than for D100 at almost all engine speeds (Fig. 7.23). The difference, however, became fairly small at 50 % partial load.

A comparison of specific fuel consumption g_e at FL and PL at various engine pump speeds shows that, compared to D100, g_e for RaBIO was higher by about 10–15 % at all engine speeds and loads (Fig. 7.24). The highest specific fuel consumptions were delivered by RaBIO at low loads and high engine speeds.

In Gumus and Kasifoglu (2010) a single-cylinder, four-stroke, air-cooled, direct injection diesel engine was tested by running it with *apricot seed kernel biodiesel*. It turned out that the effective specific fuel consumption initially decreased with increasing engine load until it reached a minimum value and then increased slightly with further increase of engine load.

In Usta et al. (2005) a four-cylinder, four-stroke, water-cooled, turbocharged, indirect injection diesel engine, Ford XLD 418 T, was tested with SuBIO and

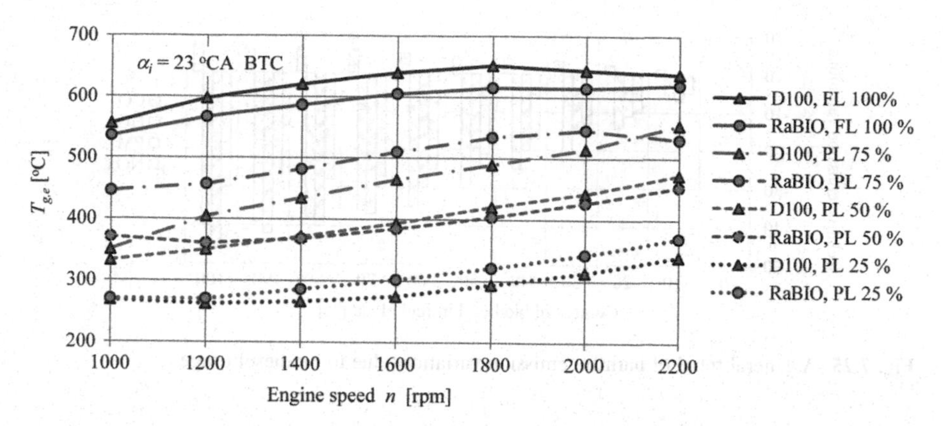

Fig. 7.23 Exhaust gas temperature at various engine regimes

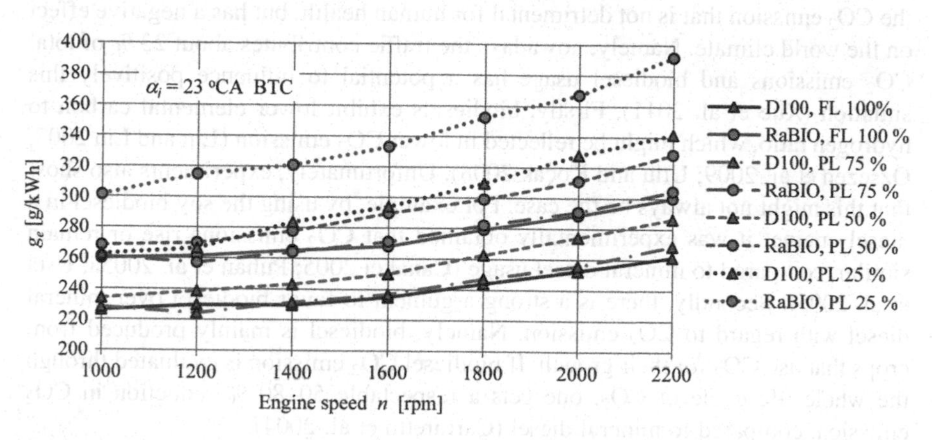

Fig. 7.24 Specific fuel consumption at various engine regimes

hazelnut biodiesel (HaBIO). The results show that a replacement of D100 by these biodiesels leads to an increase in effective specific fuel consumption. The increase at full load is higher than at partial load.

7.2 Engine Harmful Emissions

The most important diesel engine harmful emissions are NO_x, PM/smoke, CO, and unburned HC. A replacement of mineral diesel by biodiesel influences the quantities of these emissions (Xue et al. 2011). A general trend of this influence is illustrated in Fig. 7.25, showing the emission relative variations, with respect to mineral diesel (Giakoumis 2012). One can see that, in general, the situation is quite favorable for biodiesel, except for the NO_x emission.

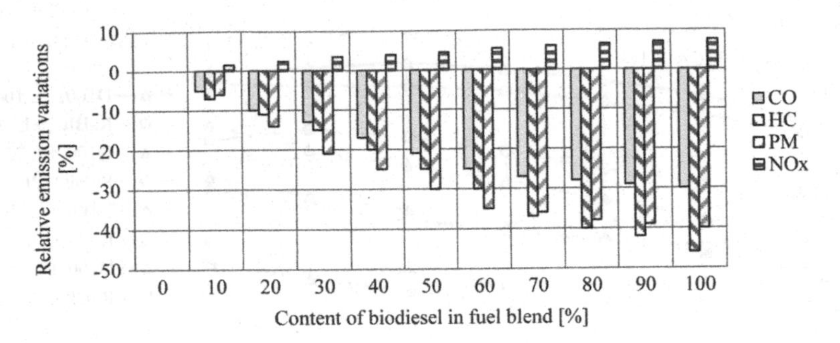

Fig. 7.25 A general trend of harmful emission variations due to biodiesel usage

Apart from the harmful emissions listed above, it may also be worth to mention the CO_2 emission that is not detrimental for human health, but has a negative effect on the world climate. Namely, nowadays the traffic contributes about 23 % of total CO_2 emissions and biodiesel usage has a potential to influence positively this situation (Xue et al. 2011). Firstly, biodiesels exhibit lower elemental carbon to hydrogen ratio, which might be reflected in lower CO_2 emission (Lin and Lin 2007; Ozsezen et al. 2009; Utlu and Koçak 2008). Unfortunately, experiments also show that this might not always be the case. For example, by using the soy biodiesel in a diesel engine, it was experimentally obtained that CO_2 emissions rise or remain similar, compared to mineral diesel usage (Canakci 2005; Puhan et al. 2005a; Usta et al. 2005). Secondly, there is a strong argument to favor biodiesel over mineral diesel with regard to CO_2 emission. Namely, biodiesel is mainly produced from crops that use CO_2 for their growth. If biodiesel CO_2 emission is evaluated through the whole life cycle of CO_2, one gets a respectable 50–80 % reduction in CO_2 emission, compared to mineral diesel (Carraretto et al. 2004).

7.2.1 Influence of Biodiesel Properties

All engine emissions depend significantly on biodiesel properties, which may vary significantly in dependence on biodiesel source (Aydin and Bayindir 2010; Buyukkaya 2010; Giakoumis et al. 2012; Kalligeros et al. 2003; Karabektas 2009; Karavalakis et al. 2009; Kegl 2008, Keskin et al. 2008; Kidoguchi et al. 2000; Koçak et al. 2007; Qi et al. 2009; Labeckas and Slavinskas 2006; Lapuerta et al. 2008b; Puhan et al. 2005b; Sayin and Gumus 2011; Usta 2005; Wu et al. 2009). The most influential properties, however, seem to be:

- *Sulfur content*: biodiesel fuels have practically zero sulfur content; note, however, that this advantage against mineral diesel gradually fades out owing to the increasingly better desulfurization of mineral diesel; lack of sulfur and aromatic compounds in biodiesels contributes to a reduction of PM emissions.

- *Oxygen content*: the relatively high oxygen concentration in biodiesel fuels, which improves the soot oxidation process, has been identified as the key contributor for the benefits related to PM emissions; high oxygen content also promotes complete combustion and thus leads to a reduction of CO emissions; since the formation of NO_x emission strongly depends on oxygen concentration and burned gas temperature, high oxygen content of biodiesel is the main reason for higher NO_x emissions; namely, due to its higher concentration, oxygen can react more easily with nitrogen during the combustion, thus causing an increase of NO_x emissions.
- *Density, bulk modulus, cetane number, iodine number*: the iodine number is closely related to density, bulk modulus, and cetane number and suggests that the observed increase in NO_x is also caused by the effects related to either injection or combustion timing; biodiesel has a higher cetane number than mineral diesel, which is reflected in a shorter ignition delay period and better autoignition capability; shorter ignition delay periods, associated with higher oxygen content of biodiesels, can also contribute significantly to lower CO emission.
- *Carbon* and *hydrogen content*: the carbon/hydrogen ratio of biodiesels is slightly lower than that of mineral diesel; this is reflected in lower CO emissions of biodiesel, compared to mineral diesel.

In Koçak et al. (2007) the effects of various biodiesel fuels on harmful emissions were tested on a Land Rover TDI 110 diesel engine (Table 7.7). The experiments were performed without engine modifications at full load and at various engine speeds. The tested fuels were CaBIO, HaBIO, *waste cooking biodiesel* (WcBIO), and D100.

In the range of engine speeds between maximal torque and maximal power, the increase of NO_x emission coincides with the increase of temperature and rise of volumetric efficiencies for all four tested fuels. Figure 7.26 shows the relative variations of NO_x emission (biodiesel, compared to mineral diesel). It can be seen that at some engine speeds the relative NO_x emission increased up to 2 %; meanwhile at some speeds it was lower up to 5 %. The reason for the decrease of relative NO_x emission for biodiesels is that the temperature at the end of combustion is somewhat lower, because the cetane numbers of biodiesel fuels are higher than those of mineral diesel (Koçak et al. 2007). Among all tested fuels, the lowest NO_x emissions were mostly obtained with CaBIO. CaBIO also delivered the lowest smoke emission at full load conditions.

In Cecrle et al. (2012) a Yammar L100V DI diesel engine with a mechanically controlled fuel injection system (Table 7.8) was used to test biodiesels at various engine loads at constant engine speed of 3,600 rpm. The tested fuels were RaBIO, PaBIO, SoBIO, WcBIO, CaBIO, *olive biodiesel* (OlBIO), and *coconut biodiesel* (CoBIO).

Figure 7.27 shows the relative NO_x emission variations for individual biodiesels with respect to D100. The results reveal that PaBIO consistently yielded the lowest

Table 7.7 Engine specifications (Koçak et al. 2007)

Engine model	Land Rover TDI 110
Engine type	Turbocharger, four stroke, four cylinders
Displacement	2,495 cm^3
Compression rate	19.5:1
Bore and stroke	90.5 mm × 97 mm
Maximal power at engine speed	82 kW at 3,850 rpm
Maximal torque at engine speed	235 Nm at 2,100 rpm
Injection model	Mechanically controlled direct injection system
Injector opening pressure	200 bar
Injection pump timing	15°CA BTC

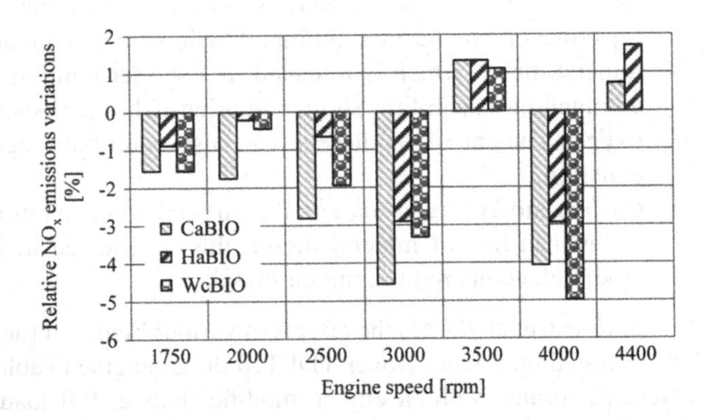

Fig. 7.26 Influence of biodiesel source on NO_x emissions at full load

Table 7.8 Engine specifications (Cecrle et al. 2012)

Engine model	Yanmar L100V
Engine type	Naturally aspirated, air cooled, four stroke, one cylinder
Displacement	435 cm^3
Compression rate	21.2:1
Bore and stroke	86 mm × 75 mm
Maximal power at engine speed	6.2 kW at 3,600 rpm
Maximal torque at engine speed	98.4 Nm at 1,440 rpm
Injection model	Mechanically controlled direct injection system
Injector opening pressure	196 bar
Injection pump timing	15.5°CA BTC

relative NO_x levels at all engine loads, while WcBIO mostly delivered the highest relative NO_x emission.

Figure 7.28 shows the relative particulate matter (PM) emission variations for various biodiesels, compared to mineral diesel. The results were obtained by running the Yanmar L100V DI diesel engine (Table 7.8) at 25 % load and at 3,600 rpm for 1 h (Cecrle et al. 2012). As expected, the PM emissions of all biodiesels were lower than those of D100. The total WcBIO emissions, which were the highest among all tested biodiesels, were approximately 20 % lower

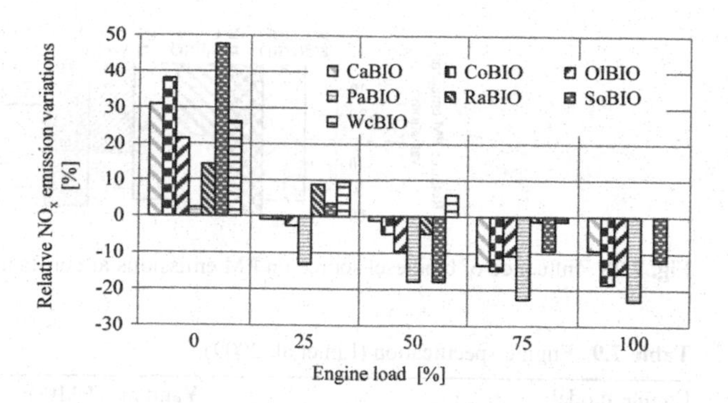

Fig. 7.27 Influence of biodiesel source on NO_x emissions at constant engine speed

than that of D100. The best result, obtained with CoBIO, produced 60 % lower emissions than D100.

In mechanically controlled injection systems, the start of injection of biodiesel fuels usually occurs earlier than that of mineral diesel due to higher density and viscosity and lower compressibility (Kegl 2006, 2007, 2008; Lapuerta et al. 2008b; Ozsezen et al. 2009).

In Banapurmath et al. (2008, 2009), Tsolakis et al. (2007) the effect of injection timing on engine performance and emissions was studied by using RaBIO and *honge biodiesel.* It turned out that the smoke emission generally increased when the injection timing was retarded. This is somewhat in contrast with the results obtained for mineral diesel. Namely, with D100 the smoke level decreases when the injection timing is advanced slightly but then increases when the injection timing is advanced further.

Biodiesel usage typically results in advance of combustion, as a result of a higher cetane number, and in advance of start of injection, due to the higher density and viscosity and lower compressibility. A high cetane number leads to a shorter ignition delay, which results in a smaller amount of heat, released at the initial combustion phase, and long combustion duration. This means that the injection timing may be somewhat retarded, which, in turn, may lead to increased PM but lower NO_x emission (Kidoguchi et al. 2000).

In Lin et al. (2009) SoBIO, SuBIO, RaBIO, PaBIO, WcBIO, and *corn biodiesel* (CrBIO) fuels were tested in a DI Yanmar TF110-F diesel engine (Table 7.9). PaBIO has the shortest carbon-chain lengths and the most saturated bonds and consequently exhibits superior ignition quality. The superior ignition quality and the higher oxygen content of this fuel allow for better combustion at lower temperatures. This is confirmed by the results, where significant reductions in exhaust gas temperature, smoke, and unburned HC emissions were observed. A lower combustion temperature also suppresses the formation of NO_x emissions. The values of the NO_x and HC emission, averaged from seven various load (12, 15, 20, 25, 30, 35, and 40 Nm) and three various engine speed (1,200, 1,800, and 2,400 rpm) regimes, are shown in Fig. 7.29.

In Wu et al. (2009) SoBIO, RaBIO, PaBIO, WcBIO, and *cottonseed biodiesel* (CtBIO) were tested in a Cummins Euro III diesel engine (Table 7.10) and compared with D100. The harmful emissions were measured at the same speed and various mean effective pressures (BMEP).

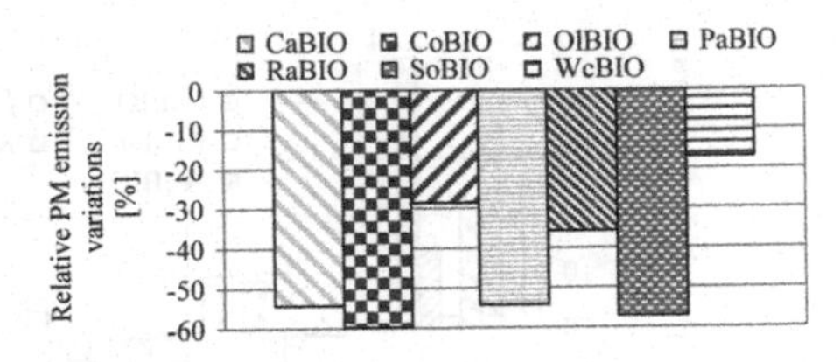

Fig. 7.28 Influence of biodiesel source on PM emissions at constant operating regime

Table 7.9 Engine specification (Lin et al. 2009)

Engine model	Yanmar TF110-F
Engine type	Water cooled, four stroke, single cylinder
Displacement	585 cm^3
Compression rate	17.9:1
Bore and stroke	88 mm × 96 mm
Maximal power at engine speed	8.1 kW at 3,600 rpm
Maximal torque at engine speed	16.7 Nm at 2,500 rpm
Injection model	Mechanically controlled direct injection system
Injector opening pressure	196 bar
Injection pump timing	17°CA BTC

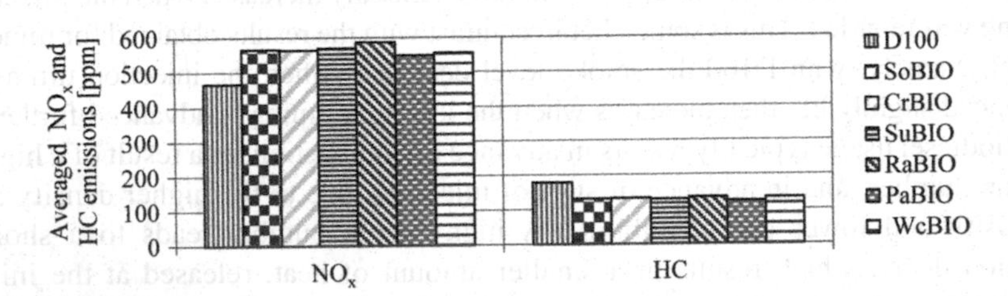

Fig. 7.29 Influence of biodiesel source on averaged NO_x and HC emissions

A comparison of NO_x emissions is shown in Fig. 7.30. All biodiesels exhibit higher NO_x emissions than D100, but the extent of the increase varies, ranging from 10 to 23 % with respect to D100. The biodiesels that result in the least NO_x, in descending order, are CtBIO, PaBIO, SoBIO, WcBIO, and RaBIO. Although PaBIO and WcBIO have almost the same oxygen content, PaBIO produces less NO_x than WcBIO. The most likely reason is the high cetane number of PaBIO. A higher cetane number reduces the ignition delay and the fuel consumed in the premixed phase and may therefore result in lower in-cylinder temperature.

A comparison of HC emissions is shown in Fig. 7.31. Compared with D100, the tested biodiesel usage reduced the average HC emission by 45–67 %. The biodiesels that reduce HC the most, in descending order, are PaBIO, WcBIO, RaBIO, CtBIO, and SoBIO. Theoretically, HC is mainly caused by misfire in a locally rich or locally lean region. The difference in HC emission of various biodiesels is likely to be a combined effect of oxygen content and cetane number.

Table 7.10 Engine specifications (Wu et al. 2009)

Engine model	Cummins ISBe6
Engine type	Turbocharged engine, 4 stroke, 6 cylinders
Displacement	5,900 cm^3
Compression rate	17.5:1
Bore and stroke	102 mm × 120 mm
Maximal power at engine speed	136 kW at 2,500 rpm
Maximal torque at engine speed	670 Nm at 1,500 rpm
Injection model	Electronically controlled direct injection system
Injection type	Common rail

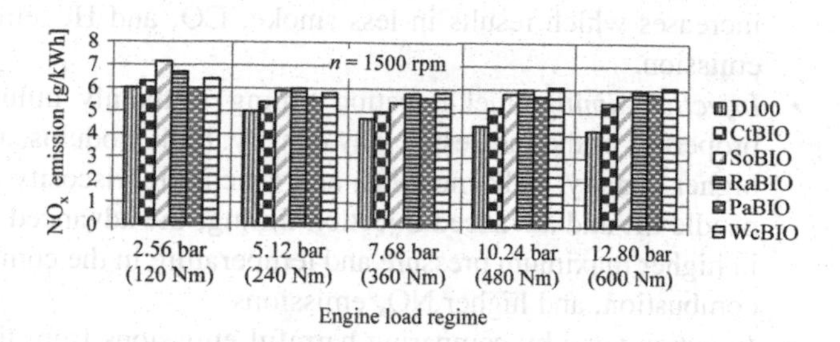

Fig. 7.30 Influence of biodiesel type on NO_x emission

From Fig. 7.31 one may also conclude that, as the cetane number increases, the HC emissions decrease consistently for biodiesel fuels. It was observed that an increase in chain length or saturation level of various biodiesels leads to a higher reduction of HC emissions. Finally, it may be worth noting that the lower HC emissions of all biofuels, compared to D100, are partly due to the oxygen presence in biodiesel molecules.

In Canakci and Van Gerpen (2003) a turbocharged DI diesel engine was tested. It was shown that there are no significant differences in HC emissions between the WCBIO and SoBIO. In Sahoo et al. (2009) a significant difference in HC emission reduction (compared to mineral diesel) was observed between the *jatropha* and *karanja biodiesels* (20.7 % and 20.6 %) and *polanga biodiesel* (6.8 %). The tests were performed on a water-cooled three-cylinder tractor diesel engine.

7.2.2 *Influence of Engine Type*

The influence of biodiesel usage on engine characteristics varies in dependence on engine type (Hazar 2009; Karabektas 2009; Kousoulidou et al. 2012, McCormick et al. 2005, Pandey et al. 2012; Sayin and Gumus 2011). The most influential parameters seem to be the following ones:

- *Compression ratio*: in general, for all compression ratios the emissions of HC, smoke, and CO, obtained with biodiesel fuels, are lower than those obtained with mineral diesel; by increasing the compression ratio, the temperature also

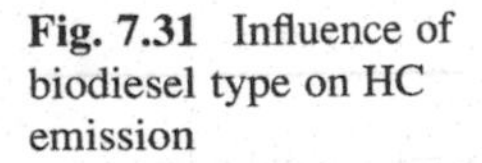
Fig. 7.31 Influence of biodiesel type on HC emission

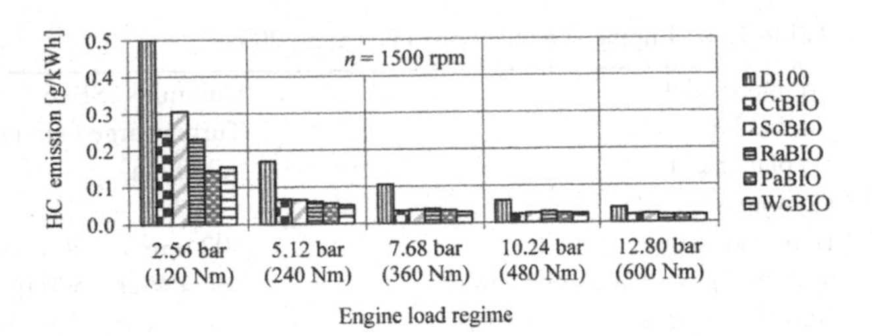

increases which results in less smoke, CO, and HC emissions, but more NO_x emission.

- *Injection timing*: fuel injection timing is mainly influenced by the physical properties of the fuel, especially density, bulk modulus, and kinematic viscosity; higher density, bulk modulus, and kinematic viscosity of a fuel lead to faster needle lift and advanced injection timing; the advanced injection timing results in higher maximum pressure and temperature in the combustion chamber, faster combustion, and higher NO_x emissions.
- *Injection type*: by comparing harmful emissions from the two different engine technologies, like common rail and unit injector injection system, one can observe that the common rail technology seems to be more compatible with biodiesels since it leads to lower NO_x and PM emissions, compared to the unit injector or to mechanically controlled fuel injection system.
- *Naturally aspirated and turbocharged operation*: a noticeable increase in the NO_x emissions can be observed in turbocharged operation for mineral diesel and biodiesel fuels when compared to the naturally aspirated operation; CO emission, on the other hand, decreases if naturally aspirated operation is replaced by turbocharged operation. In general, the performance and exhaust variations related to turbocharged operation are more significant for biodiesels than for mineral diesel.
- *Coated* and *uncoated engines*: in the coated engine the after-combustion temperature is higher as in the uncoated engine as a result of lower heat losses. This may influence positively CO and HC, but worsen NO_x emissions. This can be observed for biodiesel and mineral diesel.
- *EGR*: it seems that the use of EGR is more effective for biodiesel fuels than for mineral diesel; EGR can reduce NO_x efficiently and the accompanied smoke increase of biodiesels is lower than that observed for mineral diesel.

The experiments, performed on the *naturally aspirated DI diesel engine* (Table 7.3), show the influence of the compression ratio on harmful emissions, when using biodiesel from a commercial supplier (Sayin and Gumus 2011). By increasing the compression ratio from 17:1 to 19:1, the relative NO_x emission increased, while the relative smoke, unburned HC, and CO emissions decreased with respect to mineral diesel (Fig. 7.32). It is evident that with higher compression ratio, the difference between biodiesel and mineral diesel increases.

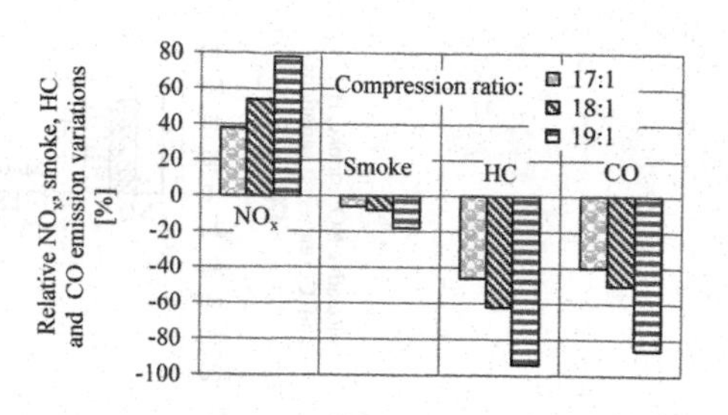

Fig. 7.32 Influence of compression ratio on relative NO_x, smoke, HC, and CO emissions

Increased compression ratio raises the density of air charge in the combustion chamber, which results in an increase of air entrainment in the spray. More air in the spray contributes to more complete combustion. From Fig. 7.32 it is evident that for all compression ratios the emissions of smoke, unburned HC, and CO of biodiesel are lower than those of mineral diesel. On the other hand, the increased compression ratio increases the in-cylinder temperature, which is one of the factors for higher NO_x emission.

In Sayin and Gumus (2011) the influence of injection timing on harmful emissions was also investigated experimentally on the engine, specified in Table 7.3. By increasing the injection timing from 15 to 25°CA BTC, the time available for carbon oxidation increases, which leads to higher in-cylinder temperatures during the expansion phase. Consequently, the relative smoke, unburned HC, and CO emissions were reduced while the relative NOx emission increased, compared to mineral diesel (Fig. 7.33). It is evident that with advanced injection timing the difference between the tested biodiesel and mineral diesel increases.

In Kegl (2006) RaBIO was tested in a *DI diesel bus* engine with *M injection system* (Table 7.1) in order to evaluate the influence of *pump injection timing* on engine emissions. For this purpose, RaBIO emissions, obtained at several pump injection timings, are compared to D100 emissions, obtained with the prescribed pump injection timing; Fig. 7.34 shows the corresponding relative emission variations. By taking into account all harmful emissions, one can say that for the tested conditions the optimal pump injection timing for RaBIO usage is $\alpha_i = 19$ °CA TDC.

It is known that the *advance* in *injection* and *combustion* for biodiesel fuels has an impact on NO_x emissions. In Carraretto et al. (2004) it was observed that NO_x emissions increase by advanced injection timing. In Tsolakis et al. (2007) the retarded injection timing resulted in reduced NO_x emissions and increased smoke, CO, and HC emissions. Furthermore, it was also found out that the variation of NO_x is a function of injection pressure and that there is a significant effect of injection pressure on NO_x emissions (Sharma et al. 2009).

In Karabektas (2009) a *naturally aspirated diesel engine* with or without installed *turbocharger system* (Table 7.4) was used to investigate the influence of RaBIO usage on CO and NO_x emissions at constant load and various *engine speeds*.

The variation of CO emission as a function of engine speed is shown in Fig. 7.35. During naturally aspirated operation, the CO emissions of RaBIO were on average about 17 % lower than those of D100. During the turbocharged operation the CO

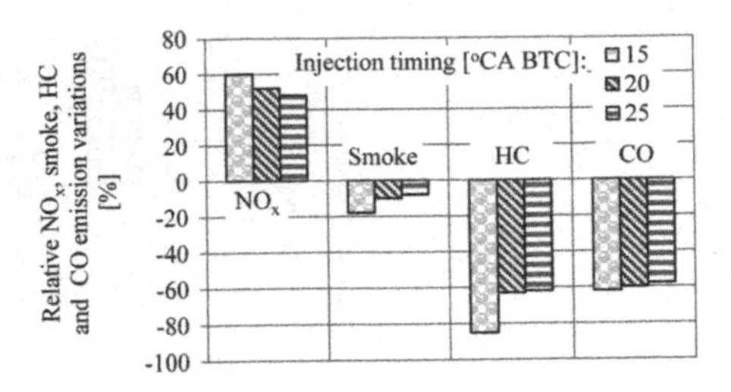

Fig. 7.33 Influence of injection timing on relative NO_x, smoke, HC, and CO emissions

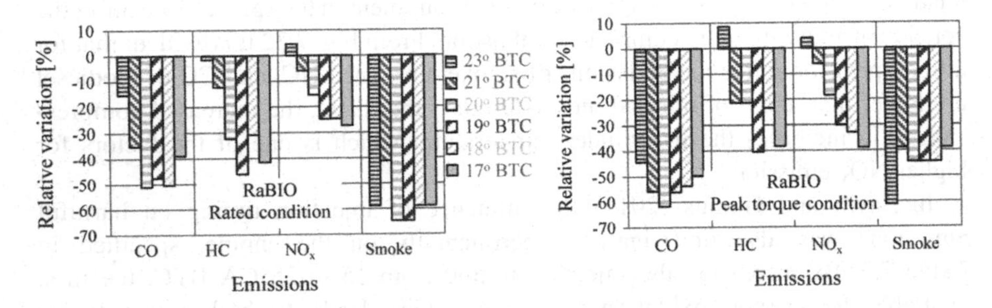

Fig. 7.34 Influence of pump injection timing on engine emissions

emissions of D100 and RaBIO were on average 47 and 52 % lower than those obtained during naturally aspirated operation. In turbocharged operation the CO emissions of RaBIO were on average 26 % lower than those of D100. The application of a turbocharger provides more air to the engine and better fuel–air mixing, thereby causing better combustion and lower CO emission values.

The variation of NO_x emission as a function of engine speed is shown in Fig. 7.36. The NO_x emissions of RaBIO were higher than those of D100 during both naturally aspirated and turbocharged operation. During naturally aspirated operation, an average NO_x emission increase of 10 % was obtained as RaBIO replaced D100. The application of a turbocharger provides more air to the engine and causes higher combustion temperatures, which yields an increase of NO_x emission. It was found out that under turbocharged operation, the NO_x emissions of D100 and RaBIO are higher on average by 27 and 21 %, respectively, compared to the naturally aspirated operation.

In Haşimoğlu et al. (2008) the influence of low heat rejection engine on engine emissions was investigated. By using SuBIO, the NO_x emission increased when the *low heat rejection engine* was used instead of the original turbocharged DI diesel engine, due to a higher combustion temperature. At the same time, the CO emission dropped when low heat rejection engine was used.

In Hazar (2009) the effects of switching between *coated* and *uncoated* version of the same engine (Table 7.5) on CO and NO_x emissions were investigated. The fuels tested were D100 and CaBIO.

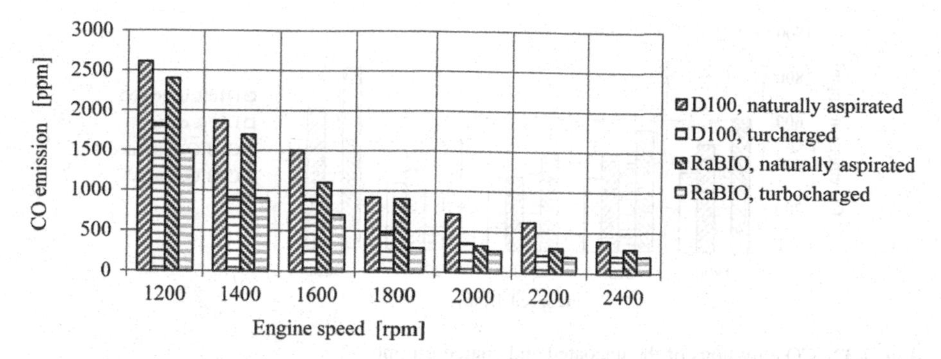

Fig. 7.35 CO emissions of the naturally aspirated and turbocharged engine

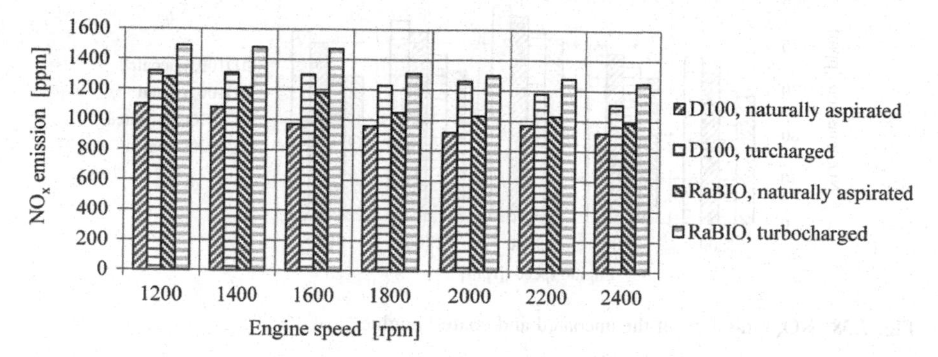

Fig. 7.36 NO_x emissions of the naturally aspirated and turbocharged engine

Figure 7.37 shows the variations of CO emissions in dependence on engine speed. By switching from the uncoated to the coated engine, the CO emission decreased by 25 % for CaBIO and by 22 % for D100.

Figure 7.38 shows the NO_x emissions in dependence on engine speed. In general, the NO_x emission initially increased from low to medium engine speeds and then decreased above medium engine speeds. The NO_x emission is low at low speeds because the combustion chamber temperature is also low. By increasing the engine speed, the temperature and consequently the NO_x emission also rise. Beyond medium speeds, however, the NO_x emission starts to decrease. This is because there is no sufficient time for the NO_x formation, despite the increase in temperature. The increases of NO_x emissions in the coated engine, compared to the uncoated version, are 4.9 % for D100 and 5.3 % for CaBIO. In general, the coated engine practically always yields higher NO_x emission than the uncoated version. This is a result of an increased after-combustion temperature due to the ceramic coating.

Another component that influences the engine emissions is *EGR*. In Agarwal et al. (2006) RaBIO was tested with EGR varying from 0 % up to 20 %. It was found out that although the smoke levels were generally low, the *smoke emission* increased by raising EGR. This is a consequence of reduced availability of oxygen, which results in relatively incomplete combustion and increased formation of *PM*.

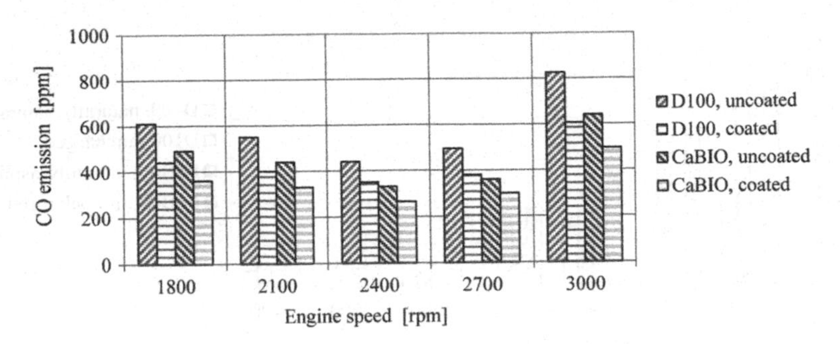

Fig. 7.37 CO emissions of the uncoated and coated engine

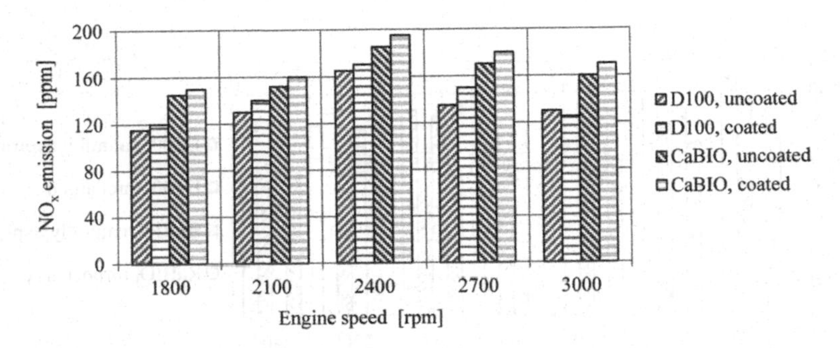

Fig. 7.38 NO_x emissions of the uncoated and coated engine

In Zheng et al. (2008) SoBIO and CaBIO fuels were tested under low temperature combustion in a single-cylinder, 4-stroke, naturally aspirated, DI diesel engine, equipped with EGR. Extensive EGR rates up to 65 % were applied in order to initiate low temperature combustion. From the results one can extract that the *smoke emission* initially increases by raising EGR, but begins to decline after a certain EGR ratio. The *NOx* emissions decreased practically monotonically with increasing EGR.

In Nabi et al. (2006) a blend of 85 % mineral diesel and 15 % *neem biodiesel* was tested in a 4-stroke naturally aspirated DI diesel engine. Compared to mineral diesel, the results showed no significant differences in engine characteristics. It is interesting to note that with neem biodiesel blend, NO_x emission was slightly lower than that of neat mineral diesel for any tested EGR rate.

In Tsolakis et al. (2007) the influence of *EGR* on engine emissions was investigated on a naturally aspirated diesel engine (Table 7.11) by using D100 and RaBIO. Figures 7.39, 7.40, 7.41, and 7.42 show the NO_x, CO, smoke, and HC emissions, obtained with D100 and RaBIO without or with 20 % EGR for two engine operating conditions. It is evident that EGR reduction of NO_x emission is larger for RaBIO than that for D100. Furthermore, EGR increases CO, smoke, and HC emissions. The differences obtained are larger for RaBIO, except for the HC emission where the effect of EGR is similar for D100 and RaBIO.

Table 7.11 Engine specifications (Tsolakis et al. 2007)

Engine model	Lister-Petter TR1 engine
Engine type	Naturally aspirated, air cooled, single cylinder
Displacement	773 cm^3
Compression rate	15.5:1
Bore and stroke	98.4 mm × 101.6 mm
Maximal power at engine speed	8.6 kW at 2,500 rpm
Maximal torque at engine speed	39.2 Nm at 1,800 rpm
Injection model	Mechanically controlled direct injection system
Injector opening pressure	180 bar
Injection pump timing	22°CA BTC

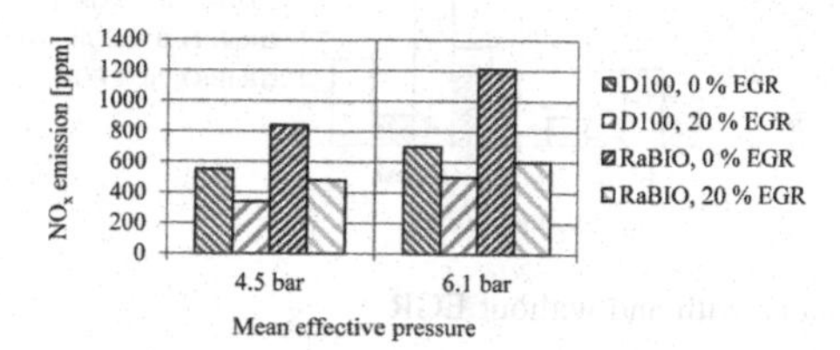

Fig. 7.39 NO_x emissions with and without EGR

7.2.3 Influence of Engine Operating Regime

The influence of biodiesel usage on engine harmful emission varies in dependence on engine operating regime (Baiju et al. 2009; Bhale et al. 2009; Buyukkaya 2010; Cheung et al. 2009; Gumus and Kasifoglu 2010; Kaplan et al. 2006; Kegl 2008, 2011; Koçak et al. 2007; Labeckas and Slavinskas 2006; Liu et al. 2009; Murillo et al. 2007; Nabi et al. 2009; Puhan et al. 2005; Sahoo et al. 2007; Song and Zhang 2008; Usta 2005):

- *Engine load:* engine load influences significantly NO_x emission; higher engine load results in higher combustion temperatures and increased NO_x formation.
- *Engine speed*: it is widely accepted that CO emission of biodiesel decreases with increasing engine speed; this is because increased engine speed results in better air–fuel mixing and/or increased fuel/air equivalence ratio (Keskin et al. 2008; Lin and Li 2009; Qi et al. 2009; Usta 2005); the oxidation converter might also influence significantly CO emission of biodiesels.

Apart from the engine load and speed, other operating conditions, such as the environment temperature, may influence significantly the effects of biodiesel usage on harmful emissions. These topics will also be addressed briefly in this section.

In Kegl (2006) a *DI diesel bus* MAN 2566 MUM engine with *M injection system* (Table 7.1) was tested in order to evaluate the influence of *engine operating conditions* on engine emissions. The pump timing was fixed to the one prescribed

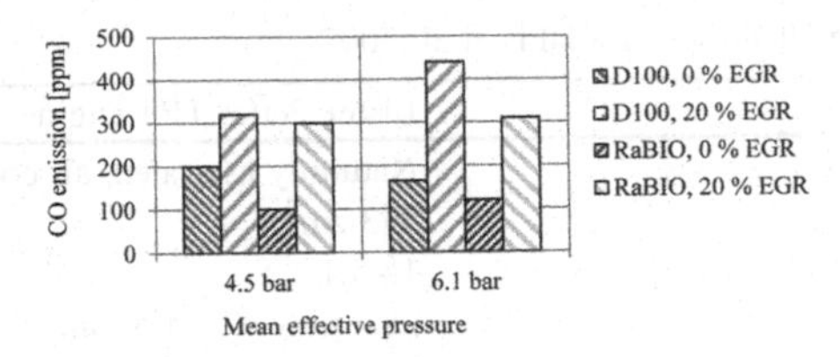

Fig. 7.40 CO emissions with and without EGR

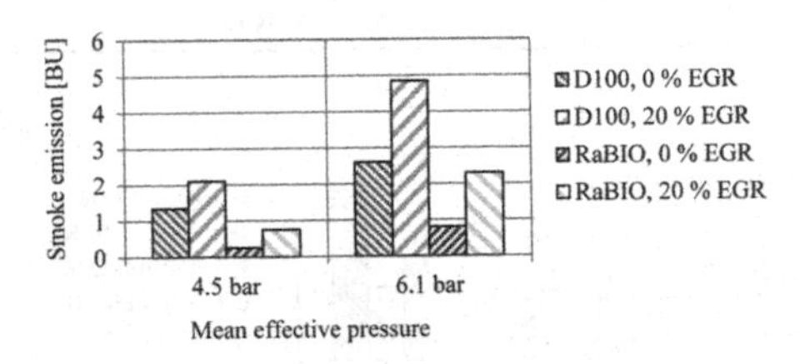

Fig. 7.41 Smoke emissions with and without EGR

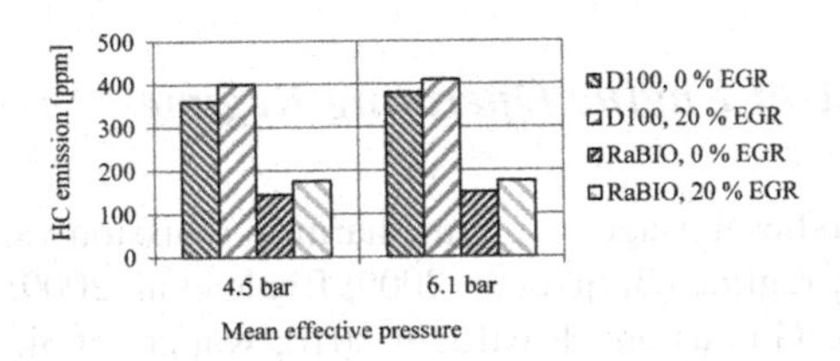

Fig. 7.42 HC emissions with and without EGR

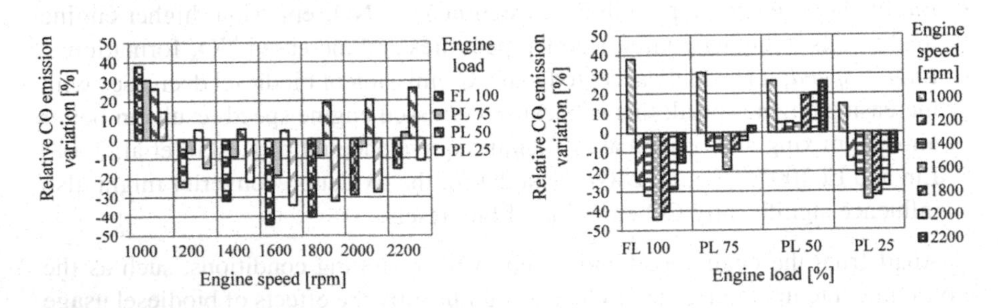

Fig. 7.43 Influence of engine operating conditions on relative CO emission

for mineral diesel. The tested fuels were D100 and RaBIO. The compared quantities are the CO, HC, NO_x, and smoke emissions.

Figure 7.43 presents the relative CO emission variations (RaBIO emission with respect to D100 emission). It is evident that RaBIO emission was higher than that of

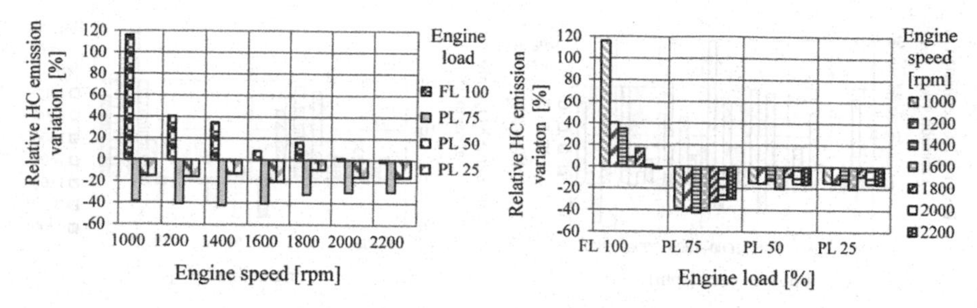

Fig. 7.44 Influence of engine operating conditions on relative HC emission

D100 at engine speed of 1,000 min^{-1} and at partial load PL 50 only. In all other tested conditions RaBIO usage resulted in lower CO emission.

Figure 7.44 presents the relative HC emission variations (RaBIO emission with respect to D100 emission). As can be seen RaBIO delivered higher HC emission at full load at all tested engine speeds. At all other tested conditions, the HC emission obtained with RaBIO was lower than the one obtained with D100.

Figure 7.45 presents the relative NO_x emission variations (RaBIO emission with respect to D100 emission). It is evident that RaBIO usage yielded higher NO_x emissions at practically all engine operating regimes. The only exception is the results obtained at middle engine speeds at partial load PL 50.

Figure 7.46 presents the relative smoke emission variations (RaBIO emission with respect to D100 emission). One can see that RaBIO usage yielded higher smoke emission at partial load PL 25 and low engine speeds only. At all other tested operation conditions, RaBIO usage resulted in lower smoke emission than D100.

In Koçak et al. (2007) TDI diesel engine (Table 7.7) was tested to evaluate the influence of *engine speed* on engine emissions, obtained at *full load* by using CaBIO, HaBIO, and WcBIO fuels.

The experimentally obtained emissions of NO_x and *CO* are shown in Fig. 7.47. For all tested biodiesels, the highest NO_x emissions were obtained within the speed range corresponding to maximum torque and maximum power. These higher NO_x emissions are related to in-cylinder temperatures and rise of volumetric efficiencies. At low engine speeds, the air–fuel mixing process is influenced by the difficulty in atomization of biodiesels due to their high viscosity. The resulting locally rich mixtures of biodiesels caused more CO to be produced during combustion. The turbocharged diesel engine used in the experiments provides more air at higher speeds, which results in an increase of turbulence intensity in the combustion chamber. This affects the air–fuel mixing process, which leads to a more complete combustion. Therefore, the CO emissions decreased at higher speeds (Fig. 7.47). The various fuel properties of tested biodiesels had only a minor effect on the CO emissions due to the dominant premixed lean combustion with a 10–11 % excess of oxygen.

The relative emissions of *smoke* and CO_2 (compared to mineral diesel) are shown in Fig. 7.48. Maximal smoke densities were measured in the range from 3,500 to 4,000 rpm. The relative values, however, mostly decrease as the engine

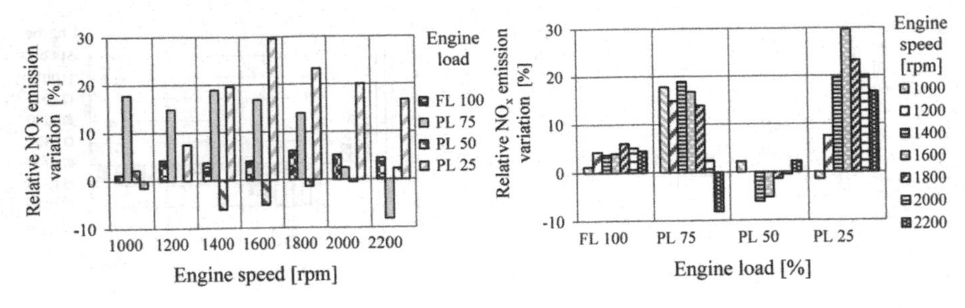

Fig. 7.45 Influence of engine operating conditions on relative NO_x emission

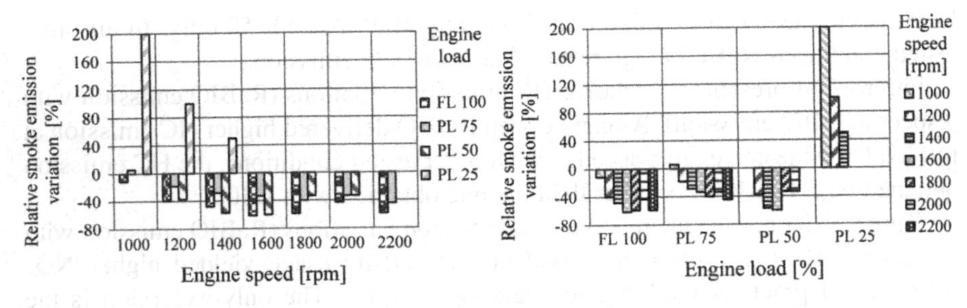

Fig. 7.46 Influence of engine operating conditions on relative smoke emission

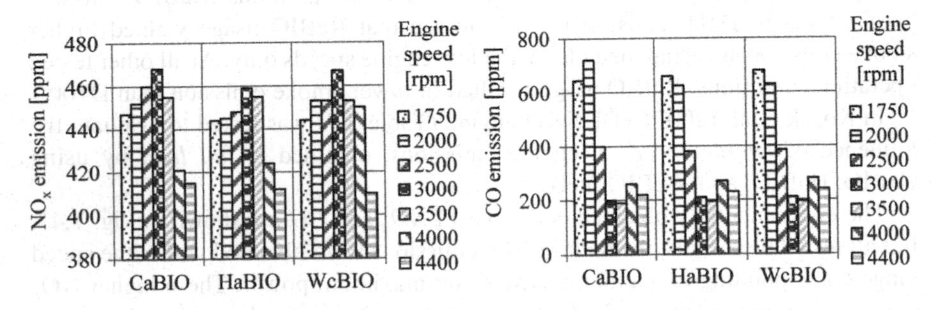

Fig. 7.47 Influence of engine operating conditions on NO_x and CO emissions

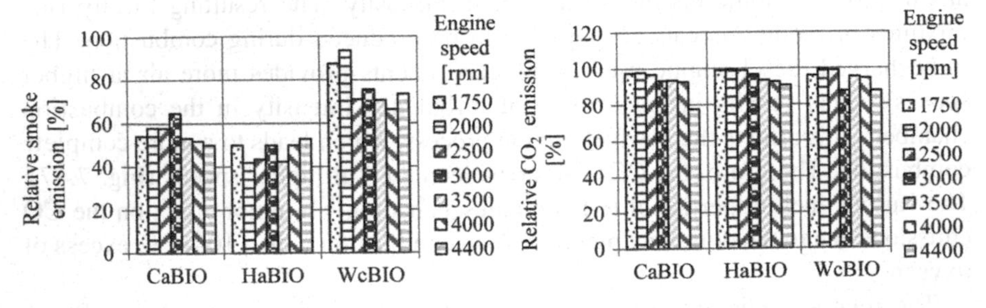

Fig. 7.48 Influence of engine operating conditions on relative smoke and CO_2 emissions

speed increases. This means that biodiesel performance, relative to mineral diesel, mostly improved as the speed increased. The only exception was the smoke emission of HaBIO.

The engine operating conditions, such as load, speed, and low temperature, play a significant role in *PM/smoke emission* production when using biodiesels. A PM emission increase, caused by an increase of engine load, was observed on various engines, for example, on a 2.5-L four-cylinder Peugeot XD3p157 engine fuelled by sunflower biodiesel (Kaplan et al. 2006), on a 6-cylinder MAN direct injection turbocharged engine fuelled with rapeseed biodiesel (Buyukkaya 2010), on a 4-stroke naturally aspirated direct injection diesel engine (Puhan et al. 2005), etc. The experiments on a high-speed single-cylinder, 4-stroke, WC Ricardo E6 engine fuelled by mahua biodiesel showed that the smoke level increased sharply with the increase of load due to the decreased air/fuel ratio; in this scenario larger quantities of fuel were injected into the combustion chamber, much of which went unburned into the exhaust (Raheman and Ghadge 2007).

The advantage of biodiesel fuels with respect to PM emissions is weakened or even reversed at *low temperature* tests. By testing SoBIO in a turbocharged direct injection diesel engine of a passenger car VW Golf 1.9 TDi, all emission levels tended to increase significantly over the cold start of the urban part of the new European driving cycle NEDC (Fontaras et al. 2009). The PM emission for biodiesel over the cold phase of the NEDC was approximately 40 times higher than at hot start of the cycle UDC. This is probably due to the fuel's higher kinematic viscosity and lower boiling point which make fuel atomization and evaporation more difficult under cold start conditions. Similar results were also obtained by testing WcBIO and SuBIO in a 4-cylinder, 4-stroke, turbocharged, intercooled, DI diesel engine (Armas et al. 2006).

In Leung et al. (2006) RaBIO was tested in a single-cylinder DI diesel engine with a mechanically controlled fuel injection system at various *load conditions*. It was observed that biodiesel-related decrease of *PM emissions* is higher at high load. SuBIO tests in a 2.5-L four-cylinder Peugeot XD3p157 engine (Kaplan et al. 2006) and in a 4-stroke, 4-cylinder, 55 kW DI diesel engine (Ulusoy et al. 2009) showed that with higher engine speeds the PM emission decreases. Probably, this is mainly because of the improved combustion efficiency that can be attributed to an increase in turbulence effects at higher engine speeds, which promotes complete combustion.

Tests performed on a 3-cylinder, 4-stroke, air-cooled, naturally aspirated DI diesel engine with *fish biodiesel* and on a Cummins 6BTA 5.9 G2-1, 158 HP-rated power, turbocharged, water-cooled DI diesel engine with *mahua biodiesel* revealed that the *NOx* concentration varies linearly with *engine load*. As the load increases, the overall fuel/air ratio also increases which results in an increase in the average in-cylinder temperature. Consequently, the NO_x formation also increases (Deshmukh and Bhuyar 2009; Godiganur et al. 2010; Raheman and Ghadge 2007; Zhu et al. 2010). On the other hand, measurements on a single-cylinder, 4-stroke, naturally aspirated DI diesel outboard engine, performed during the ISO C-3 test cycle, have shown that NO_x emission decreases as the load is increased

(Murillo et al. 2007). Presumably this could be due to the increase in turbulence inside the cylinder, which may contribute to a faster combustion and to lower residence time of reactants in the high temperature zones.

NO_x emissions are also affected by *engine speed*. By testing *fish biodiesel* in a 4-cylinder, 4-stroke, naturally aspirated DI diesel engine, the NO_x emissions decreased with increasing engine speeds (Lin and Li 2009). Probably, this is primarily due to shorter residence time available for NO_x formation, which may be the result of increased volumetric efficiency and flow velocity of the reactant mixture at higher engine speeds. In Utlu and Koçak (2011) waste cooking oil biodiesel was tested in the Land Rower TDI 110, 4-cylinder, turbocharged, inter-cooler DI diesel engine with 235 Nm torque at 2,100 rpm and 82 kW power at 3,850 rpm. It turned out that the NO_x emissions increase between maximum torque and maximum power speeds. In Keskin et al. (2008) the *tall oil biodiesel* was used at full load in a single-cylinder, 4-stroke, air-cooled, DI diesel engine Lombardini 6LD 400, which reaches the maximum power of 6.3 kW and maximum engine speed of 3,600 rpm. It turned out that the NO_x emission increased at low speeds and reached a maximum value at medium speeds. After that it decreased with further increase in engine speed.

Regarding the effects of *engine load* on *HC emissions* of biodiesel, the results are somewhat inconsistent. Some experiments indicate that HC emissions tend to increase with the increased load (Agarwal et al. 2006; Gumus and Kasifoglu 2010).

Increasing the engine load also seems to increase the *CO emission* (Agarwal et al. 2006; Gumus and Kasifoglu 2010; Ulusoy et al. 2009). The main reason for this increase is that the air–fuel ratio decreases with increased load, which is typical for diesel engines. Sometimes, however, the effect of engine load on CO emission is the opposite. In Sharma et al. (2009) and Wu et al. (2009) it was found out that in the low or intermediate load range, the CO emission may also decrease as the load increases. By testing RaBIO in 4-stroke, 4-cylinder, water-cooled, naturally aspirated DI diesel engine, the CO emission was lower at intermediate loads, but became higher in low load, heavy load, and full load (Labeckas and Slavinskas 2006; Mahanta et al. 2006).

7.3 Discussion

The determination of the effects of biodiesel usage on engine performance, economy, tribology, and harmful emissions is a complex topic. Although the use of numerical simulations increases, most of the work still has to be done by experiments. Because there are so many influential factors, experimental work is typically done under circumstances that vary significantly from one experiment to the other. The reasons for this situation are different test engines, different operating conditions or driving cycles, different biodiesels or reference mineral diesel, different measurement techniques, and so on, just to mention the most obvious ones. Consequently, the results obtained often do not show uniform trends of the influence of a particular parameter. In spite of that, some general comments can be given as follows.

- The use of biodiesel leads to the reduced *engine power* due to the lower energy content of biodiesels, compared to mineral diesel. However, high viscosity and high lubricity of biodiesels also have certain effects on engine power. For example, in some investigations the effective engine torque and power drop were less than expected because of biodiesel's higher viscosity, which enhanced fuel spray penetration and improved air–fuel mixing. On the other hand, in some investigations the higher viscosity influenced negatively the effective power, because it decreased combustion efficiency due to bad fuel injection atomization. Biodiesels typically exhibit better lubricity than mineral diesel. This results in reduced friction losses and increased effective power.
- The *effective specific fuel consumption* typically increases when biodiesel replaces mineral diesel. This increase is mainly due to its low heating value, as well as its high density and high viscosity. Various biodiesel raw materials, different production processes, and variable quality also have an impact on engine economy. In any case, the situation may be improved by using a turbocharged or a low heat release engine. Engine operating conditions, such as load and speed, also influence biodiesel engine economy, although this influence is not essential. Additives, used to improve the properties of biodiesel, may potentially improve the combustion performance and promote both engine economy and power.
- It is commonly accepted that *CO emission* decreases when biodiesel replaces mineral diesel. This is due to higher oxygen content and lower carbon to hydrogen ratio of biodiesel, compared to mineral diesel. The CO emissions of biodiesel are affected by its source and other factors such as the cetane number and advance in combustion. Engine load has also been proven to have a significant impact on CO emissions. There is a largely unanimous conclusion that CO emissions of biodiesel decrease with increasing engine speed.
- The *HC emissions* also typically decrease when biodiesel is used instead of mineral diesel. Partially, this depends on biodiesel source and its properties, especially on the chain length or saturation level of biodiesels. For mechanically controlled injection systems, the advance in injection and combustion of biodiesels also favors lower HC emissions. There are, however, inconsistent conclusions related to the effects of engine load on HC emissions.
- Because of the relatively high oxygen content and because of other physical and chemical properties of biodiesels, which lead to higher injection pressure, the *smoke emission* of biodiesel is typically lower than that of mineral diesel. Smoke emission depends on both soot formation and oxidation, which are highly dependent on local burning temperatures and local air/fuel ratios. These seem to be better when biodiesel replaces mineral diesel.
- In general, *PM emission* decreases significantly when biodiesel replaces mineral diesel. Mostly, this reduction is proportional to biodiesel content in a blend with mineral diesel. It should be noted, however, that unexpected variations of PM

emission may appear in case of certain contents of biodiesel in a blend. In any case, it seems that the most important factor, influencing the PM emission, is the higher oxygen content of biodiesel. In general, PM emissions of biodiesel increase with engine load and decrease with engine speed. The use of EGR might deteriorate PM emissions of biodiesel, although the measured levels are still much lower than those obtained with mineral diesel. It should be noted, however, that PM emissions of biodiesel might increase abnormally in the case of low temperature conditions.

- Experimental results mostly confirm that *NO_x emissions* will increase when biodiesel replaces mineral diesel. This increase is partly due to relatively high oxygen content in biodiesel. Furthermore, the higher cetane number of biodiesel, higher content of unsaturated compounds, and variations of injection characteristics also have a notable impact on NO_x emissions. The use of EGR may reduce the NO_x emissions of biodiesel, but the EGR rates must not be the same as for mineral diesel; they have to be optimized for biodiesel usage. By increasing the engine load, the level of biodiesel NO_x emissions will also typically rise. It is interesting to note that by taking into account the calculated maximum heat release rate and maximum firing temperature, the NO_x emission of biodiesel should be lower than actually measured in some experiments. Some combustion analyses indicate that for NO_x formation the maximum in-cylinder temperature and the maximum heat release rate may not be as important as the advance in the start of injection timing, caused by the biodiesel's higher bulk modulus of compressibility in a mechanically controlled injection system. The advanced injection and combustion process lead to higher in-cylinder temperature at the beginning of the combustion process. Earlier peaks prolong the period with conditions favorable for NO_x formation and consequently the total NO_x emissions.
- The measurements of *CO_2 emission* of biodiesel reveal rather inconsistent results. For example, some researches indicated that the CO_2 emission decreases when biodiesel replaces mineral diesel as a result of the low carbon to hydrocarbon ratio. On the other hand, some other researches showed that the CO_2 emission increases or stays similar because of the more effective combustion of biodiesel. Anyhow, one can agree that the CO_2 emission of biodiesel is significantly lower than that of mineral diesel, if one takes into account the whole CO_2 life cycle.
- From the quite limited literature one can conclude that, compared to mineral diesel, biodiesel mostly positively affects *carbon deposits* and *wear* of the key engine parts. At least partially, this can be attributed to good *lubricity* and *solvent action* of biodiesel. However, further studies along with biodiesel engine endurance tests would probably be necessary to clarify fully all the reasons and mechanism of wear.

On the basis of experimental and numerical investigations done so far, it should be clear that significant improvements of biodiesel engine characteristics are possible. To some extent, these improvements can be achieved by performing

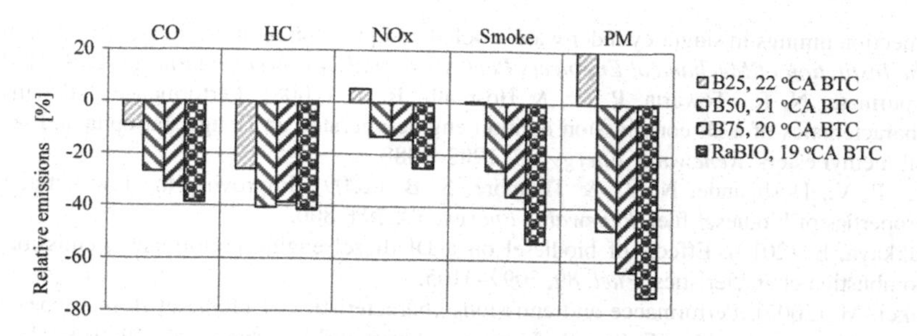

Fig. 7.49 Relative biodiesel and its blends emissions at various pump injection timings

rather inexpensive modifications on existing diesel engines. For example, for an engine equipped with a mechanically controlled fuel injection system, notable improvements can be obtained by setting optimally the pump injection timing.

As an example, it may be worth to take a look at the harmful emissions of a bus MAN diesel engine (Table 7.1) investigated at various pump injection timings (Kegl 2008). According to the ESC test, the individual characteristic points were weighted by corresponding factors that take into account the importance of individual engine regimes. RaBIO and its blends with D100 were tested at various pump injection timings and compared to D100 with the standard pump injection timing. The results clearly showed that it is possible to obtain lower CO, HC, NO_x, smoke, and PM emissions, if the injection timing is set appropriately (Fig. 7.49).

Figure 7.49 illustrates the importance of finding proper pump injection timing for each individual fuel in order to reduce harmful emissions while keeping engine performance and economy at desirable levels. Of course, the pump injection timing is only one of the many parameters which can influence the engine characteristics. In the search for the optimal values of various engine design and control parameters, a lot of experimental and numerical work needs to be done. But besides this, it may also be worth to engage systematic optimization procedures and techniques.

References

Agarwal, D., Sinha, S., & Agarwal, A. K. (2006). Experimental investigation of control of NO_x emissions in biodiesel-fueled compression ignition engine. *Renewable Energy, 31*, 2356–2369.

Armas, O., Hernández, J. J., & Cárdenas, M. D. (2006). Reduction of diesel smoke opacity from vegetable oil methyl esters during transient operation. *Fuel, 85*, 2427–2438.

Aydin, H., & Bayindir, H. (2010). Performance and emission analysis of cottonseed oil methyl ester in a diesel engine. *Renewable Energy, 35*, 588–592.

Baiju, B., Naik, M. K., & Das, L. M. (2009). A comparative evaluation of compression ignition engine characteristics using methyl and ethyl esters of Karanja oil. *Renewable Energy, 34*, 1616–1621.

Banapurmath, N. R., Tewari, P. G., & Hosmath, R. S. (2009). Effect of biodiesel derived from Honge oil and its blends with diesel when directly injected at different injection pressures and injection timings in single-cylinder water-cooled compression ignition engine. *Proceedings of the Institution of Mechanical Engineers Part A, Journal of Power and Energy, 223*, 31–40.

Banapurmath, N. R., Tewaria, P. G., & Hosmath, R. S. (2008). Performance and emission characteristics of a DI compression ignition engine operated on Honge, Jatropha and sesame oil methyl esters. *Renewable Energy, 33*, 1982–1988.

Bhale, P. V., Deshpande, N. V., & Thombre, S. B. (2009). Improving the low temperature properties of biodiesel fuel. *Renewable Energy, 34*, 794–800.

Buyukkaya, E. (2010). Effects of biodiesel on a DI diesel engine performance, emission and combustion characteristics. *Fuel, 89*, 3099–3105.

Canakci, M. (2005). Performance and emissions characteristics of biodiesel from soybean Oil. *Proceedings of the IMechE, Part D: Journal of Automobile Engineering, 219*, 915–922.

Canakci, M. (2007). Combustion characteristics of a turbocharged DI compression ignition engine fueled with petroleum diesel fuels and biodiesel. *Bioresource Technology, 98*, 1167–1175.

Canakci, M., & Van Gerpen, J. H. (2003). Comparison of engine performance and emissions for petroleum diesel fuel, yellow grease biodiesel, and soybean oil biodiesel. *Transactions of the American Society of Agricultural Engineers, 46*, 937–944.

Carraretto, C., Macor, A., Mirandola, A., Stoppato, A., & Tonon, S. (2004). Biodiesel as alternative fuel: experimental analysis and energetic evaluations. *Energy, 29*, 2195–2211.

Cecrle, E., Depcik, C., Duncan, E., Guo, J., Mangus, M., Peltier, E., Stagg-Williams, S., & Zhong, Y. (2012). Investigation of the effects of biodiesel feedstock on the performance and emissions of a single-cylinder diesel engine. *Energy & Fuels, 26*, 2331–2341.

Cheung, C. S., Zhu, L., & Huang, Z. (2009). Regulated and unregulated emissions from a diesel engine fueled with biodiesel and biodiesel blended with methanol. *Atmospheric Environment, 43*, 4865–4872.

Choi, S.-H., & Oh, Y. (2006). The emission effects by the use of biodiesel fuel. *International Journal of Modern Physics B, 20*, 4481–4486.

Demirbas, A. (2006). Biodiesel production via non-catalytic SCF method and biodiesel fuel characteristics. *Energy Conversion & Management, 47*, 2271–2282.

Demirbas, A. (2009). Progress and recent trends in biodiesel fuels. *Energy Conversion and Management, 50*, 14–34.

Deshmukh, S. J., & Bhuyar, L. B. (2009). Transesterified Hingan (Balanites) oil as a fuel for compression ignition engines. *Biomass & Bioenergy, 33*, 108–112.

Dorado, M. P., Ballesteros, E., Arnal, J. M., Gomez, J., & Lopez, F. J. (2003). Exhaust emissions from a diesel engine fueled with transesterified waste olive oil. *Fuel, 82*, 1311–1315.

Fontaras, G., Karavalakis, G., Kousoulidou, M., Tzamkiozis, T., Ntziachristos, L., Bakeas, E., et al. (2009). Effects of biodiesel on passenger car fuel consumption, regulated and non-regulated pollutant emissions over legislated and real-world driving cycles. *Fuel, 88*, 1608–1617.

Giakoumis, E. G. (2012). A statistical investigation of biodiesel effects on regulated exhaust emissions during transient cycles. *Applied Energy, 98*, 273–291.

Giakoumis, E. G., Rakopoulos, C. D., Dimaratos, A. M., & Rakopoulos, D. C. (2012). Exhaust emissions of diesel engines operating under transient conditions with biodiesel fuel blends. *Progress in Energy and Combustion Science, 38*, 691–715.

Giannelos, P. N., Sxizas, S., Lois, E., Zannikos, F., & Anastopoulos, G. (2005). Physical, chemical and fuel related properties of tomato seed oil for evaluating its direct use in diesel engines. *Industrial Crops and Products, 22*, 193–199.

Godiganur, S., Murthy, C. H. S., & Reddy, R. P. (2009). 6BTA 5.9 G2-1 Cummins engine performance and emission tests using methyl ester mahua (Madhuca indica) oil/diesel blends. *Renewable Energy, 34*, 2172–2177.

Godiganur, S., Murthy, C. H. S., & Reddy, R. P. (2010). Performance and emission characteristics of a Kirloskar HA394 diesel engine operated on fish oil methyl esters. *Renewable Energy, 35*, 355–359.

Gumus, M., & Kasifoglu, S. (2010). Performance and emission evaluation of a compression ignition engine using a biodiesel (apricot seed kernel oil methyl ester) and its blends with diesel fuel. *Biomass & Bioenergy, 34*, 134–139.

Haşimoğlu, C., Ciniviz, M., Özsert, İ., İçingürm, Y., Parlak, A., & Salman, M. S. (2008). Performance characteristics of a low heat rejection diesel engine operating with biodiesel. *Renewable Energy, 33*, 1709–1715.

Hazar, H. (2009). Effects of biodiesel on a low heat loss diesel engine. *Renewable Energy, 34*, 1533–1537.

Kalligeros, S., Zannikos, F., Stournas, S., Lois, E., Anastopoulos, G., Teas, C., et al. (2003). An investigation of using biodiesel/marine diesel blends on the performance of a stationary diesel engine. *Biomass & Bioenergy, 24*, 141–149.

Kaplan, C., Arslan, R., & Sürmen, A. (2006). Performance characteristics of sunflower methyl esters as biodiesel. *Energy Sources Part A, 28*, 751–755.

Karabektas, M. (2009). The effects of turbocharger on the performance and exhaust emissions of a diesel engine fuelled with biodiesel. *Renewable Energy, 34*, 989–993.

Karavalakis, G., Stournas, S., & Bakeas, E. (2009). Light vehicle regulated and unregulated emissions from different biodiesels. *Science of The Total Environment, 407*, 3338–3346.

Kegl, B. (2006). Experimental investigation of optimal timing of the diesel engine injection pump using biodiesel fuel. *Energy & Fuels, 20*, 1460–1470.

Kegl, B. (2007). NO_x and particulate matter (PM) emissions reduction potential by biodiesel usage. *Energy & Fuels, 21*, 3310–3316.

Kegl, B. (2008). Effects of biodiesel on emissions of a bus diesel engine. *Bioresource Technology, 99*, 863–873.

Kegl, B. (2011). Influence of biodiesel on engine combustion and emission characteristics. *Applied Energy, 88*, 1803–1812.

Keskin, A., Gürü, M., & Altıparmak, D. (2008). Influence of tall oil biodiesel with Mg and Mo based fuel additives on diesel engine performance and emission. *Bioresource Technology, 99*, 6434–6438.

Kidoguchi, Y., Yang, C., Kato, R., & Miwa, K. (2000). Effects of fuel cetane number and aromatics on combustion process and emissions of a direct injection diesel engine. *J SAE Review, 21*, 469–475.

Koçak, M. S., Ileri, E., & Utlu, Z. (2007). Experimental study of emission parameters of biodiesel fuels obtained from canola, hazelnut, and waste cooking oils. *Energy & Fuel, 21*, 3622–3626.

Kousoulidou, M., Ntziachristos, L., Fontaras, G., Martini, G., Dilara, P., & Samaras, Z. (2012). Impact of biodiesel application at various blending ratios on passenger cars of different fueling technologies. *Fuel, 98*, 88–94.

Labeckas, G., & Slavinskas, S. (2006). The effect of rapeseed oil methyl ester on direct injection diesel engine performance and exhaust emissions. *Energy Conversion and Management, 47*, 1954–1967.

Lapuerta, M., Armas, O., & Rodrígues-Fernández, J. (2008a). Effect of biodiesel fuels on diesel engine emissions. *Progress in Energy and Combustion Science, 34*, 198–223.

Lapuerta, M., Herreros, J. M., Lyons, L. L., García-Contreras, R., & Brice, Y. (2008b). Effect of the alcohol type used in the production of waste cooking oil biodiesel on diesel performance and emissions. *Fuel, 87*, 3161–3169.

Leung, D. Y. C., Luo, Y., & Chan, T. L. (2006). Optimization of exhaust emissions of a diesel engine fuelled with biodiesel. *Energy & Fuels, 20*, 1015–1023.

Lin, B. F., Huang, J. H., & Huang, D. Y. (2009). Experimental study of the effects of vegetable oil methyl ester on DI diesel engine performance characteristics and pollutant emissions. *Fuel, 88*, 1779–1785.

Lin, C. Y., & Li, R. J. (2009). Engine performance and emission characteristics of marine fish-oil biodiesel produced from the discarded parts of marine fish. *Fuel Processing Technology, 90*, 883–888.

Lin, C. Y., & Lin, H. A. (2007). Engine performance and emission characteristics of a three phase emulsion of biodiesel produced by peroxidation. *Fuel Processing Technology, 88*, 35–41.

Liu, Y. Y., Lin, T. C., Wang, Y. J., & Ho, W. L. (2009). Carbonyl compounds and toxicity assessments of emissions from a diesel engine running on biodiesels. *Journal of The Air & Waste Management Association, 59*, 163–171.

Mahanta, P., Mishra, S. C., & Kushwah, Y. S. (2006). An experimental study of Pongamia pinnata L. oil as a diesel substitute. *Proceedings of the Institution of Mechanical Engineers Part A-Journal of Power and Energy, 220*, 803–808.

McCormick, R.L., Tennant, C.J., Hayes, R.R., Black, S., Ireland, J., McDaniel, T., Williams, A., Frailey, M., Sharp, C.A. (2005). Regulated emission from biodiesel tested in heavy-duty engines. *SAE Paper*, 2005-01-2200.

Meng, X., Chen, G., & Wang, Y. (2008). Biodiesel production from waste cooking oil via alkali catalyst and its engine test. *Fuel Processing Technology, 89*, 851–857.

Murillo, S., Mıguez, J. L., Porteiro, J., Granada, E., & Moran, J. C. (2007). Performance and exhaust emissions in the use of biodiesel in outboard diesel engines. *Fuel, 86*, 1765–1771.

Nabi, M. N., Akhter, M. S., & Shahadat, M. M. Z. (2006). Improvement of engine emissions with conventional diesel fuel and diesel–biodiesel blends. *Bioresource Technology, 97*, 372–378.

Nabi, M. N., Najmul Hoque, S. M., & Akhter, M. S. (2009). Karanja (*Pongamia pinnata*) biodiesel production in Bangladesh, characterization of karanja biodiesel and its effect on diesel emissions. *Fuel Processing Technology, 90*, 1080–1086.

Oğuz, H., Öüt, H., & Eryilmaz, T. (2007). Investigation of biodiesel production, quality and performance in Turkey. *Energy Source Part A, 29*, 1529–35.

Ozsezen, A. N., Canakci, M., & Sayin, C. (2008). Effect of biodiesel from used frying palm oil on the performance, injection, and combustion characteristics of an indirect injection diesel engine. *Energy & Fuels, 22*, 1297–1305.

Ozsezen, A. N., Canakci, M., Turkcan, A., & Sayin, C. (2009). Performance and combustion characteristics of a DI diesel engine fueled with waste palm oil and canola oil methyl esters. *Fuel, 88*, 629–636.

Pandey, R. K., Rehman, A., & Sarviya, R. M. (2012). Impact of alternative fuel properties on fuel spray behavior and atomization. *Renewable and Sustainable Energy Reviews, 16*, 1762–1778.

Pehan, S., Svoljšak-Jerman, M., Kegl, M., & Kegl, B. (2009). Biodiesel influence on tribology characteristics of a diesel engines. *Fuel, 88*, 970–979.

Puhan, S., Vedaraman, N., Ram, B. V. B., Sankarnarayanan, G., & Jeychandran, K. (2005a). Mahua oil (Madhuca Indica seed oil) methyl ester as biodiesel-preparation and emission characterstics. *Biomass & Bioenergy, 28*, 87–93.

Puhan, S., Vedaraman, N., Sankaranarayanan, G., & Bharat Ram, B. V. (2005b). Performance and emission study of Mahua oil (Madhuca indica oil) ethyl ester in a 4-stroke natural aspirated direct injection diesel engine. *Renewable Energy, 30*, 1269–78.

Qi, D. H., Chen, H., Geng, L. M., & Bian, Y. Z. H. (2010). Experimental studies on the combustion characteristics and performance of a direct injection engine fueled with biodiesel/diesel blends. *Energy Conversion and Management, 51*, 2985–2992.

Qi, D. H., Geng, L. M., Chen, H., Bian, Y. Z. H., Liu, J., & Ren, X. C. H. (2009). Combustion and performance evaluation of a diesel engine fueled with biodiesel produced from soybean crude oil. *Renewable Energy, 34*, 2706–2713.

Raheman, H., & Ghadge, S. V. (2007). Performance of compression ignition engine with mahua (Madhuca indica) biodiesel. *Fuel, 86*, 2568–2573.

Raheman, H., & Phadatare, A. G. (2004). Diesel engine emissions and performance from blends of karanja methyl ester and diesel. *Biomass and Bioenergy, 27*, 393–397.

Ramadhas, A. S., Muraleedharan, C., & Jayaraj, S. (2005). Performance and emission evaluation of a diesel engine fueled with methyl esters of rubber seed oil. *Renewable Energy, 30*, 1789–1800.

Reyes, J. F., & Sepúlveda, M. A. (2006). PM-10 emissions and power of a diesel engine fueled with crude and refined biodiesel from salmon oil. *Fuel, 85*, 1714–1719.

Sahoo, P. K., Das, L. M., Babu, M. K. G., Arora, P., Singh, V. P., Kumar, N. R., et al. (2009). Comparative evaluation of performance and emission characteristics of jatropha, karanja and polanga based biodiesel as fuel in a tractor engine. *Fuel, 88*, 1698–1707.

Sahoo, P. K., Das, L. M., Babu, M. K. G., & Naik, S. N. (2007). Biodiesel development from high acid value polanga seed oil and performance evaluation in a CI engine. *Fuel, 86*, 448–454.

Sayin, C., & Gumus, M. (2011). Impact of compression ratio and injection parameters on the performance and emissions of a DI diesel engine fueled with biodiesel-blended diesel fuel. *Applied Thermal Engineering, 31*, 3182–3188.

Sharma, D., Soni, S. L., & Mathur, J. (2009). Emission reduction in a direct injection diesel engine fueled by neem-diesel blend. *Energy Sources Part A, 31*, 500–508.

Song, J. T., & Zhang, C. H. (2008). An experimental study on the performance and exhaust emissions of a diesel engine fuelled with soybean oil methyl ester. *Proceedings of the IMechE, Part D: Journal of Automobile Engineering, 222*, 2487–96.

Tsolakis, A., Megaritis, A., Wyszynski, M. L., & Theinnoi, K. (2007). Engine performance and emissions of a diesel engine operating on diesel-RME (rapeseed methyl ester) blends with EGR (exhaust gas recirculation). *Energy, 32*, 2072–2080.

Ulusoy, Y., Arslan, R., & Kaplan, C. (2009). Emission characteristics of sunflower oil methyl ester. *Energy Sources, Part A: Recovery, Utilization and Environmental Effects, 31*, 906–910.

Usta, N. (2005). Use of tobacco seed oil methyl ester in a turbocharged indirect injection diesel engine. *Biomass & Bioenergy, 28*, 77–86.

Usta, N., Öztürk, E., Can, Ö., Conkur, E. S., Nas, S., Con, A. H., et al. (2005). Combustion of biodiesel fuel produced from hazelnut soapstock/waste sunflower oil mixture in a diesel engine. *Energy Conversion and Management, 46*, 741–755.

Utlu, Z., & Koçak, M. S. (2008). The effect of biodiesel fuel obtained from waste frying oil on direct injection diesel engine performance and exhaust emissions. *Renewable Energy, 33*, 1936–1941.

Wu, F., Wang, J., Chen, W., & Shuai, S. (2009). A study on emission performance of a diesel engine fueled with five typical methyl ester biodiesels. *Atmospheric Environment, 43*, 1481–1485.

Xue, J., Grift, T. E., & Hansen, A. C. (2011). Effect of biodiesel on engine performances and emissions. *Renewable and Sustainable Energy Reviews, 15*, 1098–1116.

Zheng, M., Mulenga, M. C., Reader, G. T., Wang, M., Ting, D. S.-K., & Tjong, J. (2008). Biodiesel engine performance and emissions in low temperature combustion. *Fuel, 87*, 714–722.

Zhu, L., Zhang, W., Liu, W., & Huang, Z. (2010). Experimental study on particulate and NO_x emissions of a diesel engine fueled with ultra low sulfur diesel, RME-diesel blends and PME-diesel blends. *Science of the Total Environment, 408*, 1050–1058.

Chapter 8
Improvement of Diesel Engine Characteristics by Numeric Optimization

The term *diesel engine optimization* can be used for a variety of procedures aiming to improve the engine characteristics. In general, these procedures can be divided into two groups:

- Those, that rely on *intuitive improvements*, based mainly on past experience, experimental work, and numerical analysis
- Those, that rely on *systematic (algorithm-driven) improvements*, typically based on numerical simulation

This chapter addresses briefly the latter group, i.e., the *systematic optimization procedures* in the context of diesel engine optimization.

8.1 Systematic Optimization

Systematic optimization is a process of solving the *standard problem* of mathematical programming (standard optimization problem) by an adequate optimization method. The standard optimization problem P can be written in the following form:

$$\min f(\mathbf{b}) \tag{8.1}$$

$$g_i(\mathbf{b}) \leq 0, \ \ i = 1 \ldots K, \tag{8.2}$$

where $\mathbf{b} \in R^n$ denotes the vector of *design variables* and n is their total number. The scalar functions f and g_i are termed the *objective* and *constraint functions*, respectively. For a mechanical system, the design variables are typically geometrical and control parameters of the system that can be varied independently in order to search for the optimal solution. The objective function is related to the quality of the mechanical system, while the constraints typically reflect mechanical, technological, and other limitations. The symbol K denotes the number of constraints.

B. Kegl et al., *Green Diesel Engines*, Lecture Notes in Energy 12,
DOI 10.1007/978-1-4471-5325-2_8, © Springer-Verlag London 2013

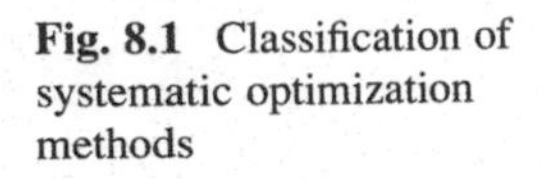

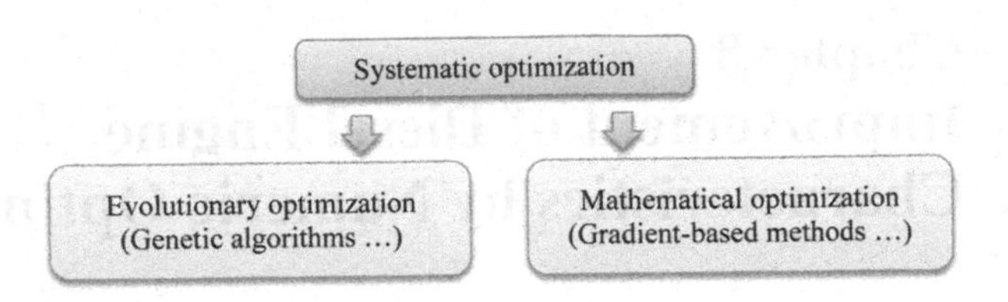

Fig. 8.1 Classification of systematic optimization methods

8.1.1 Optimization Methods

The majority of the methods employed in systematic optimization can be divided into two groups, often addressed as Fig. 8.1:

- *Mathematical optimization* (MO) methods
- *Evolutionary optimization* (EO) methods

EO methods aim to solve the standard problem by implementing algorithms that mimic various natural processes, such as, the natural selection. Perhaps the most prominent member of this group is the *genetic algorithm*, see, e.g., (Coley 2005) that comes in a wide variety of implementations. As any other numeric algorithm, the EO methods not only exhibit several attractive properties but they also have their drawbacks. Among their strengths, one can list the following:

- Rather simple implementation and usage in practice
- Relatively modest requirements regarding the involved functions (continuity, differentiability,. . .)
- Relatively good chances to find the global optimum
- No need (typically) to compute the gradients of the involved functions
- Rather easy handling of both continuous and discrete design variables

On the other side, among the drawbacks, there is only one major concern: if the function evaluation (mechanical system analysis) requires substantial computational effort, one is faced with a serious problem. Namely, evolutionary methods typically require a large number of function evaluations. This number can grow into many thousand. If the computation of the response of the mechanical system takes an hour or two (often it may take far more), the optimization could quickly take about a year of computation. This is of course unacceptable.

If the computation time, required for the mechanical system analysis, is relatively high, one has to turn to *MO methods*, which typically require a relatively low number of function evaluations. These methods usually aim to find the solution of the problem by trying to find a point that fulfills the Karush-Kuhn-Tucker optimality conditions (Bazaraa et al. 1993) and the employed solution procedures mostly rely on the first design derivatives of the involved functions. This sounds quite promising. However, it should be noted that these methods typically come with a variety of drawbacks, among which one can surely list the following:

- Rather tedious implementation and usage in practice
- Relatively strong requirements regarding the involved functions (continuity, differentiability,...)
- Relatively good chances to find only the local optimum, being the nearest to the starting point
- The need (typically) to compute the gradients of the involved functions
- Rather difficult handling of discrete design variables

There are a variety of MO methods available. However, their operation scheme to solve the standard problem P can usually be interpreted as follows:

- Set the iteration counter $k = 0$; choose initial design $\mathbf{b}^{(0)}$
- Compute f and g_i, $i = 1 \ldots K$, at the point $\mathbf{b}^{(k)}$ (*response analysis*).
- Compute $df/d\mathbf{b}$ and $dg_i/d\mathbf{b}$, $i = 1 \ldots K$, at the point $\mathbf{b}^{(k)}$ (*sensitivity analysis*).
- Use the computed function and gradient values to construct some approximation problem $P^{(k)}$ and solve the problem to get the design improvement $\Delta\mathbf{b}^{(k)}$. Depending on the method, this step may require an additional procedure to determine the proper step-size, i.e., the length of the improvement vector $\Delta\mathbf{b}^{(k)}$
- Update the design by $\mathbf{b}^{(k+1)} = \mathbf{b}^{(k)} + \Delta\mathbf{b}^{(k)}$, set $k = k + 1$, and check some appropriate convergence criteria; if not fulfilled, go back to step 2.

The methods differ mainly in the way how the approximate problem $P^{(k)}$ is constructed. For example, the well-known sequential quadratic programming (SQP) method utilizes quadratic approximation for the objective function and linear approximation for the constraint functions. Since the function approximations are rather simple, this method necessitates an additional procedure to determine the step-size. On the other hand, by the approximation method, described in (Kegl and Oblak 1997; Kegl et al. 2002), each approximate problem $P^{(k)}$ is generated by replacing the objective and constraint functions of P by strictly convex and separable nonlinear approximations. The employed approximation technique guarantees that the approximate problem $P^{(k)}$ is strictly convex and that its corresponding Lagrangian function has a relative simple form. Consequently, the solution $\Delta\mathbf{b}^{(k)}$ can be obtained directly by solving a relatively simple set of algebraic nonlinear equations, emerging from the Karush-Kuhn-Tucker conditions of $P^{(k)}$; no separate step-size determination is necessary.

8.1.2 *Transforming a Practical Optimization Problem into the Standard Form*

In engineering practice, the optimization problems are rarely defined in the standard form. The two main reasons for that are:

- The presence of an *independent parameter* (usually the time t)
- *Several objective functions* (multiobjective optimization)

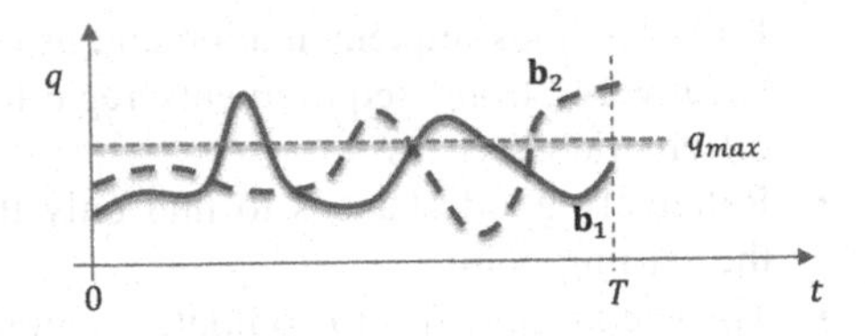

Fig. 8.2 Time-dependent constraint quantity, drawn for two different designs

The time parameter typically enters into the definition of constraint functions, if the response of the underlying mechanical system is time dependent. For example, if some time-dependent quantity $q(\mathbf{b},t)$ (Fig. 8.2) has to be lower than some specified value $q_{\max}$ in the whole time interval $[0,T]$, the corresponding constraint has to be formulated in the following form:

$$q(\mathbf{b},t) - q_{\max} \leq 0, \quad t \in [0,T] \tag{8.3}$$

This form, of course, does not fit into the standard optimization problem.

Several objective functions often appear in practical engineering problems because of the (rather frequent) requirement that several quantities have to be optimized simultaneously. For example, let $q_i(\mathbf{b})$ be the Sauter mean diameter, corresponding to the *i-th* operating regime of the engine. In order to minimize the Sauter mean diameters at all operating regimes $i = 1 \ldots N$ simultaneously, the objective should be formulated as

$$\min\{q_1(\mathbf{b}), q_2(\mathbf{b}) \ldots q_N(\mathbf{b})\} \tag{8.4}$$

This, of course, also does not fit into the standard optimization problem. Namely, the solution of such a problem is not a single point, but a set of points, called the *Pareto front*. Although there exist algorithms that try to deliver the most significant points from the Pareto front, these algorithms are rather complex and quite tedious to implement and use in practice.

Most optimization methods are well suited to solve standard optimization problems. Thus, a non-standard problem has to be transformed into the standard form somehow in order to use these methods efficiently. This transformation can be done in a variety of ways. Here, only those will be addressed that seem to be most easily implementable and exhibit good numerical stability. Anyhow, it should be noted that many of available transformations do *not preserve* exactly the mathematical properties of the original problem. Thus, the solutions of the non-standard and the transformed standard problem may not be the same.

In order to remove the independent time parameter, an attractive and efficient option is to replace the time-dependent constraint by a series of time-independent constraints, formulated at properly chosen and fixed time points. In other words, the constraint:

$$q(\mathbf{b},t) - q_{\max} \leq 0, \quad t \in [0,T] \tag{8.5}$$

is replaced by M constraints of the form:

$$q(\mathbf{b}, t_j) - q_{\max} \leq 0, \quad j = 0 \ldots M, \tag{8.6}$$

where t_j is a fixed time point and $M + 1$ is their total number. The time points t_j are typically chosen to be equally spaced through the whole time interval $[0,T]$. As a rule of the thumb, the spacing $\Delta t = t_{j-1} - t_j$ has to be chosen so that within $[t_{j-1}, t_j]$ the quantity q can be linearly approximated with reasonable accuracy.

An alternative approach to the one shown above is to use integrated quantities instead of point values. In this case, a time-dependent constraint is replaced by M constraints of the form:

$$\int_{t_{j-1}}^{t_j} q(\mathbf{b}, t_j)dt - q_{\max}\,\Delta t \leq 0, \; j = 1 \ldots M \tag{8.7}$$

This approach is very useful, if the time histories of the involved quantities are very non-smooth (jagged) functions. Namely, integration over a time subinterval can substantially reduce eventual numerical instabilities during the optimization process.

In order to transform a multiobjective optimization problem into the standard form, perhaps the most used solution is to replace the list of several objective functions by a weighted sum of these functions. In other words, a requirement like

$$\min\{q_1, q_2 \ldots q_N\} \tag{8.8}$$

is replaced by

$$\min \sum\nolimits_{i=1}^{N} w_i q_i, \tag{8.9}$$

where w_i are some properly chosen weights. This is a very simple solution. However, it should be noted that often it is very difficult to choose proper weights w_i and that the result may depend heavily on this choice. Therefore, good engineering intuition and trial and error procedures are often necessary to get a good result.

For the reason stated above, it might be reasonable to consider another approach that can be outlined as follows in order to get rid of a set of objective functions $f_1, f_2 \ldots f_N$:

- The set of design variables is increased by one artificial design variable b_{n+1}
- A new objective function is defined as $f = b_{n+1}$
- Additional constraints are defined as $w_i f_i - b_{n+1} \leq 0, \; i = 1 \ldots N,$

where w_i are some properly chosen weights. It should be noted that in this case the weights can be defined relatively easily since they merely reflect the relative importance and eventual magnitude differences of original objective functions f_i. By using this procedure, a requirement of the form

$$\min\{f_1, f_2 \ldots f_N\} \tag{8.10}$$

is replaced by

$$\min b_{n+1} \tag{8.11}$$

$$w_i f_j - b_{n+1} \leq 0, \quad i = 1 \ldots N \tag{8.12}$$

It should be noted that by minimizing the new objective function, all of the original objective functions will also be minimized.

8.1.3 The Role of the Response Equation

In practical engineering problems, the response of the system under consideration is typically described by a relatively complex system of response equations. For systems with time-dependent response, these equations are often formulated in differential form. These are the reasons that a typical objective or constraint function cannot be formulated explicitly in terms of design variables **b**, i.e., in a form

$$f = f(\mathbf{b}) \quad \text{or} \quad g_i = g_i(\mathbf{b}) \tag{8.13}$$

It is much more likely that these functions are expressed in terms of **b**, the response variables, assembled in a vector **u**, and their time derivatives $\dot{\mathbf{u}}$ and $\ddot{\mathbf{u}}$ In other words, one typically has

$$f = f(\mathbf{b}, \mathbf{u}, \dot{\mathbf{u}}, \ddot{\mathbf{u}}) \quad \text{or} \quad g_i = g_i(\mathbf{b}, \mathbf{u}, \dot{\mathbf{u}}, \ddot{\mathbf{u}}) \tag{8.14}$$

The relation between **b** and **u** is defined by the response equation, for a time-dependent system often given in the form

$$\mathbf{h}(\mathbf{b}, \mathbf{u}, \dot{\mathbf{u}}, \ddot{\mathbf{u}}) = 0, \quad \mathbf{u}(t = 0) = \mathbf{u}_0, \quad \dot{\mathbf{u}}(t = 0) = \dot{\mathbf{u}}_0, \tag{8.15}$$

where $\mathbf{u}_0$ and $\dot{\mathbf{u}}_0$ are the initial values of the response variables and their time derivatives, respectively.

In an optimization procedure, the response equation can be treated in several ways. For example, one possibility is to consider the response equation as an equality constraint of the optimization problem. In this case the response variables also become part of the total set of optimization variables, i.e., they are treated equally as the design variables. However, perhaps the most efficient and numerically stable way is to treat the response equation as an implicit dependency of the response variables on the design variables. In other words, the response equation establishes implicit dependencies:

$$\mathbf{u} = \mathbf{u}(\mathbf{b}), \quad \dot{\mathbf{u}} = \dot{\mathbf{u}}(\mathbf{b}), \quad \ddot{\mathbf{u}} = \ddot{\mathbf{u}}(\mathbf{b}) \tag{8.16}$$

In this context, the response variables and their time derivatives are merely intermediate variables. Of course, this has to be taken into account when performing the computation of design derivatives. For example, in this context the total design derivative of a function $f = f(\mathbf{b}, \mathbf{u}, \dot{\mathbf{u}}, \ddot{\mathbf{u}})$ is given by

$$\frac{\partial f}{d\mathbf{b}} = \frac{\partial f}{\partial \mathbf{b}} + \frac{\partial f}{\partial \mathbf{u}} \frac{d\mathbf{u}}{d\mathbf{b}} + \frac{\partial f}{\partial \dot{\mathbf{u}}} \frac{d\dot{\mathbf{u}}}{d\mathbf{b}} + \frac{\partial f}{\partial \ddot{\mathbf{u}}} \frac{d\ddot{\mathbf{u}}}{d\mathbf{b}} \tag{8.17}$$

which has to be taken into account, if the sensitivity analysis has to be done analytically.

8.2 Engine Optimization for Biodiesel Usage

Diesel engine is a sophisticated system. For such a system, systematic numerical optimization can usually only be employed to "fine tune" a rather minor set of its parameters. This is because a sophisticated system typically has to be defined or fixed to a large extent, if efficient numerical simulation models are needed, for example, to run an optimization process. In other words, numerical optimization can most efficiently be performed on "almost completely" defined engines.

If biodiesel usage is one of the optimization reasons, there is another aspect that is worth noting. Namely, biodiesel is often used as a more or less occasional substitute for mineral diesel on existing diesel engines. In such a scenario it is surely of interest to investigate, what can be done in order to improve engine characteristics as much as possible by making only minimal and inexpensive changes to the engine. In this context, numerical optimization can be a very useful tool.

Diesel engine characteristics are most closely related to the fuel injection system. Therefore, this chapter focuses on the diesel fuel injection system optimization. Nowadays, diesel engines are fuelled by both, mechanically and electronically controlled injection systems. Although, the underlying optimization tools and methods are essentially the same for both injection system types, the mechanically controlled system represents probably the more difficult problem. Therefore, this section will focus on optimization of a mechanical injection system with some electronic control. The underlying ideas presented, can of course be used for any injection system.

8.2.1 *Parameterization of the Injection System*

The first thing that needs to be done when a mechanical system becomes an optimization target is its proper *parameterization*. In other words, any property of the mechanical system, that is intended to be varied during optimization, has to be expressed in terms of some appropriate scalar parameters. These parameters have to be chosen in such a way that their values may be varied independently when searching for optimal design. In the context of optimization, these parameters are termed the *design variables.*

For a mechanical injection system with some electronic control, the optimization parameters can be classified into two groups:

- *Geometric* parameters
- *Control* parameters

Geometric parameters are related to the geometry of individual parts of the injection system, while the control parameters typically refer to those parameters that can be varied during operation in order to get the best response possible.

For an injection system, many of these parameters are easily determined, since they naturally appear in the description of the considered system. However, this may not be the case for a mechanical injection system, in particular, for the geometric parameters, related to the *cam profile* of the fuel pump. Since the pump cam profile is one of the most important geometrical properties of the injection system, see, e.g., (Kegl 1995, 1996, 1999), this topic needs special attention.

The cam profile of a diesel fuel injection pump can be mathematically represented by a planar curve:

$$K = \{\mathbf{r} | \mathbf{r} = \mathbf{r}(\mathbf{b}, s), \quad 0 \leq s \leq 1\}, \tag{8.18}$$

where the symbol $\mathbf{r} \in R^2$ denotes the position vector of a point on K with respect to a fixed Cartesian coordinate system, Fig. 8.3. The symbol s denotes the independent parameter, defining the position along K and $\mathbf{b}$ is the vector of design variables.

When deciding how to define the function $\mathbf{r}$, the following two aspects should be taken into account:

- It should be possible that K can take virtually any shape
- The shape of K should not exhibit excessive oscillations when the values of the design variables are varied in a reasonable range

Keeping this in mind, it seems to be a good idea to define K as a rational Bézier curve by setting:

$$\mathbf{r} = \frac{\sum_{i=1}^{k} B_i^k \psi_i \mathbf{q}_i}{\sum_{i=1}^{k} B_i^k \psi_i}, \tag{8.19}$$

where $B_i^k = B_i^k(s)$ is the Bernstein blending polynomial of the order $(k-1)$, $\mathbf{q}_i = \mathbf{q}_i(\mathbf{b})$ is the position vector of a control point, $\psi_i = \psi_i(\mathbf{b})$ is a positive weighting factor, and k is the number of all control points. According to this definition, the individual components x and y of the position vector $\mathbf{r}$ can be written as

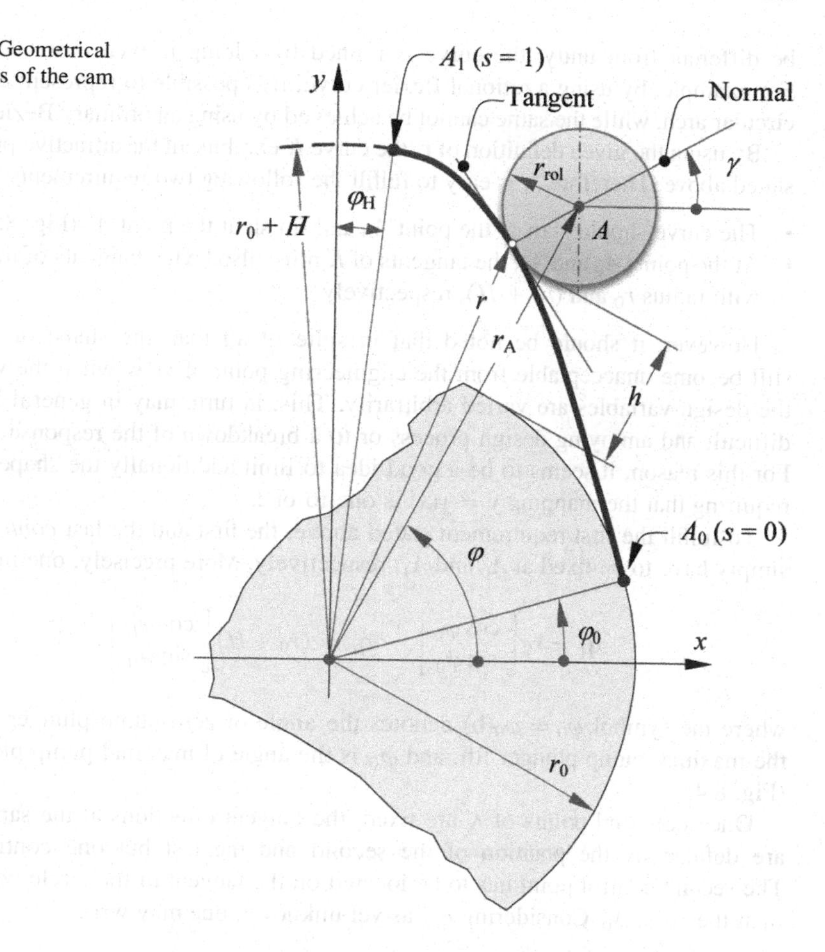

Fig. 8.3 Geometrical parameters of the cam profile

$$x = \frac{\sum_{i=1}^{k} B_i^k \psi_i q_{ix}}{\sum_{i=1}^{k} B_i^k \psi_i}, \quad y = \frac{\sum_{i=1}^{k} B_i^k \psi_i q_{iy}}{\sum_{i=1}^{k} B_i^k \psi_i} \tag{8.20}$$

where q_{ix} and q_{iy} are the components of the ith control point.

A rational Bézier curve exhibits the following attractive properties. The curve starts at the first control point and finishes at the last control point. Furthermore, it follows without excessive oscillations its defining polygon (derived by connecting successively the control points) and it is always contained within the convex hull of the defining polygon. The direction of the tangent, at the starting point of the curve, is defined by the first two control points and similar holds true for the direction of the tangent at the ending point of the curve. Finally, more flexibility of the curve can be assured by simply increasing the number of control points. All these properties are inherited from the ordinary Bézier curve that can be derived from its rational counterpart by setting $\psi_i = 1, \; i = 1 \ldots k$ which implies $\mathbf{r} = \sum_{i=1}^{k} B_i^k \mathbf{q}_i$ since $\sum_{i=1}^{k} B_i^k = 1$. However, by allowing the weighting factors to be different from unity, the curve is refined by adding it even more flexibility. For example, by using a rational Bézier curve, it is possible to represent exactly a circular arch, while the same cannot be achieved by using an ordinary Bézier curve.

By using the given definition of **r**, the curve K exhibits all the attractive properties stated above. Therefore, it is easy to fulfill the following two requirements:

- The curve should start at the point A_0 and finish at the point A_1 (Fig. 8.3)
- At the points A_0 and A_1, the tangents of K must also be the tangents of the circles with radius r_0 and $(r_0 + H)$, respectively

However, it should be noted that in spite of all that, the shape of K might still become unacceptable from the engineering point of view when the values of the design variables are varied arbitrarily. This, in turn, may in general lead to a difficult and annoying design process or to a breakdown of the response analysis. For this reason, it seems to be a good idea to limit additionally the shape of K by requiring that the mapping $y = y(x)$ is one-to-one.

To fulfill the first requirement stated above, the first and the last *control points* simply have to be fixed at A_0 and A_1, respectively. More precisely, one may write

$$\mathbf{q}_1 = r_0 \begin{bmatrix} \cos\varphi_0 \\ \sin\varphi_0 \end{bmatrix}, \quad \mathbf{q}_k = (r_0 + H) \begin{bmatrix} \cos\varphi_H \\ \sin\varphi_H \end{bmatrix}, \tag{8.21}$$

where the symbol $\varphi_0 = \varphi_0(\mathbf{b})$ denotes the angle of zero pump plunger lift, H is the maximal pump plunger lift, and φ_H is the angle of maximal pump plunger lift (Fig. 8.4).

Once both end points of K are fixed, the tangent directions at the same points are defined by the position of the second and the last but one control point. The second control point has to be located on the tangent of the circle with radius r_0 at the point A_0. Considering q_{2x} as yet unknown, one may write

$$\mathbf{q}_2 = \begin{bmatrix} q_{2x} \\ \dfrac{r_0^2 - q_{2x} q_{2x}}{q_{1y}} \end{bmatrix} \tag{8.22}$$

The last but one control point has to be located on the tangent of the circle with radius $(r_0 + H)$ at the point A_1. Considering $q_{(k-1)x}$ as yet unknown, one may write

$$\mathbf{q}_{(k-1)} = \begin{bmatrix} q_{(k-1)x} \\ \dfrac{(r_0 + H)^2 - q_{kx} q_{(k-1)x}}{q_{ky}} \end{bmatrix} \tag{8.23}$$

Finally, the one to one property of the mapping $y = y(x)$ can be assured by requiring that $q_{1x} > q_{2x} > \cdots q_{(k-1)x} > q_{kx}$ and $q_{1y} < q_{2y} < \cdots < q_{(k-1)y} < q_{ky}$.

Fig. 8.4 Control point components q_{ix} and q_{iy}, $i = 1\ldots k$

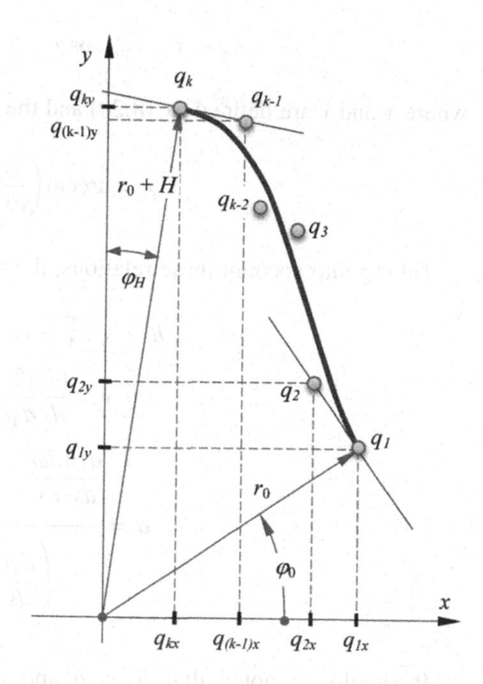

These relations are fulfilled by defining the components of the intermediate control points as

$$
\begin{aligned}
q_{ix} &= q_{(i-1)x} + c_{ix}\left(q_{kx} - q_{(i-1)x}\right), \quad i = 2\ldots(k-1) \\
q_{iy} &= q_{(i-1)y} + c_{iy}\left(q_{(k-1)y} - q_{(i-1)y}\right), \quad i = 3\ldots(k-2),
\end{aligned}
\tag{8.24}
$$

where $c_{ix} = c_{ix}(\mathbf{b}),\ i = 2\ldots(k-1)$ and $c_{iy} = c_{iy}(\mathbf{b}),\ i = 3\ldots(k-2)$ are real parameters such that $0 < c_{ix} < 1$ and $0 < c_{iy} < 1$.

The above definitions assure an acceptable shape of the curve K and, within these limits, full flexibility of a rational Bézier curve. By making the quantities Ψ_i, φ_0, c_{ix}, and c_{iy} dependent on $\mathbf{b}$, this flexibility can be fully used by the optimizer to obtain a good result.

The shape of the curve K defines the movement of the cam follower, given by the pump plunger lift h, relative pump plunger velocity $v = dh/d\varphi$, and relative pump plunger acceleration $a = d^2h/d\varphi^2$. Here, $\varphi = \arctan(y_A/x_A)$, while x_A and y_A are the components of the vector $\mathbf{r}_A$ (Fig. 8.3). The term *relative* is used to emphasize that the pump plunger lift h is differentiated with respect to the camshaft angle φ and, consequently, v and a are given in [m/rad] and [m/rad^2], respectively.

The components of the vector $\mathbf{r}_A$ may be expressed as

$$x_A = x + r_{\text{rol}} \cos\gamma, \quad y_A = y + r_{\text{rol}} \sin\gamma, \tag{8.25}$$

where x and y are defined by (8.20) and the angle γ may be written as

$$\gamma = \arctan\left(\frac{dy}{ds}\frac{ds}{dx}\right) - \frac{\pi}{2} \tag{8.26}$$

Taking into account these relations, the response quantities can be written as

$$\begin{aligned} h &= \sqrt{x_A^2 + y_A^2} - r_0 - r_{\text{rol}} \\ v &= \frac{dh}{ds}\frac{ds}{d\varphi} \\ a &= \frac{\dfrac{d^2h}{ds^2}\dfrac{d\varphi}{ds} - \dfrac{dh}{ds}\dfrac{d^2\varphi}{ds^2}}{\left(\dfrac{d\varphi}{ds}\right)^2} \end{aligned} \tag{8.27}$$

It should be noted that h, v, a and are expressed in terms of the design variables and the independent parameter s, which defines the position along K. Now, one has to remember that the response of the cam follower is needed to calculate the response of the whole injection system, where the time t represents the independent variable. Therefore, the parameter s has to take the role of a dependent parameter, expressed in terms of t. This can be done by expressing the angle φ in terms of each parameter separately. By equating both expressions, one obtains

$$\arctan\left(\frac{y_A}{x_A}\right) = \omega t, \tag{8.28}$$

where ω is the angular velocity of the cam. This relation can be regarded as a one-to-one mapping between $s \in [0,1]$ and t for a fixed $\mathbf{b}$. Therefore, the parameter s can now be viewed as a dependent variable expressed in terms of $\mathbf{b}$ and t. Taking this into account in (8.27), one may write

$$h = h(\mathbf{b}, t), \quad v = v(\mathbf{b}, t), \quad a = a(\mathbf{b}, t) \tag{8.29}$$

Considering the above dependencies, the response of the cam follower is properly parameterized and can be easily introduced into the response analysis of the whole fuel injection system and further into the optimal design procedure.

8.2.2 Optimization of Injection Rate Histories

Consider a mechanically controlled injection system. Let the objective be to determine some of its parameters so that the *actual injection rate histories* will be *as close as possible* to some *prescribed target histories*. It should be noted that a target injection rate history is associated with a corresponding engine operating regime. Thus, if N operating regimes need to be taken into account, this means that N target injection rate histories have to be defined.

Once the considered system is adequately parameterized, the first thing needing attention is the *objective function*. If the objective is to get optimal injection rate histories, the objective function must be defined as some measure of disagreement between the actual and target injection rate histories. Taking this into account, the objective function can be defined as

$$f = \max_{1 \le i \le l} \int_{t_{i-1}}^{t_i} \sum_{z=1}^{N} \frac{\left| \dot{q}^{p}_{\text{inj},z} - \dot{q}_{\text{inj},z} \right|}{w_z} dt, \tag{8.30}$$

where the symbols $\dot{q}_{\text{inj},z} = \dot{q}_{\text{inj},z}(\mathbf{b}, \mathbf{u})$ and $\dot{q}^{p}_{\text{inj},z} = \dot{q}^{p}_{\text{inj},z}(t), z = 1 \ldots N$ denote the actual and target injection rate histories, respectively, and $w_z = \max\limits_{1 \le i \le l} \int_{t_{i-1}}^{t_i} \dot{q}^{p}_{\text{inj},z}\, dt$ is a normalization constant. N denotes the number of engine operating regimes under consideration, l is the total number of subintervals of equal length on the time interval $[0, T]$, $t_0 = 0$, and $t_1 = T$. The design and response variables are assembled in the vectors $\mathbf{b}$ and $\mathbf{u}$, respectively.

It should be obvious that by minimizing the given objective function, the difference between the target and actual injection rates will also be reduced. Ideally, if $f = 0$ could be obtained, the target and actual injection rate histories would match completely.

After defining the objective function, the next important step is to define proper optimization constraints. For a mechanical injection system, at least some minimum set of *imposed constraints* should be defined. Some of these constraints have to be related to the cam follower and some to injection characteristics.

Regarding the cam profile, some attention should be focused on the local radius e. In general, the local radius of acurve defined in polar coordinates, $\rho = \rho(\varphi)$, may be expressed as $e = \frac{\left(\rho^2 + \rho'^2\right)^{1.5}}{\rho^2 + \rho'^2 - \rho\rho''}$, where $(\cdot)' = d(\cdot)/d\varphi$. By using this expression and doing some mathematical work, the *local radius of the cam profile curve* can be written as

$$e = \frac{\left[(h + r_0 + r_{\text{rol}})^2 + v^2\right]^{1.5}}{(h + r_0 + r_{\text{rol}})^2 + 2v^2 - (h + r_0 + r_{\text{rol}})a} - r_{\text{rol}} \tag{8.31}$$

For technological reasons, the minimal positive value $\xi_{\min}(+)$ and the maximal negative value $\xi_{\max}(-)$

$$\begin{aligned} \xi_{\min}(+) &= \min_{0 \le t \le T} (e_{(+)}), \quad e_{(+)} = \{e | e \ge 0\} \\ \xi_{\max}(-) &= \max_{0 \le t \le T} (e_{(-)}), \quad e_{(-)} = \{e | e < 0\} \end{aligned} \tag{8.32}$$

of the *local radius* should be constrained by

$$1 - \frac{\xi_{\min}(+)}{\xi^L_{\min}(+)} \le 0, \qquad - \frac{\xi_{\max}(-)}{\xi^U_{\min}(-)} + 1 \le 0 \tag{8.33}$$

Here and in the following text, the superscripts L and U are used to denote the lower and upper limits of a parameter or quantity.

To avoid torsion torque overloading and loosing contact between the cam and the follower, the *minimal* and *maximal pump plunger acceleration*

$$\zeta = \min_{0 \le t \le T} (a), \qquad \varsigma = \max_{0 \le t \le T} (a) \tag{8.34}$$

also have to be limited by $\zeta = \zeta^L$ and $\varsigma \le \varsigma^U$. By taking into account that $\zeta^L < 0$ these two constraints may be written as

$$\frac{\zeta}{\zeta^L} - 1 \le 0, \qquad \frac{\varsigma}{\varsigma^U} - 1 \le 0 \tag{8.35}$$

Regarding the injection performance, some attention should be focused on the *injection timing* θ and *injection durati* ϑ (both quantities are measured in degree of camshaft) and on the *fuelling* $q_{\mathrm{inj},z}$ [mm^3/stroke]. Denoting the target injection timing by θ^p_z, the target injection duration by ϑ^p_z and the target fuelling by $q^p_{\mathrm{inj},z}$ for the zth operating regime, the imposed constraints can be written as

$$\begin{aligned} &\frac{(\theta_z - \theta^p_z) - \theta^U_z}{\theta^p_z} \le 0, && \frac{\theta^L_z - (\theta_z - \theta^p_z)}{\theta^p_z} \le 0, \\ &\frac{(\vartheta_z - \vartheta^p_z) - \vartheta^U_z}{\vartheta^p_z} \le 0, && \frac{\vartheta^L_z - (\vartheta_z - \vartheta^p_z)}{\vartheta^p_z} \le 0, \\ &\frac{\left(q_{\mathrm{inj},z} - q^p_{\mathrm{inj},z}\right) - q^U_{\mathrm{inj},z}}{q^p_{\mathrm{inj},z}} \le 0, && \frac{q^L_{\mathrm{inj},z} - \left(q_{\mathrm{inj},z} - q^p_{\mathrm{inj},z}\right)}{q^p_{\mathrm{inj},z}} \le 0, \quad z = 1 \ldots N \end{aligned} \tag{8.36}$$

Finally, the values of the design variables should also be constrained by,

$$b^L_i \le b_i \le b^U_i, \quad i = 1 \ldots n \tag{8.37}$$

in order to prevent a technologically unacceptable design.

The objective function (8.30), constraints (8.33), (8.35), (8.36), (8.37), and an appropriate *response model* (for example, BKIN, Sect. 2.1.2) define completely the optimization problem. Unfortunately, the objective function and some of the constraints are in general not differentiable with respect to **b** because of the presence of the operators "min" and "max". In order to obtain a suitable form of the problem, one may proceed as follows (Kegl 1995, 1996).

In the first step, an *additional artificial design variable* $b_{n+1} \in R$ is introduced. After that one can define a *new objective function* by

$$f = b_{n+1} \tag{8.38}$$

and impose the following *additional constraints*:

$$\int_{t_{i-1}}^{t_i} \sum_{z=1}^{N} \frac{\left| \dot{q}^{p}_{\mathrm{inj},z} - \dot{q}_{\mathrm{inj},z} \right|}{w_z} dt - b_{n+1} \le 0, \quad 1 \le i \le l \tag{8.39}$$

It should be noted that the effect of minimizing the new objective function and taking into account the additional constraints, is essentially the same as the effect of minimizing the original objective function (8.30).

In the second step, one should first take a look at the time dependent quantities e and a. In the time interval $[0, T]$ these quantities may have several local minima and maxima, whose position and level changes as **b** changes. However, practical experience has shown that usually the same minimum and maximum remain global, if the design variables are varied in a reasonable range. In other words, the operators "min" and "max" in (8.32) and (8.34) always return the same extremes. Taking this into account, it is assumed that the quantities $\xi_{\min^{(+)}}, \xi_{\max^{(-)}}, \zeta$, and ς are differentiable with respect to **b** and no transformation of these constraints is necessary.

Based on the assumption stated above, the final form of the optimal design problem is given by the objective function (8.38), the constraints (8.33), (8.35), (8.36), (8.37), (8.39), and by the response equation (8.15). This problem can be solved directly by virtually any gradient based method of mathematical programming.

To illustrate the theory, one can consider the *electronic control diesel fuel injection system* (ECD FIS) *with sleeve timing controlled pump* (Needham et al. 1990) (Fig. 8.5). Its sleeve timing controlled pump offers quite a large flexibility in metering fuel quantity and injection start timing. The basic construction of a sleeve timing controlled pump is the same as that of a mechanically controlled BOSCH P size pump. The only difference is an attached sleeve that can be found at the plunger block. Consequently, the mathematical model of a conventional and ECD FIS may be the same, except that in the latter case, the sleeve-related control parameters have to be treated as independent and variable parameters rather than given constants.

By taking into account the properties of an ECD FIS, it may be worth to run an optimization procedure by adjusting both the shape of the cam profile and the

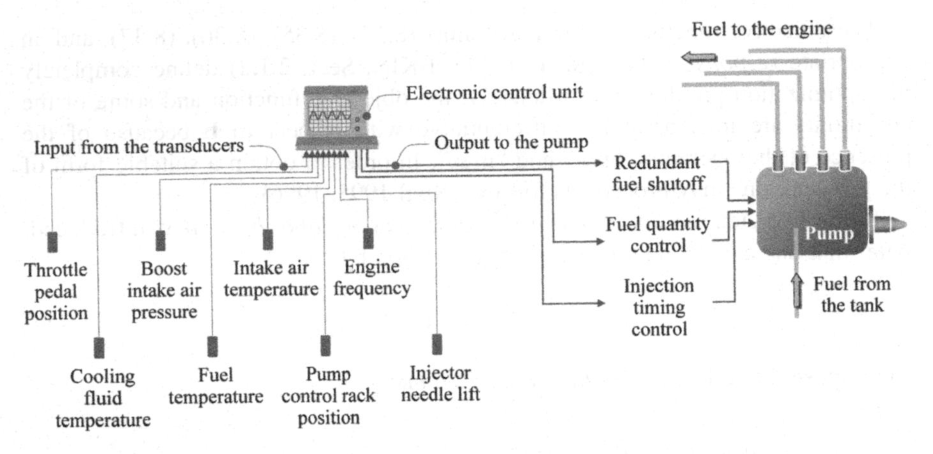

Fig. 8.5 ECD fuel injection system with sleeve timing controlled pump

values of the injection control parameters. By considering several engine operating regimes simultaneously, it might be possible to achieve near target injection rates in a wide range of engine operating regimes. The FIS input data—needed to calculate the response of the system—is taken from the numerical examples, described in (Kegl 1995, 1996).

To optimize the system, four different cases of the design problem will be considered, distinguished by the type of the cam profile curve K and by the number of engine-operating regimes that are taken into account simultaneously:

- Case A: *Bézier curve, one regime* ($z = 1$)
- Case B: *rational Bézier curve, one regime* ($z = 1$)
- Case C: *Bézier curve, five regimes* ($z = 1 \ldots 5$)
- Case D: *rational Bézier curve, five regimes* ($z = 1 \ldots 5$)

The cases A and C are taken into consideration in order to verify the benefits gained by using a rational Bézier curve rather than its nonrational counterpart.

The coordinates of six control points, which define K, are given by the following expressions:

$$
\begin{aligned}
q_{1x} &= r_0 \cos(\varphi_0), & q_{1y} &= r_0 \sin(\varphi_0), \\
q_{2x} &= q_{1x} + c_{2x}(q_{6x} - q_{1x}), & q_{2y} &= \frac{r_0^2 - q_{1x}q_{2x}}{q_{1y}} \\
q_{3x} &= q_{2x} + c_{3x}(q_{6x} - q_{2x}), & q_{3y} &= q_{2y} + c_{3y}(q_{5y} - q_{2y}) \\
q_{4x} &= q_{3x} + c_{4x}(q_{6x} - q_{3x}), & q_{4y} &= q_{3y} + c_{4y}(q_{5y} - q_{2y}) \\
q_{5x} &= q_{4x} + c_{5x}(q_{6x} - q_{4x}), & q_{5y} &= \frac{(r_0 + H)^2 - q_{6x}q_{5x}}{q_{6y}} \\
q_{6x} &= (r_0 + H)\sin(\varphi_H), & q_{6y} &= (r_0 + H)\cos(\varphi_H)
\end{aligned}
\tag{8.40}
$$

Several geometrical and control parameters are selected to serve as design variables. The geometrical design variables are the cam angle at zero pump plunger lift φ_0 [°CAM], the parameters c_{ix}, $i = 2 \ldots 5$ and c_{iy}, $i = 3, 4$ as well as the weighting factors ψ_i, $i = 2 \ldots 5$. The control design variables are the pump plunger prelifts $h_p^{(z)}$, $z = 1 \ldots N$, and the geometrical duration of delivery $t_d^{(z)}$, $z = 1 \ldots N$. The pump plunger prelift influences the injection timing and the geometrical duration of delivery influences the injection duration with the consequence of influencing also the fuelling. It should be noted that these two parameters are related to an individual operating regime. Thus, for N operating regimes $2N$ of such parameters enter the set of design variables.

For cases A, B, C, and D, the *vectors of design variables* are defined as

$$
\begin{gathered}
{}^{(A)}\mathbf{b} = \left[\varphi_0, c_{2x}, c_{3x}, c_{3y}, c_{4x}, c_{4y}, c_{5x}, h_p^{(1)}, t_d^{(1)}\right]^T \\
{}^{(B)}\mathbf{b} = \left[\varphi_0, c_{2x}, c_{3x}, c_{3y}, c_{4x}, c_{4y}, c_{5x}, \psi_2, \psi_3, \psi_4, \psi_5, h_p^{(1)}, t_d^{(1)}\right]^T \\
{}^{(C)}\mathbf{b} = \left[\varphi_0, c_{2x}, c_{3x}, c_{3y}, c_{4x}, c_{4y}, c_{5x}, \left(h_p^{(i)}, t_d^{(i)}, i = 1 \ldots 5\right)\right]^T \\
{}^{(D)}\mathbf{b} = \left[\varphi_0, c_{2x}, c_{3x}, c_{3y}, c_{4x}, c_{4y}, c_{5x}, \psi_2, \psi_3, \psi_4, \psi_5, \left(h_p^{(i)}, t_d^{(i)}, i = 1 \ldots 5\right)\right]^T
\end{gathered}
\tag{8.41}
$$

The *initial values of all design variables* as well as their *lower and upper limits*, given in Table 8.1, are in all cases the same.

The following *lower* and *upper limits* of the *constrained quantities* have also been taken to be the same in all cases:

$$
\begin{gathered}
\xi_{\min}^L(+) = 2\,\text{mm}, \xi_{\max}^U(-) = -60\,\text{mm}, \varsigma^L = -100\,\text{mm}, \varsigma^U = -100\,\text{mm}, \\
q_{\text{inj},z}^L = -5\,\text{mm}^3/\text{stroke}, q_{\text{inj},z}^U = 5\,\text{mm}^3/\text{stroke}, \\
\theta_z^L = -0.5°\,\text{CAM}, \theta_z^U = 0.5°\,\text{CAM}, \vartheta_z^L = -0.5°\,\text{CAM, and } \vartheta_z^U = 0.5°\,\text{CAM}.
\end{gathered}
$$

The *target injection rate histories* are chosen in accordance with the guidelines discussed in (Kegl 1999). The chosen injection rate histories (Fig. 8.6) define the values of the following parameters: the target start of injection θ_z^p, the target fuelling $q_{\text{inj},z}^p$, and the target injection duration ϑ_z^p. Their values are given in Table 8.2.

The interval of calculation in degree of camshaft is [20, 80] for all operating regimes. The corresponding time intervals $[0,T]$ in seconds are [0, 0.01111] and [0, 0.01667] for operating regimes $z = 1, 2$, respectively, and [0, 0.02] for $z = 3, 4, 5$. For all operating regimes, the total number of sub intervals is set to $l = 10$.

In cases A and B only the first engine-operating regime is considered. At this regime, the after injection is present at the starting design. By optimizing the fuel injection system, the after injection was removed (Fig. 8.7). The agreement

Table 8.1 Initial and limit values of the design variables

Design variables	Initial value	Lower limit	Upper limit
φ_0 [°CAM]	35	20	50
c_{ix}, c_{jy} [mm], $i = 2 \ldots 5, j = 3, 4$	0.25	0	0.5
ψ_i [−], $i = 2 \ldots 5$	1.0	0	2.0
$h_p^{(z)}$ [mm], $z = 1 \ldots N$	5	3	7
$t_d^{(z)}$ [mm], $z = 1 \ldots N$	2	0.1	3

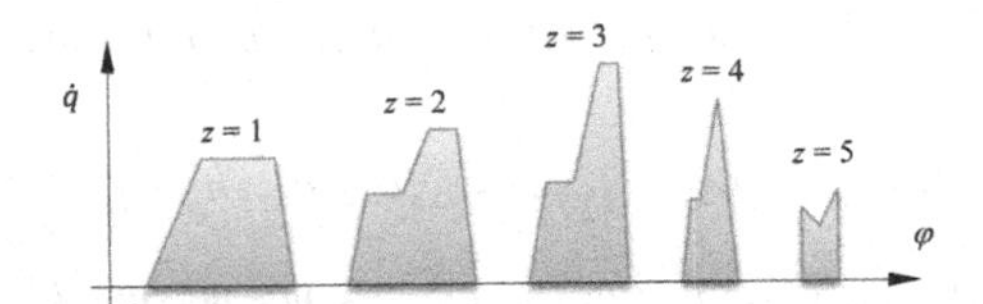

Fig. 8.6 Target injection rate histories

Table 8.2 Pump rotational speed and target data for all engine-operating regimes

	Operating regime				
Target data	$z = 1$	$z = 2$	$z = 3$	$z = 4$	$z = 5$
n_p [min^{-1}]	900	600	500	500	500
θ_z^p [°CAM]	65	64	63	62	60
$q_{\text{inj},z}^p$ [mm^3/stroke]	150	150	150	50	25
ϑ_z^p [°CAM]	10	8	7	3.5	2.5

between the target and optimal injection rate histories is satisfactory in both cases. As expected, the optimal design in case B is slightly better than the optimal design in case A. The *optimal values* of the *design variables* are as follows:

$$^{(A)}\mathbf{b} = [28.947, 0.255, 0.216, 0.305, 0.248, 0.270, 0.240, 6.873, 1.898]^T$$
$$^{(B)}\mathbf{b} = [29.718, 0.262, 0.231, 0.294, 0.261, 0.262, 0.253, 1.056, 0.885, 1.006, 1.087, 6.747, 1.905]^T$$

The corresponding values of the objective functions are $^{(A)}b_{n+1} = 0.0754$ and $^{(B)}b_{n+1} = 0.0655$.

In cases C and D all five engine-operating regimes have been included into the optimal design problem. Again, the agreement between the target and optimal injection rate histories is slightly better in case D than in case C. Optimal values of the design variables are as follows:

$$^{(C)}\mathbf{b} = \begin{bmatrix} 30.068, 0.263, 0.235, 0.278, 0.250, 0.263, 0.244, \\ 6.717, 1.876, 6.418, 1.635, 6.240, 1.5907, 5.907, 0.343, 4.646, 0.227 \end{bmatrix}^T$$
$$^{(D)}\mathbf{b} = \begin{bmatrix} 30.854, 0.255, 0.231, 0.283, 0.252, 0.262, 0.246, 1.041, 0.917, 0.972, 1.128, \\ 6.780, 1.900, 6.460, 1.652, 6.214, 1.598, 5.780, 0.401, 4.686, 0.231 \end{bmatrix}^T$$

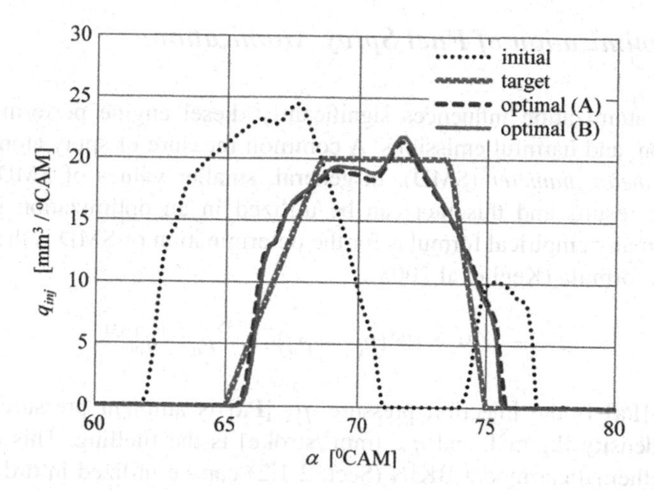

Fig. 8.7 Injection rate histories for cases A (Bézier curve) and B (rational Bézier curve)

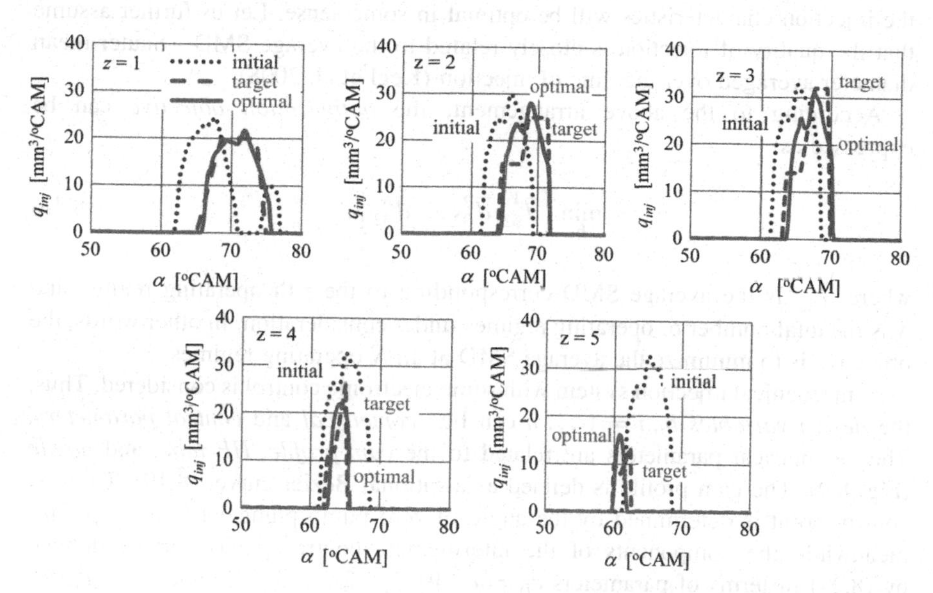

Fig. 8.8 Injection rate histories for case D (rational Bézier curve) for 5 operating regimes

The corresponding values of the objective functions are $^{(C)}b_{b+1} = 1.3813$ and $^{(D)}b_{b+1} = 1.3674$. For case D, the target, initial, and optimal injection rate histories are compared in Fig. 8.8, where the injection rate $\dot{q}_{\text{inj}}$ is given in [mm^3/stroke] and the camshaft angle φ in [°CAM].

8.2.3 Optimization of Fuel Spray Atomization

Fuel spray atomization influences significantly diesel engine performance, fuel consumption, and harmful emissions. A common measure of spray atomization is the *Sauter mean diameter* (SMD). In general, smaller values of SMD typically yield better results and this fact can be utilized in an optimization procedure. One of the many empirical formulas for the determination of SMD is the Hiroyasu and Kadota formula (Kegl et al 2008):

$$d_{32} = 2.39 \times 10^3 \left(p_{\text{inj}} - p_a\right)^{-0.135} \rho_a^{0.121} q_{\text{inj}}^{0.131}, \tag{8.42}$$

where p_{inj} [Pa] is the injection pressure, p_a [Pa] is ambient pressure, ρ_a is air (ambient) density [kg/m^3], and q_{inj} [mm^3/stroke] is the fuelling. This expression and the mathematical model BKIN (Sect. 2.1.2) can be utilized in order to solve numerically the following optimization problem.

Let the objective be to determine such values of the design variables **b** that the injection characteristics will be optimal in some sense. Let us further assume that the quality of injection is closely related to the average SMD—Sauter mean diameter averaged over the time of injection (Kegl et al. 2008).

According to the above arrangement, the *optimization objective* can be expressed as

$$\min_{\mathbf{b}} \left\{ \overline{d}_{32}^{\,1}, \overline{d}_{32}^{\,2} \ldots \overline{d}_{32}^{\,N} \right\}, \tag{8.43}$$

where $\overline{d}_{32}^{\,z}$ is the average SMD corresponding to the z-th operating regime and N is the total number of operating regimes under consideration. In other words, the objective is to minimize the average SMD at all N operating regimes.

A mechanical injection system with some electronic control is considered. Thus, the *design variables* b_i, $i = 1 \ldots n$ can be *geometrical* and *control parameters.* The geometrical parameters are related to the *cam profile*, *HP tube,* and *nozzle* (Fig. 8.9). The cam profile is defined as a rational Bézier curve, (8.19). The first control point is determined by the angle of zero pump plunger lift $\varphi_0 = \varphi_0\,(\mathbf{b})$; meanwhile the components of the intermediate control points can be defined by (8.24) in terms of parameters $c_{ix} = c_{ix}(\mathbf{b})$, $i = 2 \ldots (k - 1)$ and $c_{iy} = c_{iy}(\mathbf{b})$, $i = 3 \ldots (k - 2)$. The last two parameters are the high pressure tube length L_t and the nozzle hole diameter d_n.

The *control parameters* are the *pump plunger pre lift* h_p and the *geometrical duration of fuel delivery* t_d. The pump plunger pre lift influences the injection

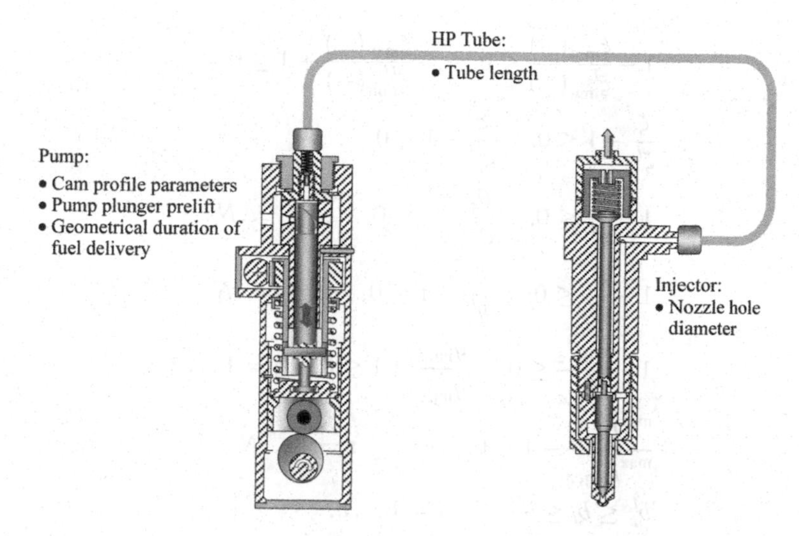

Fig. 8.9 Geometrical design variables of the inline fuel injection system

timing, while the geometrical duration of delivery influences the injection duration.

For technological reasons, the *minimal positive* $\xi_{\min}(+)$ and *minimal negative* $\xi_{\max}(-)$ *value* of the *local radius* of the *cam profile* should be constrained properly. To avoid torsion torque overloading and loosing contact between the cam and the follower, the *minimal* ζ and *maximal* ς *pump plunger acceleration* has to be limited also.

Additionally, proper *constraints* also have to be imposed on the *injection timing* θ and *injection duration* ϑ (both quantities are measured in degree of camshaft), on the *fuelling* q_{inj} (measured in mm^3/stroke), and on the *maximal injection pressure* ${}^{\max}p_{\mathrm{inj}}$.

Finally, to prevent a technologically unacceptable design, constraints on *lower* and *upper limits* of the *design variables* have to be imposed as well.

All of the above constraints can be written in a normalized form as

$$
\begin{aligned}
& 1-\frac{\xi_{\min}(+)}{\xi^{L}_{\min}(+)} \leq 0, \quad -\frac{\xi_{\max}(-)}{\xi^{U}_{\min}(-)}+1 \leq 0 \\
& \frac{\zeta}{\zeta^{L}}-1 \leq 0, \quad \frac{\varsigma}{\varsigma^{U}}-1 \leq 0 \\
& 1-\frac{\theta_z}{\theta^{L}_z} \leq 0, \quad \frac{\theta_z}{\theta^{U}_z}-1 \leq 0, \quad z=1\ldots N \\
& 1-\frac{\vartheta_z}{\vartheta^{L}_z} \leq 0, \quad \frac{\vartheta_z}{\vartheta^{U}_z}-1 \leq 0, \quad z=1\ldots N \\
& 1-\frac{q_{\mathrm{inj},z}}{q^{L}_{\mathrm{inj},z}} \leq 0, \quad \frac{q_{\mathrm{inj},z}}{q^{U}_{\mathrm{inj},z}}-1 \leq 0, \quad z=1\ldots N \\
& \frac{{}^{\max}p_{\mathrm{inj},z}}{{}^{\max}p^{U}_{\mathrm{inj},z}}-1 \leq 0, \quad z=1\ldots N \\
& b^{L}_i \leq b_i \leq b^{U}_i, \quad i=1\ldots n,
\end{aligned}
\tag{8.44}
$$

where the right superscripts L and U denote the prescribed lower and upper limits.

The objective (8.43) to minimize N functions, the selection of the design variables, the constraints (8.44), and an appropriate *response model* (for example, BKIN, Sect. 2.1.2) define completely the optimization problem. Unfortunately, this problem does not have the standard form since it is a multiobjective one. In order to solve it efficiently, one has to transform it into the standard form, i.e., into a form with one objective function.

This transformation can be done in several different ways that typically yield different results—usually a point belonging to the Pareto front. In any case, the requirement to minimize N independent functions has to be replaced by minimization of a substitute scalar function. This can be done in a satisfactory manner only if the relative importance (weighting) factors of original objective functions are known. In general, these factors cannot be determined easily; in fact, observation of the optimization progress and good intuition are needed to get a satisfactory result.

Fortunately, in our case the original objectives are related to individual operating regimes, which in turn are already weighted adequately by the ESC test weighting factors χ^z. Therefore, it seems to be a good choice to take the ESC weights to be the factors of the original objective functions. Thus, by multiplying each $\overline{d}^{\,z}_{32}$ by the corresponding χ^z and summing up, one gets a substitute scalar objective function of the form

$$
f=\sum\nolimits_{z=1}^{N}\chi^z\overline{d}^{\,z}_{32} \tag{8.45}
$$

It is expected that minimization of the above sum will yield a Pareto point, being satisfactory from the engineering point of view.

The purpose of this optimization is to see what the needed design differences are, if *mineral diesel* (D100) is replaced by *rapeseed biodiesel* (RaBIO). Furthermore, it

Table 8.3 Mineral diesel and rapeseed biodiesel properties

Target data	D100	RaBIO
Kinematic viscosity @ 30 °C [mm²/s]	3.34	5.51
Surface tension @ 30 °C [N/m]	0.0255	0.028
Calorific value [kJ/kg]	43,800	38,177
Cetane number	45–55	>51

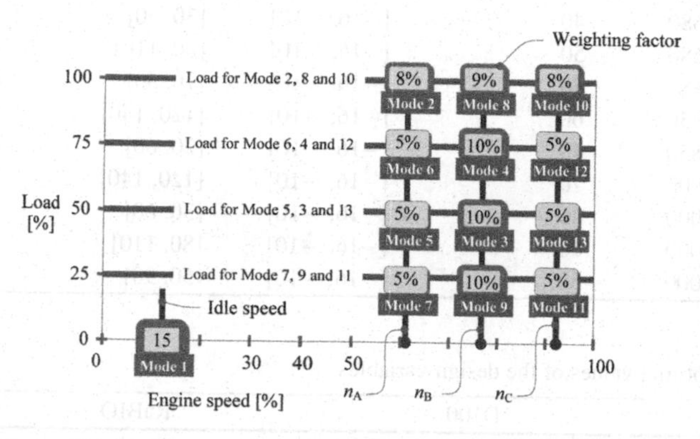

Fig. 8.10 ESC test, 13 mode cycle, weighting factors

Table 8.4 Initial and limit values of the design variables

Design variables	Initial value	Lower limit	Upper limit
φ_0 [°CAM]	35	20	50
c_{ix}, c_{jy} [mm], $i = 2 \ldots 5, j = 3, 4$	0.25	0	0.5
ψ_i [–], $i = 2 \ldots 5$	1.0	0	2.0
L_t [mm]	1,000	500	1,500
d_n [mm]	0.68	0.5	0.9
$h_p^{(z)}$ [mm], $z = 1 \ldots N$	1.87	1	5
$t_d^{(z)}$ [mm], $z = 1 \ldots N$	0.5, 1.5, 0.5, 1.0, 0.5, 1.0, 0.5, 1.5, 0.5, 1.5, 0.5, 1.0, 0.5	0.1	3

would be interesting to see, if by the optimized injection and fuel spray characteristics, the harmful emissions of biodiesel can be expected to be within the range of those of mineral diesel (or even lower).

The tested fuels are neat D100, conforming to the European standard EN 590 and neat RaBIO, conforming to the European standard EN 14214. Some measured properties of these fuels are given in Table 8.3.

The measurements and computations of injection and engine characteristics were performed at the 13 characteristic mode steady-state regimes of the *ESC test*. The importance of an individual mode is determined by the corresponding weighting factor χ^z, $z = 1 \ldots N$ in the objective function (8.45). The weighting factors for all modes in percent (%) are given in Fig. 8.10.

Table 8.5 Pump rotational speed, upper limit of maximal injection pressure ${}^{\max}p_{\text{inj}}$, lower and upper limits (intervals) for fuelling q_{inj}, injection timing θ, and injection duration ϑ

Mode	n [rpm]	${}^{\max}p_{\text{inj}}$ [MPa]	θ [°CAM]	q_{inj} [mm^3/stroke]	ϑ [°CAM]
1	275	30	[−16, −10]	[5, 15]	[1, 4]
2	680	50	[−16, −10]	[110, 130]	[10, 14]
3	850	30	[−16, −10]	[50, 80]	[8, 12]
4	850	50	[−16, −10]	[80, 110]	[10, 14]
5	680	40	[−16, −10]	[30, 60]	[5, 9]
6	680	50	[−16, −10]	[80, 110]	[8, 12]
7	680	40	[−16, −10]	[10, 50]	[3, 7]
8	850	60	[−16, −10]	[120, 140]	[13, 17]
9	850	40	[−16, −10]	[20, 60]	[6, 10]
10	1,000	70	[−16, −10]	[120, 140]	[14, 18]
11	1,000	40	[−16, −10]	[30, 60]	[9, 13]
12	1,000	60	[−16, −10]	[80, 110]	[11, 15]
13	1,000	50	[−16, −10]	[50, 90]	[10, 14]

Table 8.6 Optimal values of the design variables

Case	D100	RaBIO
φ_0 [°CAM]	28.58	22.55
c_{ix}, c_{jy} [mm] $i = 2 \ldots 5, j = 3, 4$	0.30352, 0.25692, 0.36866, 0.28845, 0.32365, 0.28776	0.25566, 0.17621, 0.29047, 0.25302, 0.26798, 0.23985
ψ_i [−], $i = 2 \ldots 5$	1.26038, 0.88739, 1.13024, 1.39873	1.05785, 0.83403, 0.93552, 1.25995
L_{t} [mm]	1.01332	0.95790
d_{n} [mm]	0.61247	0.61600
$(h_p^{(z)}, t_d^{(z)})$ [mm] $z = 1 \ldots N$	1.97414, 0.54523, 1.75976, 1.39697, 1.86833, 0.35384, 1.95489, 0.95997, 1.79904, 0.25808, 1.90468, 0.95529, 1.76308, 0.10000, 1.91244, 1.57345, 1.72530, 0.16213, 1.84628, 1.55342, 1.65557, 0.10016, 2.01694, 0.97043, 1.93086, 0.47963.	1.20393, 0.71998, 2.07246, 1.44619, 1.87820, 0.29018, 2.30993, 0.80478, 1.64899, 0.30884, 1.85066, 0.85077, 1.62637, 0.13611, 2.05228, 1.69821, 1.85486, 0.13186, 1.79030, 1.73583, 1.59724, 0.11141, 1.92101, 0.81730, 1.88366, 0.39181.

According to the ESC test, the pump speeds to be tested in this paper are $n_{\text{A}} = 680$ rpm, $n_{\text{B}} = 850$ rpm, $n_{\text{C}} = 1,000$ rpm and at the idle $n_i = 275$ rpm.

The following geometrical and control parameters are selected to be the *design variables*:

- The parameters φ_0, c_{2x}, c_{3x}, c_{3y}, c_{4x}, c_{4y}, c_{5x} , related to the cam profile
- The weighting factors ψ_2, ψ_3, ψ_4, ψ_5 , related to the rational Bezier curve of the cam profile
- The HP tube length L_{t} and the nozzle hole diameter d_{n}
- The control parameters $(h_p^{(i)}, t_d^{(i)}, i = 1 \ldots N)$ related to individual engine operating regimes

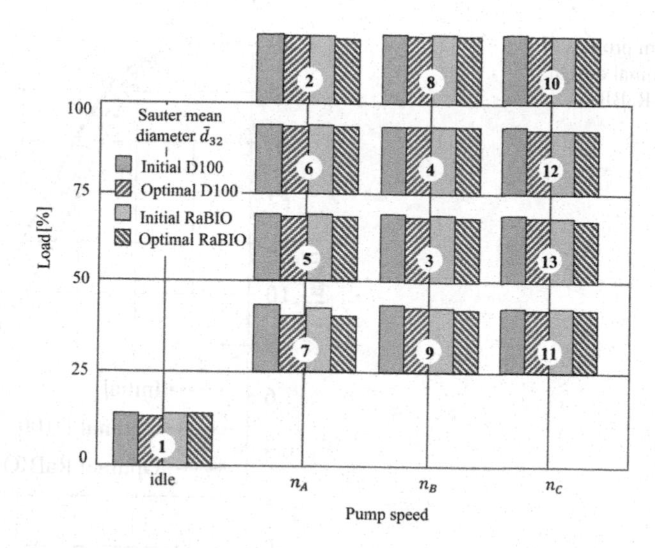

Fig. 8.11 Average SMD at initial and optimal designs for D100 and RaBIO

Since $N = 13$, there are a total of $n = 39$ design variables. The initial values of the design variables as well as their lower and upper limits are given in Table 8.4.

The following *lower* and *upper limits* of the constrained quantities are taken to be the same in all operating regimes: $\xi^L_{\min}(+) = 2$ mm, $\xi^U_{\max}(-) = -60$ mm, $\zeta^L = -100$ mm, and $\varsigma^U = 100$ mm. The limits of the maximal injection pressure $^{\max}p_{\text{inj}}$, fuelling q_{inj}, injection timing θ, and injection duration ϑ,differ for individual operating regimes (Table 8.5).

Two optimization cases were considered; neat D100 was used in the first case and neat RaBIO in the second case. Both cases have been solved successfully within 8 iterations. The optimal values of design variables are given in Table 8.6.

The average Sauter mean diameters $\overline{d}_{32}$ [μm] for $z = 1 \ldots N$ at *initial design* for D100 are 22.15, 31.94, 30.49, 30.73, 30.53, 31.33, 30.53, 31.59, 30.49, 31.23, 30.19, 30.47, 30.19 and for RaBIO 23.65, 31.74, 30.51, 30.93, 30.46, 31.42, 30.46, 31.45, 30.51, 31.17, 30.22, 30.39, 30.22, respectively. The *optimal values* for D100 are 21.56, 31.16, 29.58, 30.42, 28.86, 30.83, 25.92, 30.84, 29.25, 30.59, 29.25, 29.89, 29.62 and for RaBIO 24.55, 31.10, 29.50, 30.44, 28.79, 30.79, 27.23, 30.86, 29.26, 30.69, 29.32, 30.22, 29.47.

Figure 8.11 shows the differences between the average SMDat *initial* and *optimal designs*. It can be seen that the improvement at higher pump speeds and higher loads is not significant. However, it must be pointed out that at these regimes the constraints were not fulfilled at the initial design, whereas at optimal design all of the imposed constraints are fulfilled. The weighted sums $\sum_{z=1}^{N} \chi^z \overline{d}^{\,z}_{32}$ are 29.55 μm and 28.59 μm at initial and optimal design for mineral diesel and 29.76 μm and 29.11 μm at initial and optimal design for rapeseed biodiesel. As can be seen from Fig. 8.9, the optimized designs offer atomization

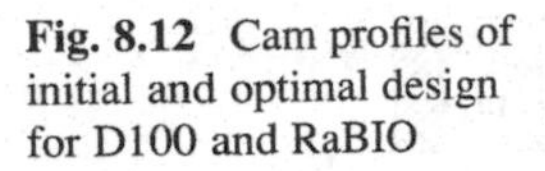

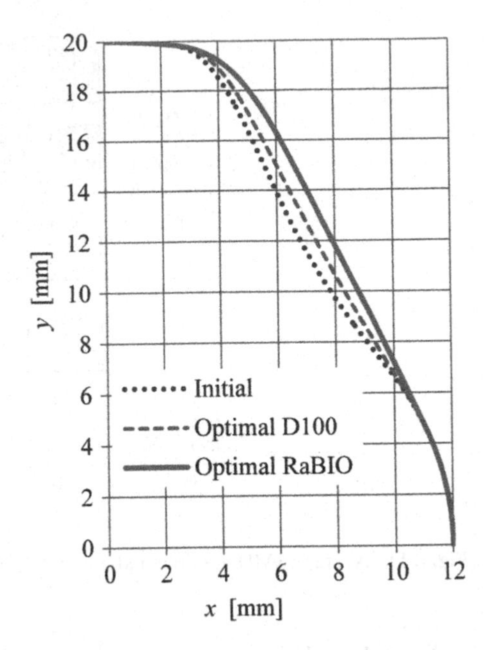

Fig. 8.12 Cam profiles of initial and optimal design for D100 and RaBIO

improvements practically at all engine-operating regimes without increasing the maximal injection pressure beyond the prescribed limits. This result might offer a possibility to improve NO_x–PM trade-off.

The *cam profiles* of *initial* and *optimal design* are compared in Fig. 8.12, where x and y denote the Cartesian coordinates of the profile. The optimal design differs from the initial one significantly. One can also observe that optimal designs for D100 and RaBIO are quite different.

Clearly, during the optimization procedure, the injection characteristics also change. For some operating regimes, the injection pressure, needle lift, and injection rate histories at the initial and optimal designs are compared for RaBIO and D100 (Fig. 8.13). As expected, at the same initial design, the injection pressure and injection rate are higher, and the needle lift opens earlier for RaBIO. After optimization the opposite is observed. The injection pressure of RaBIO is lower at all operating regimes. Furthermore, at some operating regimes the needle opens later and the injection rate history is shifted accordingly. This actually means that injection timing of RaBIO is retarded compared to D100. This could potentially mean lower NO_x emission.

8.3 Implementation of the Optimization Process

Nowadays, systematic optimization in engineering practice is still a rather challenging task. A common reason for this is that optimization procedures are often not supported/available directly by the simulation software used. Consequently, the

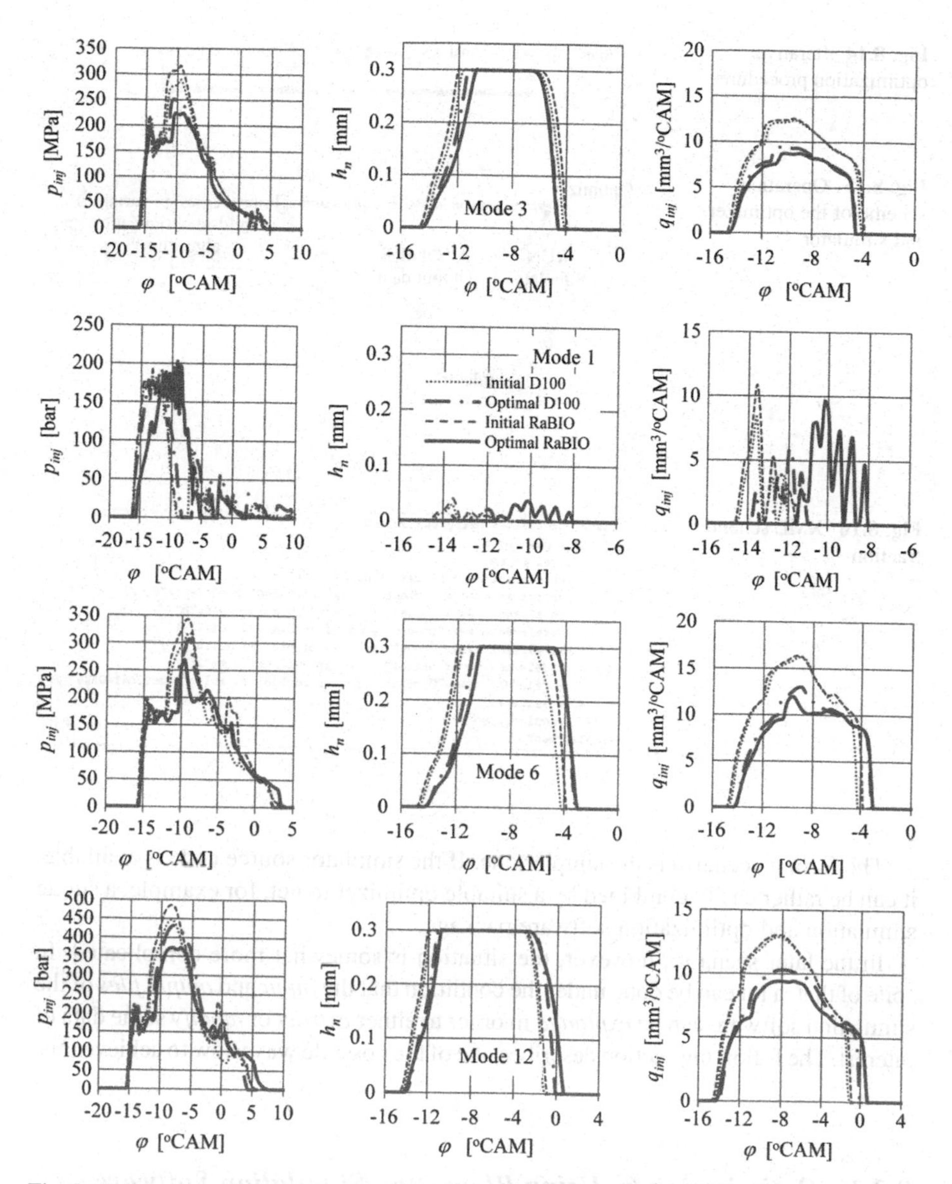

Fig. 8.13 Some injection characteristics for D100 and RaBIO at initial and optimal design

engineer is faced with the problem of combining the simulation software with a suitable optimizer. From that point on, one should distinguish between two scenarios:

- The simulation software *source code* is *known/available*
- The simulation software is a *black-box program*

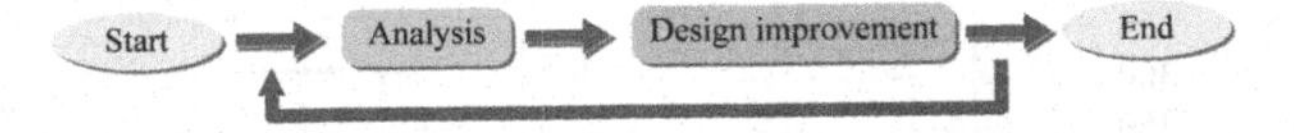

Fig. 8.14 Iterative optimization procedure

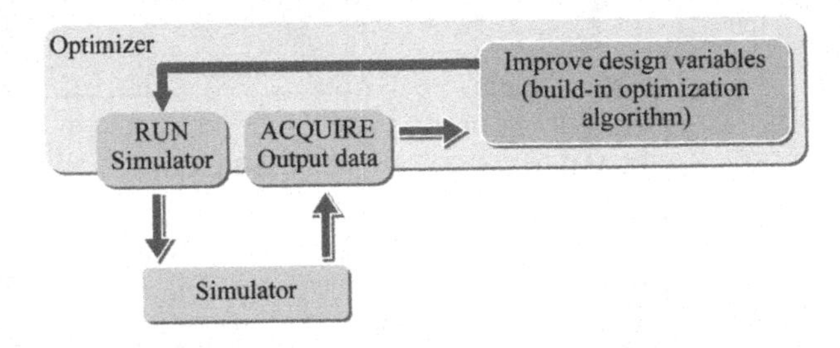

Fig. 8.15 Operation scheme of the optimizer and simulator

Fig. 8.16 XML scheme fraction

```
<xs:element name="DesignVariables">
  <xs:complexType>
    <xs:sequence>
      <xs:element name="Index" type="xs:int" />
      <xs:element name="Comment" type="xs:string" minOccurs="0" />
      <xs:element name="Value" type="xs:double" default="0" />
      <xs:element name="Lower" type="xs:double" default="-1" />
      <xs:element name="Upper" type="xs:double" default="1" />
      <xs:element name="Status" type="xs:int" default="1" />
      <xs:element name="DerEps" type="xs:double" default="0.000244140625" />
    </xs:sequence>
  </xs:complexType>
</xs:element>
```

The former scenario is the simpler one. If the simulator source code is available, it can be rather easily combined by a suitable optimizer to get, for example, a single simulation and optimization software package.

In the later scenario, however, the situation is somewhat more complicated. In spite of that, a lot can be done under the condition that the *input* and *output files* of the simulation software can be *exploited* in order to either *extract* or *modify* some data of interest. The following section describes one of the possible ways how to achieve this.

8.3.1 Optimization by Using Black-Box Simulation Software

In general, the implementation of mathematical optimization methods is more difficult than that of evolutionary methods. This is especially true, if a gradient-based optimizer has to be implemented. This section addresses such a situation.

By employing a gradient-based optimizer, the *solution procedure* layout is well known and can graphically be represented as shown in Fig. 8.14.

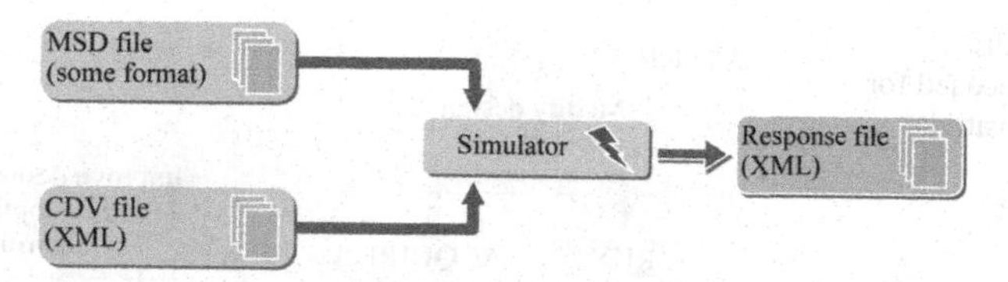

Fig. 8.17 Expected functionality of a simulator

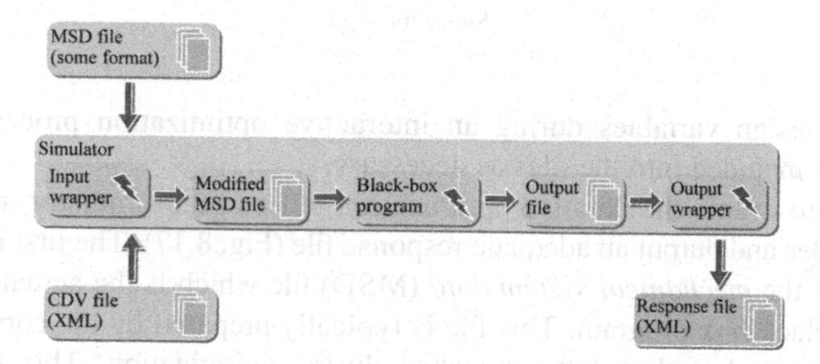

Fig. 8.18 Getting the expected simulator functionality by employing wrapper programs

In our case the analysis part (the calculation of functions) has to be done by a *black-box program,* while the design improvement part will be done by a *gradient-based* optimization algorithm. Of course, in order to run the whole procedure, one additionally needs some *driving program* that will execute the corresponding processes as needed.

From the practical point of view, one may note that the design improvement part as well as the driver is completely independent of the actual optimization problem. Thus, it is probably most appropriate to combine the design improvement part and the driver into a single (general-purpose) program. For the sake of convenience we will use here the term *optimizer* to denote this program. On the other hand, the black-box analysis program will be termed the *simulator* (Fig. 8.15).

By adopting the layout shown on Fig. 8.15, one actually has a situation of two stand-alone programs. Between both programs there obviously needs to be some data exchange. In order to reduce the platform dependency, this communication can be done through adequate data files. The format of these files should preferably be platform independent and self-descriptive. A good choice is that all data exchange files have an XML conforming syntax based on an appropriate *XML scheme*. An example fraction of such a scheme, describing the design variable data format, is shown in Fig. 8.16.

Employing XML syntax might appear to be an exaggeration since the data files are relative large due to unavoidable presence of XML tags. It should be noted, however, that the input/output operations in the procedure described above, represent a really negligible fraction of total computer time. On the other side, the benefit gained is that optional data, needed for example for activation/deactivation of

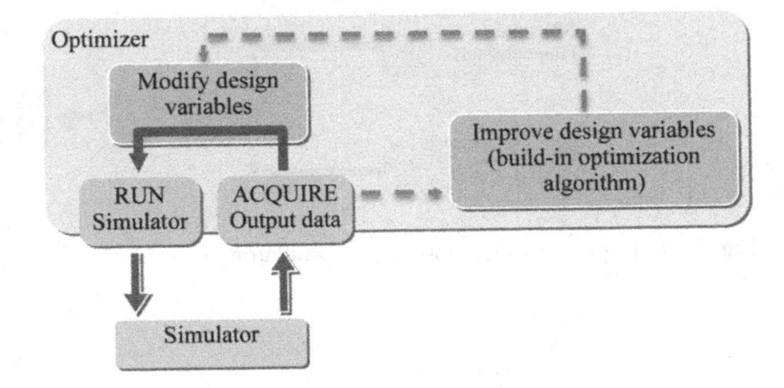

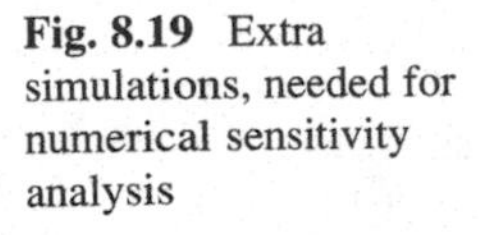
Fig. 8.19 Extra simulations, needed for numerical sensitivity analysis

individual design variables during an interactive optimization process, can be consistently included into the files as necessary.

In order to fit into the iterative optimization process, the simulator must accept two input files and output an adequate response file (Fig. 8.17). The first input file is termed here the *mechanical system data* (MSD) file which is the actual input data file of the black-box program. This file is typically prepared by the corresponding pre-processor and is then left untouched during optimization. This file is also typically written in some program-dependent format. The second input file is the *current design variables* (CDV) file which contains current design variable values. This file is written by the optimizer in each iteration just before the simulator is run. In our case the CDV file is expected to have the appropriate format in accordance with the corresponding XML scheme. The simulator should output an adequate response file according to the underlying XML scheme. The response file contains data about function values evaluated at the current design.

A commercial black-box program obviously does not exhibit the functionality shown in Fig. 8.17. Thus, in order to circumvent this problem, it seems to be an attractive solution to code two (usually rather simple) ad hoc *wrapper programs* and to define the simulator as a combination of three stand-alone programs: the *input wrapper*, the black-box program, and the *output wrapper* (Fig. 8.18). This can be achieved, for example, by defining the simulator as a command file which takes two input parameters (the names of both input files) and then runs all the required processes. The purpose of the input wrapper is to replace the original design variables data in the MSD file with the current values of the design variables and to output the modified MSD file, which then represents the actual input to the commercial black-box program. The output of the black-box program is then processed by the output wrapper in order to get the response file with the expected XML format.

Although the use of wrapper programs might appear to be somewhat tedious to implement, one should emphasize that the wrappers are usually really simple programs and can be coded very quickly. Especially, if the same black-box program is employed several times, virtually the same wrappers may be employed, containing perhaps only minor changes of the code.

A gradient-based optimizer requires that the response and sensitivity analysis have to be done in each iteration. Theoretically, the sensitivity analysis can be done in several ways. However, in the case of using a commercial black-box and response-analysis-only program one actually does not have many options left. In fact, one has to do the sensitivity analysis by using numerical finite (preferably central) differences. At this point it might be worth noting that the sensitivity analysis by using finite differences can be implemented completely in the optimization program.

By implementing finite differences, the optimizer simply has to run the simulator more than once. To be more precise, if the number of design variables is n, the total required number of simulations is $n + 1$ in case of using simple finite differences. In case that the central finite differences are preferred, the total number of simulations within each iteration raises to $2n + 1$. Of course, when computing the finite differences, the driver has to make appropriate changes to the design variables, as required by the corresponding formulas (Fig. 8.19).

8.4 Discussion

Optimization may lead to significant improvements of the engine characteristics regardless of the fuel used. One must note, however, that the optimal design variable values may depend significantly on the fuel type. This is true for both, geometrical and control, design variables. It follows that, in general, a diesel engine, equipped with a mechanically controlled injection system, cannot run very efficiently, for example, with D100 and RaBIO. It has to be adjusted to the actual fuel used.

In spite of that, optimization results reveal that it is possible to employ a relative small number of design variables in order to obtain good engine characteristics at several engine operating regimes simultaneously. Thus, use of an appropriate optimal design procedure and careful choice of design variables offer a possibility to adjust a diesel engine for biodiesel usage by making only minor and low cost modifications on its design.

Using the *injection rate history* in the objective function seems to be an attractive approach to optimization of the injection system. In accordance with the numerical results, obtained for an ECD fuel injection system, the following conclusions can be made:

- It seems that quite good injection rate histories of the ECD fuel injection system may be obtained by performing adequate adjustments of the cam profile and some control parameters only
- The rational Bézier curve gives better results than its nonrational counterpart; although the differences are not dramatically, it seems to be worth to include the weighting factors into the set of design variables

Furthermore one can say that the *Sauter mean diameter* can also be successfully employed in the formulation of an optimization problem. According to the results, the following conclusions can be made:

- Improvements of the fuel spray characteristics may be obtained by performing adequate adjustments of the cam profile, the nozzle hole diameter, high pressure tube length, and control parameters like as pump plunger prelift and duration of fuel delivery
- The values of the Sauter mean diameter can potentially be reduced at several engine-operating regimes simultaneously.

Regarding the optimization methods, one can say that the *gradient-based algorithms* may well represent an attractive choice when addressing engine optimization problems. These algorithms are somewhat tedious to implement because of the need to compute the design derivatives of the involved functions. The situation becomes even trickier, if one employs a commercial black-box program in order to analyze the considered system. It seems, however, that the solution of using two stand-alone programs—the optimizer and the simulator—works quite well. The optimizer can be coded as a completely independent general-purpose program by agreeing on suitable data exchange protocols. XML data exchange files, based on adequate XML schemes, represent a very good platform-independent choice. In order to fit the commercial simulator program into the optimization process, simple input/output wrapper programs may be utilized. These ad hoc programs have to be coded for every particular optimization problem considered. However, this can typically be done very quickly.

Finally, in order to make the optimization process as efficient as possible and to reduce the possibility to get trapped into unsatisfactory local optima, a careful selection of design variables is necessary. A *study* of the *influencing parameters* is therefore always recommended.

References

Bazaraa, S., Sherali, H. D., & Shetty, C. M. (1993). *Nonlinear programming: theory and algorithms*. New York: Wiley.

Coley, D. A. (2005). *An introduction to genetic algorithms for scientists and engineers*. Singapore: World Scientific Publishing.

Kegl, B., Kegl, M., & Pehan, S. (2008). Optimization of a fuel injection system for diesel and biodiesel usage. *Energy & Fuels, 22*, 1046–1054.

Kegl, B. (1996). Successive optimal design procedure applied on conventional fuel injection equipment. *ASME Journal of Mechanical Design, 118*, 490–493.

Kegl, B. (1995). Optimal design of conventional in line fuel injection equipment. *Proceedings of the Institution of Mechanical Engineers, Part D: Journal of Automobile Engineering, 209*, 135–141.

Kegl, M., Butinar, B. J., & Kegl, B. (2002). An efficient gradient-based optimization algorithm for mechanical systems. *Communications in Numerical Methods in Engineering, 18*, 363–371.

Kegl, B. (1999). A procedure for upgrading an electronic control diesel fuel injection system by considering several engine operating regimes simultaneously. *Journal of Mechanical Design, Transactions of the ASME, 121*, 159–165.

Kegl, M., & Oblak, M. M. (1997). Optimization of mechanical systems: On non-linear first-order approximation with an additive convex term". *Communications in Numerical Methods in Engineering, 13*, 13–20.

Needham, J. R., May, M. P., Doyle, D. M., Faulkner, S. A. (1990). Injection timing and rate control—a Solution for low emissions. *SAE paper 900854*

Chapter 9
Concluding Remarks

The *ultimate green diesel engine*—the one with *zero harmful emissions*, with *theoretically minimal fuel consumption*, and running on *environment-friendly renewable fuels*, is something that will always remain out of reach. However, reaching the exact theoretical limits is not as important as is the question: how close to these limits can we push a modern diesel engine. By looking back into the last decade, one can only be astonished what a tremendous progress was achieved by diesel engine developers and producers. Fuel consumption decreased substantially, exhausts became much cleaner, and the engines became lighter and smaller due to downsizing in piston displacement volume and number of cylinders. From that point of view one can be very optimistic for the future. In spite of that, one should be aware of several hard nuts still waiting to be cracked in the coming years. Two of the perhaps mostly exposed are briefly addressed in the following.

The first serious problem is related to the introduction of alternative fuels. A modern diesel engine is developed for mineral diesel. Mineral diesels are quite well regulated in terms of quality, purity, and so on. In spite of this good regulation, the usage of modern diesel engines revealed several problems related to diesel fuel quality and purity. On the other hand, these problems were practically negligible in traditional engines with mechanically controlled fuel injection systems. The point here is that, as the diesel engine technology advances, the fuel quality and purity problems rise. In this context, the introduction of alternative fuels is a very problematic scenario. Namely, alternative fuels come in such a variety in terms of their composition, quality, and purity, that it seems practically impossible that a modern diesel engine would be able to consume at least some of these fuels efficiently, without risking any harm or damage. At a first glance the situation looks quite frustrating. Namely, we know from engineering practice that the more we improve some system/technology, the more precisely the input data specification needs to be. For diesel engine development, fuel characteristics are probably the most important input data. But the introduction of alternative fuels scatters these data in an extremely wide range. The consequences can also easily be seen from the experimental results, assembled in this book; the measured quantities may vary dramatically in dependence on, e.g., type of biodiesel. For this reason, it would be

B. Kegl et al., *Green Diesel Engines*, Lecture Notes in Energy 12,
DOI 10.1007/978-1-4471-5325-2_9, © Springer-Verlag London 2013

of great help to engine developers, if the governments would do every effort to bring out good classification, appropriate standards, and regulation of alternative fuels. Unfortunately, this will also be very hard to do since alternative fuels production technologies are still developing. And we know that one cannot write good standards and regulation, if the technologies under consideration are still under intensive development.

The second important problem is related to modeling and computation. During the whole history, engineers improved and optimized their designs or products. At first, this optimization was mainly driven by intuition and experiments. But, as the modeling and computation methods developed, numerical simulation became an ever more important factor. For a system/technology at the development stage being not very close to the limits, rather simple numerical models can be of great benefit and can significantly contribute to the improvement of the product. But, the more the development stage of some system approaches the theoretical limits, the more sophisticated and accurate numerical models are needed in order to achieve at least some improvement. A diesel engine is such a complex system, that we are far from being able to simulate it as a whole with the accuracy, expected nowadays. If one just looks at the most important process in the engine: fuel starting from within the injector, traveling through the injector nozzle orifice, entering the combustion chamber, atomizing and mixing with air, evaporating, and eventually burning, one must recognize that full 3D modeling and computation of this process still represent a major problem. The difficulties range from unacceptably long computation times, over numerical instabilities, to numerous uncertainties, related to the modeling. In practice this means that even if we can manage to get some results, we cannot be very sure what we have actually got. To mitigate the situation, partial modeling is often applied in practice, e.g., by modeling the spray development from the nozzle orifice exit downstream. So far so good, but setting the boundary conditions at the nozzle orifice exit manually, introduces another source of error, which has to be taken into account properly. Once we will manage to run numerical simulation of diesel engine processes efficiently and accurately, this will set the base for the use of another important tool—systematic numerical optimization. If efficient computational models are available, the only exposed problem in this step is the additional computational effort. Namely, one must recognize that the computational effort, needed for the optimization of a considered system, may be one, two, or even three orders of magnitude larger than the computational effort, needed for the response analysis of the system. In other words, if the response computation takes, for example, one day, optimization may potentially take a completely unacceptable time span.

At the bottom line, pushing a diesel engine toward its theoretical limits, is not and will not be an easy task. However, by considering the immense efforts put into the diesel engine-related research at the universities and companies, one can be confident that the future of the green diesel engine looks pretty bright.

Index

B. Kegl et al., *Green Diesel Engines*, Lecture Notes in Energy 12,
DOI 10.1007/978-1-4471-5325-2, © Springer-Verlag London 2013

Printed in the United States
By Bookmasters